# Advances in Clean Energy

# Advances in Clean Energy

## Production and Application

Anand Ramanathan, Babu Dharmalingam,
and Vinoth Thangarasu

CRC Press
Taylor & Francis Group
Boca Raton London New York

CRC Press is an imprint of the
Taylor & Francis Group, an **informa** business

First edition published 2021
by CRC Press
6000 Broken Sound Parkway NW, Suite 300, Boca Raton, FL 33487-2742

and by CRC Press
2 Park Square, Milton Park, Abingdon, Oxon, OX14 4RN

ISBN: 978-0-367-51912-4 (hbk)
ISBN: 978-1-003-05568-6 (ebk)

Typeset in Times
by Deanta Global Publishing Services, Chennai, India

# Contents

# Preface

Fossil fuels are depleting due to enormous usage in power generation and transportation sectors. The energy sector has faced several challenges over the past decades; the critical problem is to meet the increasing demand for energy in affordable, environmentally friendly, efficient, and safer ways. Among the various possibilities, biodiesels offer to meet energy demand across the world, and they remain a promising alternative energy source. They possess environmental advantages relative to fossil fuels, as well as previously mentioned attributes.

Crude oil fuels supply about 96% of the world's transportation energy demand. The petroleum reserves, on the other hand, are small, and will eventually run out. Different studies placed the date of a global peak in oil production from 1996 to 2035. Decreasing reserves of fossil fuels, such as oil, coal, and natural gas, and the high cost of reliance on their imports are the main challenges facing nations worldwide. Future energy requirements have encouraged research interest in the area of alternative fuels. Over the years, non-edible seed crops have been used as feedstock for biodiesel production to substitute for conventional fossil fuels. This book focuses primarily on innovations for scholars, researchers, practicing engineers, technologists, and students in the fields of energy technology, renewables, low-emission and sustainable energy, energy conservation, and energy and environmental sustainability.

Biodiesel is one of the promising alternative fuels capable of replacing conventional fossil fuels to meet future energy demands. Chapters 1–3 provide complete coverage of the utilization of biomass sources to produce eco-friendly green fuels. Chapter 1 provides an overview of global energy sources and the present energy scenario. Chapters 2 and 3 present state-of-the-art information on extraction technologies such as microwave energy and ultrasonic sonication to extract the oil efficiently from seeds. Also, it offers a timely reference on the recent advanced technologies for biodiesel production and the physicochemical and thermal behavior of biodiesel.

Fossil fuel combustion remains a leading source of various emissions such as CO, NOx, and smoke. Moreover, promising carbon-free alternatives are not yet readily available. The above-stated emissions are the main contributors to climate change, ozone layer depletion, and environmental pollution. Advanced combustion techniques offer the opportunity to minimize carbon emissions compared to fossil fuel combustion, but more needs to be done to enhance our research and development on the future use of lean-burn, including carbon-neutral fuels such as biodiesel. Chapters 4–6 discuss the pros and cons of currently available techniques such as metal and alcohol additives on diesel engine combustion, novel advanced injection strategy, and low-temperature combustion technology.

Lignocellulosic biomass is a possible source of biofuels, easily and readily usable around the world. At the same time, significant demand for liquid biofuels is found in the transport and industrial sectors. Hence this book focuses in particular on the processing of lignocellulosic biomass to liquid biofuels. An increasing number of liquid biofuels, such as bioethanol, biodiesel, methanol, and reformulated gasoline components, are being evaluated for commercial use now. The production of biofuels from lignocellulosic biomass still poses some technical hurdles, but robust work is underway to resolve those obstacles.

Chapter 7 presents the various technologies for waste reduction and also recycling of the products. The chapter provides the possible solution for solid waste–generated health hazards and environmental pollution in the global community, which requires the safe handling of the wastes with less pollution to the environment. Also, it presents information about the various processes and considerations, including future policies for effective waste management. Chapter 8 discusses the characterization of biomass, such as the physical and chemical properties of biomass, and the technologies used to measure desired properties. This chapter discusses the economic sustainability of biomass power generation. Chapter 9 focuses on the conversion technology of biomass gasification, which consists of the chemistry of gasification and the effect of process parameters. Also, this chapter presents the production technologies of higher alcohol biobutanol from biomass. Chapter 10 provides a general idea about microbiology associated with biogas generation, environmental factors, selection of substrates, pre-treatment, and monitoring involved in biogas production processes. Chapter 11 presents an overview of the preparation of feedstock and anaerobic processes for biogas processing and its conversion to biofuel.

Numerous simple, concise diagrams and tables supplement the text in each chapter. This book aims to support sustainable clean energy technology and green fuel for clean combustion. For beginners or experts, we hope the book will offer a solid framework for current and potential biofuel enthusiasts to help their research journey toward a sustainable clean energy production to meet future energy demand.

# Authors

**Anand Ramanathan, PhD,** Associate Professor, Department of Mechanical Engineering in National Institute of Technology, Tiruchirappalli, Tamil Nadu, India, is a recipient of an Endeavour Fellowship from the Australian Government in 2015 at Australian National University, Canberra, Australia. He has received awards from various reputed organizations such as the N.K. Iyengar Memorial Prize (IEI, India, 2009), an Endeavour Executive Fellowship (Australia, 2015), Outstanding Scientist in Mechanical Engineering (Venus International Foundation, 2017), Outstanding Reviewer Award in Elsevier (2014–2018), and the Dr Radhakrishnan Excellence Award (India, 2018). His current research focuses on alternative fuels, internal combustion engines, fuel cells and waste-to-energy conversion. He is a principal investigator in various research projects in collaboration with several countries, including Brazil, the United Kingdom, and Russia. He has received many projects from national and international funding agencies such as DST-BRICS, MHRD-SPARC, GTRE-DRDO, DST-UKERI, DST India, and IEI-India. His research-oriented journey has involved the publication of more than 52 publications in Science Citation (SCI)/ Scopus Indexed research journals, 7 patents, 1 book, and 12 book chapters from renowned publishers such as Elsevier and Springer. He has delivered many lectures globally and served as a member of many research committees.

**Babu Dharmalingam, PhD,** completed his Doctor of Philosophy at the Department of Mechanical Engineering, National Institute of Technology, Tiruchirappalli, Tamil Nadu, India, in 2019. He earned a BE degree in Mechanical Engineering in 2007 and an ME (Mechanical-Thermal Engineering) in 2012. His area of interest is alternative fuels for internal combustion engines and ecology. He has published three international papers, three book chapters, and two international conference papers.

**Vinoth Thangarasu, ME,** is a Research Scholar at the Department of Mechanical Engineering, National Institute of Technology, Tiruchirappalli, Tamil Nadu, India. He earned his Master's degree in Thermal Engineering from the Government College of Technology, Coimbatore, Tamil Nadu. He earned his Bachelor's degree in Mechanical Engineering from the Institute of Road and Transport Technology (affiliated with Anna University, Chennai). He joined the National Institute of Technology, Tiruchirappalli, as a Junior Research Fellow, under the DST project (YSS/2015/000429) and since has been promoted to a Senior Research Fellow. He has published three international research papers, three patents, four book chapters, and two international conference papers.

# 1 Global Energy Sources and Present Energy Scenario

## 1.1 INTRODUCTION

Total primary energy consumption (TPEC) in the world is rising daily due to growing population and industrialization. In 2015, the world's TPEC amounted to more than 150,000,000 GW h and is projected to increase by 57% by 2050 (Hajjari et al. 2017). This rapid rise in energy consumption could potentially lead to more greenhouse gas (GHG) emissions and more environmental problems (Hosseini et al. 2015). Today, fossil fuels supply more than 80% of the overall energy used in the world, resulting in a significant contribution to environmental and health issues. Because of these issues, significant attempts have been made to identify the best alternative energy sources to address the economic and environmental impacts of fossil fuel use worldwide (Aghbashlo et al. 2015; Dadak et al. 2016).

The United States and China are the topmost energy users in the world. India ranks as the fourth largest energy user worldwide. Since March 2013, India's gross consumption of electricity is about 917.2 kWh per capita. India's electricity energy consumption is projected to be around 2280 BkWh and 4500 BkWh by 2021–2022 and 2031–2032, respectively. India's energy usage has rapidly grown due to population growth and increased standards of living. India's current integrated energy planning is primarily based on thermal power plants to meet energy demands and its percentage of total installed power plant capacity is close to 70%. This over-reliance generates pressure on fossil fuel. The key question emerges about how to conserve the petroleum-based fuels for our next generation while at the same time using different energy resources for fast and sustainable economic development.

Moreover, thermal power plants also adversely impact the natural climate. It should also be remembered that very high emissions of $CO_2$ (0.9–0.95 kg/kWh), $SO_x$, and $NO_x$ from thermal power plants lead to an increase in global temperature leading to climate change. Studies and literature have shown in the past century that since the industrial revolution, $CO_2$ emission rose by 28%. The global surface temperature has increased by 0.3°C to 0.6°C, and sea level has increased by 10–15 cm over the last 100 years. Scientists expect that if GHG emissions increase and no successful environmental mitigation measures are taken, the average temperature may rise by 1–3.5°C, and sea levels may increase by 15–95 cm. The continuously increasing energy demand and the associated negative effect of fossil fuels on the environment have pushed India towards a deliberate policy of renewables.

India's government is also trying to meet the country's demand for electricity while simultaneously protecting the environment from the pollution generated by utilizing fossil fuel energy sources such as coal for electricity generation. Hence, it is necessary to find new energy sources. In these circumstances, renewable energy is the principal alternative. The potential of renewable energy can be explored and exploited to meet electricity needs. Our survival requires numerous renewable energy sources such as hydro, solar, wind, and biomass to address this energy crisis. These renewables could meet future energy demand with great power capacity.

The main aims of this chapter are to explore possible ways of, first, supplying energy to all, even in rural parts of the globe; second, reducing the burden on fossil resources and preserving them for the next generation; and third, protecting the atmosphere from global warming and ultimately avoiding environmental disasters.

## 1.2 BRIEF HISTORY OF FOSSIL FUELS

### 1.2.1 COAL

Coal is a strong material, typically darker brown or black in color, with a significant portion of carbon and hydrocarbons formed from peat under higher pressures. It is one of the most significant and abundant natural resources of energy that can be used as fuel in various applications such as power plants, gasification, coal-powered engines, and liquefaction. Different types of coal, namely, lignite, sub-bituminous, bituminous, steam coal, anthracite, and graphite, are available. The types may vary with plant materials, impurities, and degree of coalification. Even though the carbon concentration in the Earth's crust is not greater than 0.1% by weight, carbon is necessary for life, and is also the primary source of human energy (*Coal – Simple English Wikipedia, the Free Encyclopedia* 2020).

Coal is the most abundant fuel in the fossil fuel family and has the most extensive and diverse history. Coal was used for heating food and homes since the days of the caveman. Archaeologists have found that Romans in England were familiar with the usage of coal by 100–200 AD. In the thirteenth century, Hopi Indians used coal to cook food and heat buildings. Later, coal was rediscovered and usage increased exponentially due to the industrial revolution (*History of Coal Mining – Wikipedia* 2020).

Coal is the leading fuel source for electricity generation—it accounts for a 38% share—and the international coal business increased by 4% in 2018. The world coal need has increased by 1.2% due to a 2% rise in coal power generation in 2018. The coal consumption of the Organization for Economic Co-operation and Development (OECD) countries has decreased by 3.2%, but consumption has increased by 2.5% for non-OECD countries. China is the topmost coal producer in the world, contributing 3550 Mt in 2018. India is the second-largest coal producer in the world, accounting for 771 Mt of the world total in 2018 (Schlissel, Feaster, and Wamstead 2019). The other three countries in the top five producers are the United States, Indonesia, and Australia. World coal reserves were 1055 billion metric tons in 2018, and they are projected to last for 132 years as per the current reserves/production ratio (*BP Statistical Review of World Energy* 2019). The distribution of coal reserves in 1998, 2008, and 2018 is depicted in Figure 1.1.

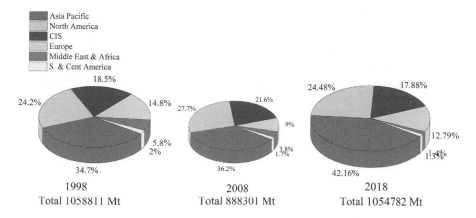

**FIGURE 1.1** Distribution of coal reserves in 1998, 2008, and 2018.

India had proven coal reserves of 101 billion metric tons at the end of 2018. The majority of these reserves are located in Jharkhand, Odisha, Chhattisgarh, and West Bengal. Over the last two decades, the consumption of coal has rapidly increased at an average rate of 6.3%. In 2017, coal consumption was 942 billion metric tons: 806, 47, and 89 million metric tons of coal account for thermal, brown, and metallurgical coal, respectively. This is due to the exponential use of thermal coal for electricity generation, which is 70% of the total (Philalay, Drahos, and Thurtell 2019).

## 1.2.2 PETROL AND DIESEL

Petrol and diesel are significant sources of energy for various sectors but are also used as a source for chemical and synthetic materials. These fuels can be derived from crude oil by refining. Dead animals and other organisms sank into the seabed, which formed the crude oil. Over time, these deposited bodies turned to black oil under higher pressures and temperatures of 60–120°C. Crude oil can be found in the seabed and underground (*Petroleum – Wikipedia* 2020). In 1853, the first crude oil refinery was established in Baku for the production of kerosene, which was used for lamp heating. Also, people started to use crude oil as a fuel for transportation in 1858 (Jeffery 2020). Later, the modern crude oil industry was developed by John D. Rockefeller around 1870. The global petrol price decreased from 2.56 USD in 1876 to 0.56 USD in 1892 due to rapid production in the United States and Russia (*History Of Crude Oil | Events That Drove The Oil Price History | IG UK* 2020).

The global oil demand rose by 0.9% in 2018 because of the need for petrochemicals and notably increased fuel consumption in the transport sector. In 2018, the use of oil was increased by 2.5% in the United States (leading consumer globally), followed by China (6%), and India (2.7%). At the same time, fuel consumption in Latin America and the European Union decreased by 3.3% and 0.6%, respectively. At the end of 2018, the total oil reserves of North America, South and Central America, Europe, the CIS, the Middle East, Africa, and the Asian Pacific were 35.4, 51.1, 1.9, 19.6, 113.2, 16.6, and 6.3 Mt, respectively. The world reserves/production ratio shows that oil reserves in 2018 represented 50 years of current production. The top oil reserve countries in the world are Venezuela, Saudi Arabia, Canada, Iran, and Iraq; they account for 17.5%, 17.2%, 9.7%, 9%, and 8.5%, respectively, of total oil reserves (*BP Statistical Review of World Energy* 2019).

India is the largest oil consumer in the world and its projected energy demand for 2035 is 1516 Mt. India's oil demand was 5.156 Mt barrels per day in 2018 and it is projected to be 5.05 Mt barrels per day in 2020. According to the BP Statistical Review of World Energy 2019, India's oil reserves are 4500 Mt and production is 39.5 Mt at the end of 2018. The diesel and petrol consumption during 2018–2019 rose to 83.5 Mt and 28.3 Mt, respectively. In order to meet this demand, USD5.615 billion worth of petrol and diesel was imported during 2018–2019 (*BP Statistical Review of World Energy* 2019).

### 1.2.3 NATURAL GAS

Natural gas is a hydrocarbon mixture composed largely of methane. However, it also composed of alkenes and small amounts of $CO_2$, $N_2$, and He. It occurs when layers of decaying plant and animal matter are exposed to extreme pressure and temperature below the soil surface for millions of years. Natural plants convert the sunlight into gas, in which the solar energy is stored in the form of chemical bonds (*Natural Gas – Wikipedia* 2020). Natural gas is a non-renewable source of energy that is used for cooking, heating buildings, and producing power. Since ancient times, natural gas has been known, but its commercial uses are comparatively recent (*History of Natural Gas* 2020). The Chinese used bamboo pipelines to transport natural gas to convert seawater to drinking water in 500 BC. In 1785, natural gas was produced from coal and used for commercial purposes. The first natural gas well was found in the United States in 1821.

Nowadays, natural gas is an essential part of the world's energy supply. Natural gas currently contributes 50% of the world's energy consumption for residential and commercial purposes. Approximately 41% of natural gas is consumed by US industry. Natural gas is a clean, safe, and widely used source of energy as compared to other sources (*A Brief History of Natural Gas – APGA* 2020). In 2018, world natural gas reserves were increased from 0.7 to 198.9 trillion cubic meters (Tcm). This is due to gas reserves in Azerbaijan, Russia, Iran, and Qatar increasing by 0.8, 38.9, 31.9, and 24.7 Tcm, respectively.

India's total natural gas reserves at the end of 2018 were 1.3 Tcm, which is 0.7% of the entire world gas reserves. The current Indian reserves/production ratio shows that natural gas reserves in 2018 accounted for 46.9 years of current production. The natural gas production was decreased from 29.4 Tcm in 2008 to 27.5 Tcm in 2018. Hence, the production rate was reduced by 0.7%. On the other hand, the natural gas consumption rate rose to 8.1% in 2018 as compared to 2008, which accounted for 1.5% of total world consumption (*BP Statistical Review of World Energy* 2019).

### 1.2.4 OTHER FOSSIL FUELS

Uranium is another kind of fossil fuel. Uranium is available in various concentrations in different sources, namely, granite (20 ppm), coal (20 ppm), and seawater

(0.0005 ppm). Uranium content and quality increase with decreased ore grade. The primary source of uranium is Precambrian granitic rock that was buried and subjected to extreme pressure and temperature (*List of Fossil Fuels | Environment* 2020). Kazakhstan, Canada, and Australia are the world's largest uranium producers. They contributed 41%, 13%, and 12% of the total world uranium share, respectively. Fifty percent of uranium is produced through *in situ* leaching. Uranium production from mines was increased from 80% in 2009 to 83% in 2018. In 2009, India's uranium production was 290 metric tons, and this rapidly rose to 423 metric tons in 2018 (*World Uranium Mining – World Nuclear Association* 2020).

Nuclear energy comes from the rearrangement of the nuclei of the atoms, and its high energy release is due to the fact that the forces between them are much higher than the forces between the outer electrons of the atom. The energy released from nuclear is 100 times higher than the energy released from chemical reactions. The first fully commercialized nuclear power plant with 250 MWe was established in 1992 at Westinghouse, United States. Around the same time, the Argonne National Laboratory developed the first boiling water reactor (BWR) (*Nuclear Power in India – Wikipedia* 2020). The world's consumption of nuclear energy increased by 2.4% in 2018. The United States, France, and China are the top three nuclear energy consumers around the globe, at 192.2, 93.5, and 66.6 Mt oil equivalent, respectively, at the end of 2018. India's fifth primary source of energy is nuclear power. As of 2018, India has 22 reactors with a capacity of 6780 MW; in these, seven nuclear power plants are successfully running. The nuclear energy share for electricity was about 3.22% in 2017. The consumption of nuclear energy has increased two-fold over the last ten years (*BP Statistical Review of World Energy* 2019).

## 1.3 RENEWABLE ENERGY SOURCES – BIOMASS

### 1.3.1 WHAT IS BIOMASS?

Biomass is a biological material derived from plants and other living organisms. Biomass is one of the primary renewable, eco-friendly, and abundantly available energy sources. Biomass can be directly used as a sole fuel or converted into biofuels such as biogas or biodiesel. The estimation of bioenergy potential is the assessment of the amount of energy available in the biomass sources. The utilization of bioenergy could supplement fossil fuels and reduce global greenhouse gas emissions. But the question is how much bioenergy could help. Research to assess bioenergy potential can provide a complete understanding of biomass types and quantities, and potential bioenergy distribution. Hence, such data will also help to identify how much bioenergy will substitute for fossil fuel and fulfill energy demands (Long et al. 2013).

### 1.3.2 BIOMASS FEEDSTOCKS

The European Commission defines biomass types as agriculture residues, crop residues, forestry, and municipal solid waste. Crop residues are widely accepted biomass sources and are used for energy production (Erb, Haberl, and Plutzar 2012). A great deal of work has been carried out on alternative biofuel crops to generate biodiesel worldwide (Kumar and Sharma 2011). But choosing a low-cost feedstock is a critical problem to minimize the overall production cost. The feedstock should be selected based on availability and raw material cost (Atabani et al. 2012). The land availability, soil conditions, and local climate determine the feedstock availability (Silitonga et al. 2013). Biodiesel feedstocks are categorized into food-based feedstock and non-edible feedstock.

However, researchers are focusing on the non-edible crops for energy production due to food versus fuel and land degradation issues. The non-edible feedstock, waste frying oil, and animal fat are considered second-generation feedstock. Many researchers have studied algae as non-edible feedstock, but few researchers have categorized it as a separate feedstock (No 2011). However, it has to considered as an emerging source to produce biodiesel because it has more oil content than vegetable oil feedstock and it can be cultivated on a large scale in a short time (Bhuiya et al. 2016).

### 1.3.2.1   Edible Crops
The main source of biodiesel production is edible vegetable oil. At present, over 95% of the world's biodiesel is made from edible vegetable oils. Among these, rapeseed is the predominant source for biodiesel; it accounted for 84%, followed by sunflower oil (13%), palm oil (1%), soybean oil, and others (2%). Nevertheless, the utilization of edible oils is related to important habitat issues such as deforestation, depletion of soil resources, and the use of certain agricultural land. However, the price of edible oil has increased significantly, which will eventually affect the sustainability of edible oil-based fuels. At the beginning of the biodiesel era the use of edible feedstock for biodiesel production was very widespread. The key benefits of first-generation feedstocks are the quality of crops and the comparatively simple conversion process (Singh et al. 2020). The risk of food supply constraint is the major drawback to the use of such feedstocks, which raises the cost of food products (Aransiola et al. 2014). Moreover, edible oils are not produced in adequate amounts in many countries of the world to satisfy local human food needs and thus must be imported from other places (Bhuiya et al. 2016). Barriers to the production of biodiesel from edible feedstock also include adaptability to environmental conditions, high costs, and restricted area of cultivation. These shortcomings have forced consumers to turn to alternative sources for the production of biodiesel (Tariq, Ali, and Khalid 2012).

### 1.3.2.2   Non-Edible Crops
Non-edible oils are promising potential alternative sources of energy to meet future energy demand. Researchers are focusing on non-edible oil resources because they are readily available in the world and can be cultivated on wasteland. Moreover, they are not suitable for human use, and by-products are more eco-friendly and more economical compared to edible oils (Demirbas 2009). All over the world, an enormous variety of non-edible plant trees are available. Azam, Waris, and Nahar (2005) found that 75 non-edible feedstocks have more than 30% oil content. Among these, 26 feedstocks are identified as a promising source to produce biodiesel.

The use of non-edible vegetable oils instead of edible oils is the best choice for biodiesel production because it is not suitable for human consumption, owing to the presence of toxins (Banković-Ilić, Stamenković, and Veljković 2012). The potential of non-edible feedstocks can be utilized according to their availability to produce biodiesel. Widely used non-edible feedstock sources are jatropha, pongamia, castor, mahua, rubber seed, *Calophyllum inophyllum*, neem, *Simarouba glauca*, jojoba, *Thevetia peruviana* Merrill, etc. These feedstocks are abundantly available all over the world and are low-cost as compared to edible feedstocks (Atabani et al. 2013). Detailed descriptions and biological aspects of a few non-edible feedstocks are presented below.

The jatropha plant is widely available in India, South America, Southeast Asia, and Africa. It is capable of growing in semi-arid and low-rainfall regions (Kibazohi and Sangwan 2011). The productivity data for these plant species are, to some extent, limited and indefinite. However, the seed productivity of the jatropha plant varies from 0.1 to 15 t/ha/year (Kumar and Sharma 2011). Many researchers have tried to increase jatropha productivity under various excess salt and drought conditions (Achten et al. 2010; Kumar, Sharma, and Mishra 2010). The maximum oil content that can be extracted from the seed is 40–60%. Seed oil is used in many applications such as making soap, lighting, biodiesel production, and lubricants. Jatropha oil has attracted attention in the alternative energy field due to its higher hydrocarbon levels. Jatropha seed oil is largely comprised of oleic acid (42%), linoleic acid (35%), palmitic acid (14%), and stearic acid (6%). The percentage of the chemical composition may vary with region and soil conditions (Kumar and Sharma 2008).

The pongamia tree is a short tree with a curved trunk and a wide crown of expanding or hanging branches. The plant is native to India and Myanmar but it also grows in Sri Lanka, Pakistan, Nepal, Bangladesh, and Southeast Asia (Atabani et al. 2013). The seed productivity of a single tree is 9–90 kg. The pongamia plant is considered as a nitrogen-fixing tree (Karmee and Chadha 2005). The seed contains 30–40 wt.% of oil and is of 0.77 to 1.11 cm in length and 0.69 to 0.92 cm in width (Mukta, Murthy, and Sripal 2009). This seed oil has many medicinal properties. Moreover, it has been getting attention in the biodiesel industry. The chemical composition of pongamia oil/feedstock may vary by region (Kumar and Sharma 2011).

*Calophyllum inophyllum*, also known as polanga, is an ornamental, evergreen tree from Indian, Malaysian, Indonesian, and Philippine tropical regions (Pinzi et al. 2009). It grows to a height of 25 m and produces a slightly toxic fruit containing a single, large seed. A shell (endocarp) and a thin, 3- to 5-mm layer of pulp covers the single, large seed. Oil yields were recorded at 2000 kg/ha per unit land area (Hathurusingha, Ashwath, and Subedi 2011; Venkanna and Venkataramana Reddy 2009). The seed productivity of a single tree is 100–200 fruits/kg. The seed is not suitable for human use because of its high acidity, i.e., 44 mg KOH/g of oil. The seed oil is mainly comprised of unsaturated fatty acids such as 38.3 wt.% of linoleic acid and 34.1 wt.% of oleic acids (No 2011).

*Simmondsia* (also known as jojoba) is a perennial shrub, native to Mexico, California, and Arizona's Mojave and Sonoran Deserts. Jojoba seed has a peculiar lipid content in the seeds which varies between 45 and 55 wt.%, as opposed to triglycerides in the form of long-chain esters of fatty acids and alcohols (wax esters) (Canoira et al. 2006). Because of the peculiar structure of jojoba wax, methanolysis produces a product comprised of a mixture of fatty acid methyl ester and long-chain alcohols, since removing such materials is troublesome. A few physiochemical properties of jojoba biodiesel are poor as compared with biodiesel prepared from other feedstocks because the kinematic viscosity is 11.82 mm²/s at 40°C and the cold filter plugging point value is 4°C. The cold filter plugging point value can be improved to –14°C by removing the alcohol content (Kumar and Sharma 2011; Bouaid et al. 2007).

Mahua is a 20 m-tall tropical tree primarily found in the northern and eastern plains and forests of India (Demirbas 2009; Jena et al. 2010). It is fast growing and has evergreen or semi-eternal leaves and is suited for tropical areas. In India, *indica* and *longifolia* are the two main species of the genus *Mahua*. *Mahua indica* is one of India's non-edible, forest-based feedstocks with a high annual production capacity of almost 60 million metric tons. Mahua fruit comprises one or two kernels, which can be harvested from a 4- to 7-year matured tree, and the trees can live for up to 60 years. The *Mahua indica* seed yield differs from 5 to 200 kg/tree based on the size and age of the tree. Seventy percent of the seed is in the kernel and it has 50% oil content (Atabani et al. 2013). The chemical composition is 17.8 wt.% of palmitic acid, 14 wt.% of stearic acid, 46.3 wt.% of oleic acid, and 17.9 wt.% linoleic acid. Hence, it has a higher percentage of saturated fatty acids as compared to other non-edible oils. Accordingly, this seed oil has poor cold flow properties. Mahua oil contains around 20% of free fatty acid and it requires a large amount of acid catalyst to produce biodiesel from oil (Kumar and Sharma 2011).

The neem tree can adjust to a wide temperature range (0–49°C). The neem tree is found in a variety of regions, including in South America, Africa, and Asia. The neem tree grows in almost all soil types, like clay, sandy, alkaline, acidic, stony, and shallow soils and even solid soils with heavy calcareous soil (No 2011). A well-grown tree can yield 30–50 kg of fruit, and tree life is up to 150–200 years. So, it can produce 540,000, 107,000 and 425,000 metric tons of seed, oil, and cake, respectively. The neem seed oil is dark brown and has a bitter taste. It could be a promising alternative source of energy to replace other fossil fuel sources. This oil is primarily used in cosmetic, ayurvedic, and biopesticide preparations. The raw neem oil has a higher free fatty acid of 21.6 mg of KOH/g of oil and it requires more acid catalysts to convert oil into biodiesel (Atabani et al. 2013).

The rubber seed tree, referred to as *Hevea brasiliensis*, is part of the Euphorbiaceae family. This rubber tree is native to the Amazon (Brazil). The rubber tree is the predominant natural rubber source, supplying 99% of natural rubber in the world. It is also possible to obtain and use the sap-like extract of the tree (known as latex) for multiple purposes. It is found primarily in India, Indonesia, Sarawak, Malaysia, Sri Lanka, Thailand, and Liberia. The tree can grow up to a height of 34 m; the plant requires heavy rainfall and yields seeds ranging from 2 to 4 g, which are not used for commercial purposes nowadays (Tomes, Lakshmanan, and Songstad 2011).

Typically, 37% of the seed weight is the shell and the rest is the kernel. Rubber seed oil is a non-edible oil having 50–60 wt.% oil and 40–50 wt.% of brown-colored oil in the kernel. The seed oil has a higher percentage of unsaturated fatty acids such as 36.3 wt.% of linoleic acid, 24.6 wt.% of oleic acid, and 16.3 wt.% of linolenic acid (Kumar and Sharma 2011).

### 1.3.2.3 Lignocellulosic Biomass Waste

Lignocellulosic biomass is the expression used for the combination of lignin, cellulose, and hemicellulose polymers interlinked in a heterogeneous matrix as biomass from woody or fibrous plant material. The total mass of cellulose and hemicellulose content may vary with plants, but usually the biomass consists of 30% of lignin and 50–70% of cellulose and hemicellulose combined. Via a sequence of thermochemical and biological processes, cellulose and hemicellulose can be processed into sugars and finally fermented to bioethanol. Lignocellulosic biomass is a readily available and sustainable resource that consists of cereal, straw, wheat chaff, rice husks, maize cobs, corn stove, sugarcane bagasse, nut shells, forest harvest, residues from wood processing, and marginal and degraded land energy crops. According to an EIA report (International Energy Agency 2010), there will be more than 1500 EJ of technological capacity for bioenergy by 2050. The estimated potential of biofuel and bioenergy from agricultural and forestry residues would be 85 EJ in 2050 (Hoogwijk et al. 2005).

Lignocellulosic feedstocks are typically categorized into three types: agricultural residues (e.g., crop residues, sugarcane bagasse), forest residues, and crops with herbaceous and woody biomass.

A wide variety of agricultural residues, namely corn stove, grass, wheat grass, bagasse, etc., could be used to produce second-generation biofuels. These fuels are usually regarded as renewable because they are produced from food crop waste materials and do not compete for land with food crops. The crop residues can be converted into value-added products using different conversion technologies. Ethanol can be produced from agricultural residues through biochemical processes such as anaerobic digestion and fermentation. Also, gasification can be used to produce syngas from residues; subsequently, syngas can be converted to liquid fuels with the aid of catalysts.

The second-generation biofuels are mostly produced from two main types of forest residues. The residues of wood cutting, such as branches, leaves, roots, etc., can be used for biofuel production. The demand for cut wood includes the difference between the average amount of wood that can be harvested and the actual amount of wood required to satisfy the demand. The thermochemical route is the one of the best ways to convert residues from forests into biofuels. In Sweden, the world's first BioDME plant was started, using black liquor from forest residues through gasification.

In order to provide feedstocks for the production of biofuels, there are a number of energy crops that could be cultivated on land which is not appropriate for agricultural farming. Hence, new energy crops, especially perennial grasses such as miscanthus, switch grass, prairie grass, and forest species, namely eucalyptus, poplar, and *Robinia*, can be used for biofuel production. Drawbacks associated with forest residues are that they are used to produce many other products, and that energy production on waste lands is problematic due to high energy demands and the low ability to adapt the new crops to these soil conditions (Ullah et al. 2015).

## 1.4  GLOBAL ENERGY NEEDS – PRESENT AND FUTURE

The development of a social economy requires enormous natural resources, especially energy, which are considered the most significant growth drivers. Global energy demand grew by 2.9% last 2019 This was the fastest pace of the decade, owing to a stable global economy and increased heating and cooling needs. The non-OECD countries consume more energy than the OECD countries due to the rapid growth of population and healthy economic development. A 2019 international energy outlook report predicted that the increase in energy consumption of OECD countries would be 15% between 2018 to 2050 but that of non-OECD countries would be 70%. Worldwide, the need for all types of fuels was increased but growth was higher for gaseous and renewable fuels compared to other fuels. Natural gas is the fuel of choice with the most considerable revenues and increases in global energy consumption of 45%. Consumption of natural gas was increased to 195 billion cubic meters, which was the highest growth since 1984. The United States and China are the topmost countries where the demand for gaseous fuels increased sharply in 2018. US gaseous fuel consumption increased faster in 30 years. Similarly, China's gaseous fuels demand was increased by 18% (*International Energy Outlook* 2019).

Oil is one of the primary energy sources, accounting for 30% of the global energy need. Oil consumption peaked in the OECD countries in the last decade. In 2018, global oil consumption increased from 90 million barrels per day to 122 million barrels per day, and it is forecast to be 107 million barrels per day in 2050. This is because of the robust oil consumption of non-OECD countries. China will be the major oil consumer in the world in 2030, and this trend will continue up to 2040, declining after that. India will become the second major oil consumer, surpassing the United States in 2050 (*IEEJ Outlook* 2018).

According to the British Petroleum 2019 report, renewable energy is a fast-growing energy source that contributes 14% of the world's energy consumption, which is projected to reach 50%. Renewable energy sources are getting attention in the power generation sector, since they accounted for nearly 45% of the energy growth in 2018. China is the top leader in the production of energy from renewable sources, such as solar and wind. The production of energy from renewable sources such as solar and wind increased at a two-digit rate (*BP Statistical Review of World Energy* 2019). However, this was not enough to meet the growing global electricity demand. Energy from biomass was also considered as a potential source for future energy demand. Hence, using energy from untapped biomass will provide a cost-effective and sustainable supply of energy.

The United States and Brazil are the world's largest biodiesel producers, with a share of 87% of global biodiesel production, while the remaining share of biodiesel production is evenly distributed between Asia, the Americas, and Europe. The European Union is the most significant biodiesel producer in the world, producing 37% of global output. Rapeseed is the predominant feedstock in the EU. Recently, the biodiesel market has been getting more attention in Southeast Asia. Palm oil is the principal source to produce biodiesel in Indonesia and Malaysia. The world biodiesel production was 34.08 million metric tons in 2016. In 2016, 2.15 million metric tons of biodiesel were produced in Germany (*UFOP Report on Global Market Supply 2017/2018* 2019). According to the REN21 report, global biodiesel production increased by nearly 5%, which is mainly due to US biodiesel growth (*Renewables Global Status Report* 2019).

The United States produced 159 million gallons of biodiesel in July 2019, which is 17 million gallons higher than the previous month. Of this 68% came from the Midwest region. Ninety-five biodiesel feedstocks with an annual capacity of 2.5 billion gallons have been used to produce biodiesel. Soybean oil is the largest source, with 709 million pounds being used to produce biodiesel in 2019. The annual biodiesel production capacity of the United States was 1596 million gallons in 2018, which is two times higher than in 2017. Sales of B100 and biodiesel blends were 765 and 1093 million gallons, respectively, in 2019 (*Monthly Biodiesel Production Report* 2019).

Biodiesel is one of the leading biofuels used in the transport sector. Worldwide, many state government fleets such as military vehicles, buses, trains, and trucks use biodiesel and its blends (B20) as fuel. The consumption of biodiesel in the United States has increased significantly since 2001. The consumption of biodiesel increased from about 10 million gallons in 2001 to about 2 billion gallons in 2016. In 2017 and 2018, there was a decline in the consumption of biodiesel due to imposed import duty taxes on biodiesel from Indonesia and Argentina; these two countries are the primary import source of biodiesel to the United States. Nowadays, many countries are giving priority to the use of biodiesel. In 2016, 9.3 billion gallons of biodiesel were consumed the entire world. In this, the United States, Brazil, France, Indonesia, and Germany are the top five countries, consuming 58% of biodiesel produced in 2016, with the remaining 42% consumed in all other countries (*U.S. EIA Monthly Energy Review* 2019).

## 1.5  INDIAN ENERGY NEEDS – PRESENT AND FUTURE

Non-edible seeds oils are second-generation feedstock, gaining more attention than edible oils because they are not used for food purposes. The problems associated with edible oils such as food versus fuel, and economic and environmental issues, can be overcome by using non-edible feedstocks to produce biodiesel (Roschat et al. 2017). Besides, non-edible biodiesel crops are expected to use predominantly non-productive and affected areas and degraded forests. Public areas like irrigation canals, along with national highways and railway tracks, can be used to plant non-edible plants (Reddy, Sharma, and Agarwal 2016). The development of non-edible oil–based biodiesel could become a significant part of the poverty programs for poor rural populations, including providing energy to rural areas and modernizing the rural non-agricultural sector (Wu et al. 2016). All these problems have a significant impact on the sustainability of biodiesel production. All over the world, more research has been done on alternative feedstocks to produce biodiesel, and it has been shown that these could be promising sources. Moreover, there are 350 oil-bearing feedstocks around the world. Hence, non-edible can be considered as a sustainable alternative source for biodiesel production (Singh et al. 2017).

India has a high potential of 3.0–3.5 million metric tons of tree-borne oilseeds (TBO), but each year, the collection is about 0.5 million metric tons. Currently, 67% of Indian oil demand is fulfilled by imports of vegetable oil due to untapped hidden sources of potential TBO. Hence, using edible oil to produce biodiesel is not possible. Under Indian conditions, only non-edible sources, which can be grown on a large scale on wastelands, can be considered to produce biodiesel (Kukreti 2018). Several non-edible feedstocks are available in India, but only 77 species were identified as a source for biodiesel production. Among these, only 12 species were selected as promising non-edible feedstocks to produce biodiesel on an industrial scale. Jatropha, karanja, neem, mahua, and rubber are some proven non-edible feedstocks (Karmakar, Karmakar, and Mukherjee 2012). Such proven non-edible feedstocks are not adequate to meet the huge energy demands of a country like India. This is where the value of discovering new non-edible feedstocks for biodiesel production comes in.

## 1.6  CLIMATE CHANGE

Human-made global climate change is severe not just because of the extent of the destruction but also because of its irreversibility (Solomon et al. 2009). The main GHG emission causing climate change is $CO_2$. More than 80% of the energy produced in industrial society is from fossil fuels, which leads to more $CO_2$ emissions. The world energy demand is growing exponentially due to sophisticated lifestyle and global population expansion (Bose 2010). The transport industry and commodities account for a fifth of global energy demand (Creutzig et al. 2015). Due to worldwide rapid motorization and immense reliance on fossil fuels, the reduction of carbon emissions in the transport industry is now more challenging than in other sectors (Loo and Banister 2016).

Future energy consumption and $CO_2$ emissions related to transport will continue to rise without clear and successful policy action. Moreover, the transport sector will continue to rely heavily on fossil fuels. Hence, adopting advanced vehicle techniques and alternate fuels with different options for behavioral modification will bring about a major $CO_2$ emissions reduction in the transport sector (McCollum and Yang 2009). Reducing activity growth and modal changes to low-carbon technology may have a marginal impact on reducing emissions, but advanced techniques are likely to be more useful in reducing emissions (Edelenbosch et al. 2017).

In particular, electric vehicles may be seen as promising alternative technology for reducing carbon emissions (Dhar, Pathak, and Shukla 2017). On the other hand, there is no strong evidence from empirical studies to support alternative energy for carbon emission reduction (Neves, Marques, and Fuinhas 2018). The stringent emissions norms cannot be achieved only through technological advancements without major improvements in behavioral action such as greater usage of rail, reduced travel, greater public transport usage, and higher occupancy (Bristow et al. 2008). The transport industry has unique and powerful relationships with other sectors, which influences the energy system and the economy via several dimensions. The transport of electrical supply and electrification have intersectoral trade, and consistency between energy and transport policies is required (Kannan and Hirschberg 2016). Therefore, there is a need for policymaking to model cross-sectoral interaction and to detect how carbon emissions reduction policies can alter energy and emissions profiles as well as economic growth.

## 1.7 CONCLUSION

This chapter presents a summary of the available energy sources and the current energy scenario with respect to global and Indian perspectives based on research findings and critical descriptive statistics gathered from the available literature on the topic. Fossil fuels, such as oil, coal, and natural gas, will remain important major sources of energy for human life at least for the next 50 years. But, as mentioned above, all of these fossil fuels have different issues, and their future in the long term is highly questionable. Moreover, overuse of fossil fuels drastically increases $CO_2$ concentration and global temperature. Significant analysis of biomass resource properties and availability was presented. India has a huge amount of biomass available in its forests that can be used for energy generation. The global community needs to support the transition from fossil fuels to renewable and alternative energy supplies in order to slow down climate change and create sustainable energy supplies.

## REFERENCES

*A Brief History of Natural Gas – APGA*. 2020. Accessed February 26. https://www.apga.org/apgamainsite/aboutus/facts/history-of-natural-gas.

Achten, W. M. J., W. H. Maes, B. Reubens, E. Mathijs, V. P. Singh, L. Verchot, and B. Muys. 2010. "Biomass Production and Allocation in Jatropha Curcas L. Seedlings under Different Levels of Drought Stress." *Biomass and Bioenergy* 34 (5): 667–76. doi:10.1016/j.biombioe.2010.01.010.

Aghbashlo, Mortaza, Meisam Tabatabaei, Pouya Mohammadi, Navid Pourvosoughi, Ali M. Nikbakht, and Sayed Amir Hossein Goli. 2015. "Improving Exergetic and Sustainability Parameters of a Di Diesel Engine Using Polymer Waste Dissolved in Biodiesel as a Novel Diesel Additive." *Energy Conversion and Management*. doi:10.1016/j.enconman.2015.07.075.

Aransiola, E. F., T. V. Ojumu, O. O. Oyekola, T. F. Madzimbamuto, and D. I. O. Ikhu-Omoregbe. 2014. "A Review of Current Technology for Biodiesel Production: State of the Art." *Biomass and Bioenergy*. doi:10.1016/j.biombioe.2013.11.014.

Atabani, A. E., A. S. Silitonga, H. C. Ong, T. M. I. Mahlia, H. H. Masjuki, Irfan Anjum Badruddin, and H. Fayaz. 2013. "Non-Edible Vegetable Oils: A Critical Evaluation of Oil Extraction, Fatty Acid Compositions, Biodiesel Production, Characteristics, Engine Performance and Emissions Production." *Renewable and Sustainable Energy Reviews* 18. Elsevier: 211–45. doi:10.1016/j.rser.2012.10.013.

Atabani, A. E., A. S. Silitonga, Irfan Anjum Badruddin, T. M. I. Mahlia, H. H. Masjuki, and S. Mekhilef. 2012. "A Comprehensive Review on Biodiesel as an Alternative Energy Resource and Its Characteristics." *Renewable and Sustainable Energy Reviews* 16 (4). Elsevier Ltd: 2070–93. doi:10.1016/j.rser.2012.01.003.

Azam, M. Mohibbe, Amtul Waris, and N. M. Nahar. 2005. "Prospects and Potential of Fatty Acid Methyl Esters of Some Non-Traditional Seed Oils for Use as Biodiesel in India." *Biomass and Bioenergy* 29 (4): 293–302. doi:10.1016/j.biombioe.2005.05.001.

Banković-Ilić, Ivana B., Olivera S. Stamenković, and Vlada B. Veljković. 2012. "Biodiesel Production from Non-Edible Plant Oils." *Renewable and Sustainable Energy Reviews.* doi:10.1016/j.rser.2012.03.002.

Bhuiya, M. M. K., M. G. Rasul, M. M. K. Khan, N. Ashwath, and A. K. Azad. 2016. "Prospects of 2nd Generation Biodiesel as a Sustainable Fuel – Part: 1 Selection of Feedstocks, Oil Extraction Techniques and Conversion Technologies." *Renewable and Sustainable Energy Reviews* 55. Elsevier: 1109–28. doi:10.1016/j.rser.2015.04.163.

Bose, Bimal K. 2010. "Global Warming: Energy, Environmental Pollution, and the Impact of Power Electronics." *IEEE Industrial Electronics Magazine.* doi:10.1109/MIE.2010.935860.

Bouaid, A., L. Bajo, M. Martinez, and J. Aracil. 2007. "Optimization of Biodiesel Production from Jojoba Oil." *Process Safety and Environmental Protection.* doi:10.1205/psep07004.

*BP Statistical Review of World Energy.* 2019. https://www.bp.com/content/dam/bp/busine ss-sites/en/global/corporate/pdfs/energy-economics/statistical-review/bp-stats-review -2019-full-report.pdf.

Bristow, Abigail L., Miles Tight, Alison Pridmore, and Anthony D. May. 2008. "Developing Pathways to Low Carbon Land-Based Passenger Transport in Great Britain by 2050." *Energy Policy.* doi:10.1016/j.enpol.2008.04.029.

Canoira, Laureano, Ramón Alcántara, Ma Jesús García-Martínez, and Jesús Carrasco. 2006. "Biodiesel from Jojoba Oil-Wax: Transesterification with Methanol and Properties as a Fuel." *Biomass and Bioenergy.* doi:10.1016/j.biombioe.2005.07.002.

*Coal – Simple English Wikipedia, the Free Encyclopedia.* 2020. Accessed February 26. https://simple.wikipedia.org/wiki/Coal.

Creutzig, Felix, Patrick Jochem, Linus Mattauch Oreane Y. Edelenbosch, David McCollum Detlef P. van Vuuren, and Jan Minx. 2015. "Transport: A Roadblock to Climate Change Mitigation?" *Science* 350 (6263): 911–12.

Dadak, Ali, Mortaza Aghbashlo, Meisam Tabatabaei, Habibollah Younesi, and Ghasem Najafpour. 2016. "Exergy-Based Sustainability Assessment of Continuous Photobiological Hydrogen Production Using Anaerobic Bacterium Rhodospirillum Rubrum." *Journal of Cleaner Production.* doi:10.1016/j.jclepro.2016.08.020.

Demirbas, A. 2009. "Potential Resources of Non-Edible Oils for Biodiesel." *Energy Sources, Part B: Economics, Planning and Policy.* doi:10.1080/15567240701621166.

Demirbas, Ayhan. 2009. "Progress and Recent Trends in Biodiesel Fuels." *Energy Conversion and Management* 50 (1): 14–34. doi:10.1016/j.enconman.2008.09.001.

Dhar, Subash, Minal Pathak, and Priyadarshi R. Shukla. 2017. "Electric Vehicles and India's Low Carbon Passenger Transport: A Long-Term Co-Benefits Assessment." *Journal of Cleaner Production.* doi:10.1016/j.jclepro.2016.05.111.

Edelenbosch, O. Y., D. L. McCollum, D. P. van Vuuren, C. Bertram, S. Carrara, H. Daly, S. Fujimori, et al. 2017. "Decomposing Passenger Transport Futures: Comparing Results of Global Integrated Assessment Models." *Transportation Research Part D: Transport and Environment.* doi:10.1016/j.trd.2016.07.003.

Erb, Karl Heinz, Helmut Haberl, and Christoph Plutzar. 2012. "Dependency of Global Primary Bioenergy Crop Potentials in 2050 on Food Systems, Yields, Biodiversity Conservation and Political Stability." *Energy Policy.* doi:10.1016/j.enpol.2012.04.066.

Hajjari, Masoumeh, Meisam Tabatabaei, Mortaza Aghbashlo, and Hossein Ghanavati. 2017. "A Review on the Prospects of Sustainable Biodiesel Production: A Global Scenario with an Emphasis on Waste-Oil Biodiesel Utilization." *Renewable and Sustainable Energy Reviews* 72 (November 2016). Elsevier Ltd: 445–64. doi:10.1016/j.rser.2017.01.034.

Hathurusingha, Subhash, Nanjappa Ashwath, and Phul Subedi. 2011. "Variation in Oil Content and Fatty Acid Profile of Calophyllum Inophyllum L. with Fruit Maturity and Its Implications on Resultant Biodiesel Quality." *Industrial Crops and Products.* doi:10.1016/j.indcrop.2010.12.026.

*History of Coal Mining – Wikipedia.* 2020. Accessed February 26. https://en.wikipedia.org/wiki/History_of_coal_mining.

*History Of Crude Oil | Events That Drove The Oil Price History | IG UK.* 2020. Accessed February 26. https://www.ig.com/uk/commodities/oil/history-of-crude-oil-price.

*History of Natural Gas.* 2020. Accessed February 26. http://naturalgas.org/overview/history/.

Hoogwijk, Monique, André Faaij, Bas Eickhout, Bert De Vries, and Wim Turkenburg. 2005. "Potential of Biomass Energy out to 2100, for Four IPCC SRES Land-Use Scenarios." *Biomass and Bioenergy.* doi:10.1016/j.biombioe.2005.05.002.

Hosseini, Seyed Sina, Mortaza Aghbashlo, Meisam Tabatabaei, Ghasem Najafpour, and Habibollah Younesi. 2015. "Thermodynamic Evaluation of a Photobioreactor for Hydrogen Production from Syngas via a Locally Isolated Rhodopseudomonas Palustris PT." *International Journal of Hydrogen Energy.* doi:10.1016/j.ijhydene.2015.08.092.

Institute of Energy Economics, Japan. 2018. "IEEJ Outlook – Prospects and Challenges Until 2050 – Energy, Environment and Economy".

International Energy Agency. 2010. *Energy Technology Perspectives 2010: Scenarios & Strategies to 2050. Strategies.* doi:10.1049/et:20060114.

*International Energy Outlook.* 2019. doi:10.1080/01636609609550217.

Jeffery, Brian R. 2020. *Energy – Petrol and Diesel | Young People's Trust For the Environment.* Accessed February 26. https://ypte.org.uk/factsheets/energy/petrol-and-diesel?hide_donation_prompt=1.

Jena, Prakash C., Hifjur Raheman, G. V. Prasanna Kumar, and Rajendra Machavaram. 2010. "Biodiesel Production from Mixture of Mahua and Simarouba Oils with High Free Fatty Acids." *Biomass and Bioenergy* 34 (8). Elsevier Ltd: 1108–16. doi:10.1016/j.biombioe.2010.02.019.

Kannan, Ramachandran, and Stefan Hirschberg. 2016. "Interplay between Electricity and Transport Sectors – Integrating the Swiss Car Fleet and Electricity System." *Transportation Research Part A: Policy and Practice.* doi:10.1016/j.tra.2016.10.007.

Karmakar, Anindita, Subrata Karmakar, and Souti Mukherjee. 2012. "Biodiesel Production from Neem towards Feedstock Diversification: Indian Perspective." *Renewable and Sustainable Energy Reviews* 16 (1). Elsevier Ltd: 1050–60. doi:10.1016/j.rser.2011.10.001.

Karmee, Sanjib Kumar, and Anju Chadha. 2005. "Preparation of Biodiesel from Crude Oil of Pongamia Pinnata." *Bioresource Technology* 96 (13): 1425–29. doi:10.1016/j.biortech.2004.12.011.

Kibazohi, O., and R. S. Sangwan. 2011. "Vegetable Oil Production Potential from Jatropha Curcas, Croton Megalocarpus, Aleurites Moluccana, Moringa Oleifera and Pachira Glabra: Assessment of Renewable Energy Resources for Bio-Energy Production in Africa." *Biomass and Bioenergy* 35 (3): 1352–56. doi:10.1016/j.biombioe.2010.12.048.

Kukreti, Ishan. 2018. "Tree-Borne Oilseeds Can Reduce India's Import Dependency." *Down To Earth.* https://www.downtoearth.org.in/news/forests/oil-grows-on-trees-60255.

Kumar, Ashwani, and Satyawati Sharma. 2008. "An Evaluation of Multipurpose Oil Seed Crop for Industrial Uses (Jatropha Curcas L.): A Review." *Industrial Crops and Products.* doi:10.1016/j.indcrop.2008.01.001.

Kumar, Ashwani, and Satyawati Sharma. 2011. "Potential Non-Edible Oil Resources as Biodiesel Feedstock: An Indian Perspective." *Renewable and Sustainable Energy Reviews* 15 (4). Elsevier Ltd: 1791–1800. doi:10.1016/j.rser.2010.11.020.

Kumar, Ashwani, Satyawati Sharma, and Saroj Mishra. 2010. "Influence of Arbuscular Mycorrhizal (AM) Fungi and Salinity on Seedling Growth, Solute Accumulation, and Mycorrhizal Dependency of Jatropha Curcas L." *Journal of Plant Growth Regulation* 29 (3): 297–306. doi:10.1007/s00344-009-9136-1.

*List of Fossil Fuels | Environment.* 2020. Accessed February 27. http://www.environmentalpollution.in/environment/list-of-fossil-fuels-environment/3509.

Long, Huiling, Xiaobing Li, Hong Wang, and Jingdun Jia. 2013. "Biomass Resources and Their Bioenergy Potential Estimation: A Review." *Renewable and Sustainable Energy Reviews* 26. Elsevier: 344–52. doi:10.1016/j.rser.2013.05.035.

Loo, Becky P. Y., and David Banister. 2016. "Decoupling Transport from Economic Growth: Extending the Debate to Include Environmental and Social Externalities." *Journal of Transport Geography.* doi:10.1016/j.jtrangeo.2016.10.006.

McCollum, David, and Christopher Yang. 2009. "Achieving Deep Reductions in US Transport Greenhouse Gas Emissions: Scenario Analysis and Policy Implications." *Energy Policy.* doi:10.1016/j.enpol.2009.08.038.

*Monthly Biodiesel Production Report.* 2019.

Mukta, N., I. Y. L. N. Murthy, and P. Sripal. 2009. "Variability Assessment in Pongamia Pinnata (L.) Pierre Germplasm for Biodiesel Traits." *Industrial Crops and Products.* doi:10.1016/j.indcrop.2008.10.002.

*Natural Gas – Wikipedia.* 2020. Accessed February 26. https://en.wikipedia.org/wiki/Natur al_gas.

Neves, Sónia Almeida, António Cardoso Marques, and José Alberto Fuinhas. 2018. "Could Alternative Energy Sources in the Transport Sector Decarbonise the Economy without Compromising Economic Growth?" *Environment, Development and Sustainability.* doi:10.1007/s10668-018-0153-8.

No, Soo Young. 2011. "Inedible Vegetable Oils and Their Derivatives for Alternative Diesel Fuels in CI Engines: A Review." *Renewable and Sustainable Energy Reviews* 15 (1). Elsevier Ltd: 131–49. doi:10.1016/j.rser.2010.08.012.

*Nuclear Power in India – Wikipedia.* 2020. Accessed February 27. https://en.wikipedia.org/ wiki/Nuclear_power_in_India.

*Petroleum – Wikipedia.* 2020. Accessed February 26. https://en.wikipedia.org/wiki/ Petroleum.

Philalay, Monica, Nikolai Drahos, and David Thurtell. 2019. "Coal in India 2019." *Commonwealth of Australia.* doi:10.1017/CBO9781107415324.004.

Pinzi, S., I. L. Garcia, F. J. Lopez-Gimenez, M. D. Luque DeCastro, G. Dorado, and M. P. Dorado. 2009. "The Ideal Vegetable Oil-Based Biodiesel Composition: A Review of Social, Economical and Technical Implications." *Energy and Fuels.* doi:10.1021/ ef801098a.

Reddy, M. Sarveshwar, Nikhil Sharma, and Avinash Kumar Agarwal. 2016. "Effect of Straight Vegetable Oil Blends and Biodiesel Blends on Wear of Mechanical Fuel Injection Equipment of a Constant Speed Diesel Engine." *Renewable Energy* 99 (December 2017). Elsevier Ltd: 1008–18. doi:10.1016/j.renene.2016.07.072.

*Renewables Global Status Report.* 2019. https://wedocs.unep.org/bitstream/handle/20.500.11 822/28496/REN2019.pdf?sequence=1&isAllowed=y%0Ahttp://www.ren21.net/cities/ wp-content/uploads/2019/05/REC-GSR-Low-Res.pdf.

Roschat, Wuttichai, Theeranun Siritanon, Boonyawan Yoosuk, Taweesak Sudyoadsuk, and Vinich Promarak. 2017. "Rubber Seed Oil as Potential Non-Edible Feedstock for Biodiesel Production Using Heterogeneous Catalyst in Thailand." *Renewable Energy* 101. Elsevier Ltd: 937–44. doi:10.1016/j.renene.2016.09.057.

Schlissel, David, Seth Feaster, and Dennis Wamstead. 2019. *Coal Outlook 2019* (March): 1–30. http://ieefa.org/wp-content/uploads/2019/03/Coal-Outlook-2019_March-2019.pdf.

Silitonga, A. S., H. H. Masjuki, T. M. I. Mahlia, H. C. Ong, W. T. Chong, and M. H. Boosroh. 2013. "Overview Properties of Biodiesel Diesel Blends from Edible and Non-Edible Feedstock." *Renewable and Sustainable Energy Reviews* 22. Elsevier: 346–60. doi:10.1016/j.rser.2013.01.055.

Singh, Digambar, Dilip Sharma, S. L. Soni, Sumit Sharma, Pushpendra Kumar Sharma, and Amit Jhalani. 2020. "A Review on Feedstocks, Production Processes, and Yield for Different Generations of Biodiesel." *Fuel* 262 (February). Elsevier: 116553. doi:10.1016/J.FUEL.2019.116553.

Singh, Yashvir, Abid Farooq, Aamir Raza, Muhammad Arif Mahmood, and Surbhi Jain. 2017. "Sustainability of a Non-Edible Vegetable Oil Based Bio-Lubricant for Automotive Applications: A Review." *Process Safety and Environmental Protection* 111. Institution of Chemical Engineers: 701–13. doi:10.1016/j.psep.2017.08.041.

Solomon, Susan, Gian Kasper Plattner, Reto Knutti, and Pierre Friedlingstein. 2009. "Irreversible Climate Change Due to Carbon Dioxide Emissions." *Proceedings of the National Academy of Sciences of the United States of America* 106 (6): 1704–09. doi:10.1073/pnas.0812721106.

Tariq, Muhammad, Saqib Ali, and Nasir Khalid. 2012. "Activity of Homogeneous and Heterogeneous Catalysts, Spectroscopic and Chromatographic Characterization of Biodiesel: A Review." *Renewable and Sustainable Energy Reviews* 16: 6303–16.

Tomes, Dwight, Prakash Lakshmanan, and David Songstad. 2011. *Biofuels: Global Impact on Renewable Energy, Production Agriculture, and Technological Advancements. Biofuels: Global Impact on Renewable Energy, Production Agriculture, and Technological Advancements.* doi:10.1007/978-1-4419-7145-6.

*UFOP Report on Global Market Supply 2017/2018.* 2019.

Ullah, Kifayat, Vinod Kumar Sharma, Sunil Dhingra, Giacobbe Braccio, Mushtaq Ahmad, and Sofia Sofia. 2015. "Assessing the Lignocellulosic Biomass Resources Potential in Developing Countries: A Critical Review." *Renewable and Sustainable Energy Reviews* 51: 682–98. doi:10.1016/j.rser.2015.06.044.

*U.S. EIA Monthly Energy Review.* 2019. Vol. 24. doi:10.1044/leader.ppl.24062019.28.

Venkanna, B. K., and C. Venkataramana Reddy. 2009. "Biodiesel Production and Optimization from Calophyllum Inophyllum Linn Oil (Honne Oil) – A Three Stage Method." *Bioresource Technology.* doi:10.1016/j.biortech.2009.05.023.

*World Uranium Mining* – World Nuclear Association. 2020. Accessed February 27. https://www.world-nuclear.org/information-library/nuclear-fuel-cycle/mining-of-uranium/world-uranium-mining-production.aspx.

Wu, Lian, Tengyou Wei, Zijun Lin, Yun Zou, Zhangfa Tong, and Jianhua Sun. 2016. "Bentonite-Enhanced Biodiesel Production by NaOH-Catalyzed Transesterification: Process Optimization and Kinetics and Thermodynamic Analysis." *Fuel* 182 (May). Elsevier Ltd: 920–27. doi:10.1016/j.fuel.2016.05.065.

# 2 Biodiesel Production Techniques – The State of the Art

## 2.1 INTRODUCTION

The increasing everyday demand for energy and the threat posed by the crisis linked with the depletion of fossil fuels have made researchers all over the world switch over to alternative sources of energy such as biofuels. Biodiesels, which are alkyl esters of long-chain fatty acids derived from lipids such as animal fats, vegetable oils, alcohols, etc., have properties similar to those of diesel and can be a potential replacement for diesel in the near future. They are known as biodiesels because they are biologically renewable, contrary to conventional diesel, which is non-renewable (Ahmad et al. 2011; Kafuku and Mbarawa 2010; Atabani and César 2014; Canakci and Sanli 2008). Biodiesels are considered as an eco-friendly and sustainable source of energy to reduce dependence on conventional fossil fuels (Kumar and Ali 2013). The demand for fossil fuels, especially high-speed diesel, increases every year at a very steady rate (Mofijur et al. 2013).

Apart from the non-renewability factor of fossil fuels, regular usage of fossil energy has contributed to 25% of the world's toxic emissions and the greenhouse effect. Since the industrial revolution, many scientists all over the world have presented pieces of evidence about the abnormal increase in the temperature of earth registered due to human actions. This is because of high levels of $CO_2$ in the atmosphere due to the burning of fossil fuels. As people are getting more aware of pollution-free fuels and ongoing climate change, there is a rising demand for sustainable production of energy (Sadeghinezhad et al. 2013; Mat Yasin et al. 2017; Tarabet et al. 2014; Varatharajan, Cheralathan, and Velraj 2011; Utlu and Koçak 2008; Kojima, Imazu, and Nishida 2014; Millo et al. 2015; Armas et al. 2005; Maiboom and Tauzia 2011). While biodiesel is not known to suffer from any drawbacks, elements of the process behind producing the feedstock, such as the agricultural input in the form of raw materials and the changes in the land being used for growing the feedstock, have been known to produce ill effects due to the usage of artificial fertilizers and chemicals.

Experimental studies have proved that direct usage of petroleum diesel also causes similar ill effects to the environment. The usage of fertilizers and chemicals can cause harm to waterways in the form of nitrification. Though biodiesel has direct advantages over conventional diesel for a local environment, it does have many drawbacks from the global environment point of view (Devi, Das, and Deka 2017). Biodiesel is a promising fuel for the future due to its extraordinary behavior when compared to other alternatives like non-toxicity, renewability, eco-friendliness, inherent lubricity, high flash and fire point, high cetane number rating, and reduced emissions (Rashid, Anwar, Moser, and Knothe 2008). The entire biodiesel feedstocks at present cannot produce a sufficient amount of biodiesel to meet the global demand. Hence it can be blended with diesel at any proportion and can be used in diesel engines without any modification (Shahabuddin et al. 2012; Jain and Sharma 2011). Biodiesel production is promoted all over the world because of the high availability of a wide range of feedstocks in the world (Janaun and Ellis 2010; Atadashi, Aroua, and Aziz 2010). Around 300 crops have been identified as potential feedstocks for the synthesis of biodiesel (Karmakar, Karmakar, and Mukherjee 2010; Atabani et al. 2012). It has been found that the cost of feedstock constitutes 75% of the entire biodiesel production cost. Hence selection of feedstock is essential and has to be economical to take care of the entire biodiesel cost production. Edible oils have been considered as the primary feedstocks for biodiesel production for a very long time. Nevertheless, this has raised a lot of concerns related to the food crisis due to the usage of these edible feedstocks for biodiesel production. This has also marked a rise in deforestation and destruction of important arable lands with fertile soil. As a result of these, edible oil prices have gone up tremendously (Balat and Balat 2010; Atabani et al. 2013, 2012; Lin et al. 2011; Lim and Teong 2010).

The global production of biodiesels has increased at a steady rate every year. Apart from transportation fuel, biodiesel has several other uses, but it becomes a problem when quantity becomes huge. It can be used as heating oil, lubricant, power-generating oil, solvents in various applications, and a remedy for oil spills (Fernández-Álvarez et al. 2007; Ng et al. 2015; DeMello et al. 2007; Mudge and Pereira 1999; Mudge 1999; Prince, Haitmanek, and Lee 2008). The feasibility of biodiesel production has always been a critical factor in deciding its superiority over petroleum diesel. New innovative technologies used for synthesizing biodiesels are adopted at a much slower rate all over the globe due to the high investment of capital. On the other hand, the limited availability of fossil fuels has reduced the affordability of these fuels for a normal person. Thus, one cannot completely say that biodiesel is a perfect alternative for petroleum diesel because it is yet to be widely accepted. Thus biodiesel, despite being adopted as a sustainable source of energy, has yet to be put into practical implementation in many places (Knothe and Razon 2017).

## 2.2 CATALYST IN TRANSESTERIFICATION

Transesterification is the best known and most widely used technology to convert triglycerides into biodiesel. Compared to other technologies, it is a cost-effective and promising method to reduce the viscosity of vegetable oil. Besides, due to its higher conversion efficiency in a shorter time and low cost, this technique is commonly used for industrial biodiesel production (Bhuiya et al. 2016). The transesterification process requires a catalyst to reduce the kinetic energy and hasten the reaction forward. Usually, homogeneous or heterogeneous catalysts are employed to break down the triglycerides into esters. Moreover, many assisted catalytic techniques are available to drive the transesterification reaction, namely (i) conventional heating with mechanical stirring, (ii) ultrasound irradiation, and (iii) microwave irradiation (Jain et al. 2018). The overall transesterification reaction mechanism is presented in Figure 2.1.

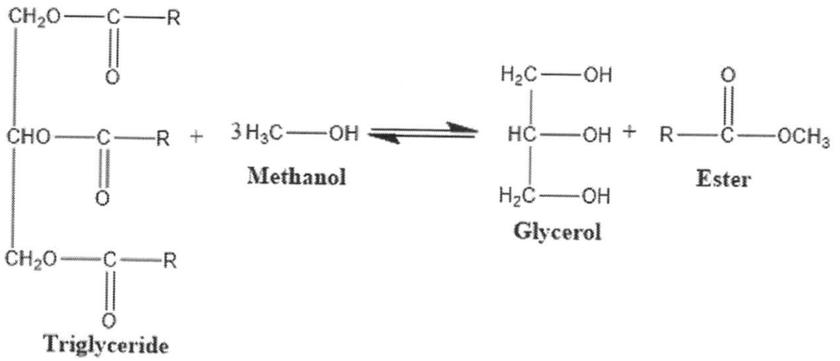

**FIGURE 2.1** The overall transesterification reaction mechanism.

## 2.2.1 HOMOGENEOUS CATALYST

### 2.2.1.1 Acid Catalyst

Esterification of soybean oil with hydrochloric acid (HCl) gives 98.19% of ester yield at an optimum condition of oil to methanol molar ratio of 1:7.9 and a reaction temperature of 77°C for 104 min. The acid-catalyzed transesterification of beef tallow using an HCl catalyst concentration of 0.5 wt.% and a methanol to beef tallow volume ratio of 6:1 for a 1.5-h reaction time at 60°C yields 96.30% of biodiesel yield (Ehiri, Ikelle, and Ozoaku 2014). Sunflower oil acid-catalyzed transesterification with 1.5 wt.% of HCl and a reaction temperature of 100°C produced a 95.2% biodiesel yield (Sagiroglu et al. 2011). An optimum biodiesel yield of 92.5% was obtained by acid-catalyzed transesterification of *Chlorella pyrenoidosa* oil using $H_2SO_4$ catalyst with a concentration of 0.5 wt.% (Cao et al. 2013). The transesterification of jatropha oil with $H_2SO_4$ catalyst resulted in a 99% biodiesel yield with a methanol to oil ratio of 0.28 by volume, a reaction temperature of 60°C, and a catalyst concentration of 1.4 v/v for an 88-min reaction time (Sarin et al. 2007). A maximum of 99.7% of oleic methyl ester was obtained with an $H_2SO_4$ catalyst concentration of 1 wt.% and a methanol to oil molar ratio of 1:3 at 100°C (Lucena, Silva, and Fernandes 2008).

Homogeneous acid catalyst has advantages over homogeneous alkaline catalysts, like the production of biodiesel from low-grade feedstock which has a higher amount of free fatty acid (FFA) than the EN 14104 limits. Moreover, the acid catalyst is not sensitive to water content in the feedstock, hence there is a chance to produce soap in reaction (Thanh et al. 2012). On either side, the homogeneous acid catalyst needs an elevated reaction temperature, a higher methanol to oil ratio, and a longer reaction time for conversion of feedstock into biodiesel than a homogeneous alkaline catalyst (Tariq, Ali, and Khalid 2012). An acid catalyst has a smaller number of active sites and fewer micropores than a base catalyst, which leads to a slow reaction rate in biodiesel production using a homogeneous acid catalyst (Dias, Alvim-Ferraz, and Almeida 2008).

### 2.2.1.2 Base Catalyst

Alkali homogeneous catalysts hasten the transesterification reaction by 4000 times as compared to acid catalyst. Among homogeneous catalysts, potassium hydroxide, sodium hydroxide, and sodium methoxide are the most widely used homogeneous catalysts to produce biodiesel, at lower costs and with excellent catalytic activity (Buasri et al. 2012; Fabiano et al. 2010). Figure 2.2 illustrates the reaction mechanism of alkali homogenous catalyzed transesterification. The optimum biodiesel yield of 98.2% was achieved from waste cooking oil using a KOH catalyst of 1 wt.% and an oil to methanol ratio 1:6 for 1 h of reaction time at 70°C (Agarwal et al. 2012). Thanh et al. (2014) achieved 98.1% of fatty acid methyl esters with a KOH catalyst concentration of 1 wt.%, and a 1:6 ratio of oil-methanol for 25 min by an ultrasonic sonicator method (Thanh et al. 2014). Microwave-assisted transesterification of castor oil with a KOH alkaline catalyst concentration of 1.4 wt.%, an 8 w/w ethanol to oil molar ratio, 275 W power, and 7 min of reaction time gives 95% biodiesel yield (Thirugnanasambandham et al. 2016). Rapeseed oil methanolysis with a KOH alkaline catalyst amount of 0.89–1.33 wt.% gives better yield at a temperature range of 45 to 75°C, a methanol to oil molar ratio of 1:6 to 1:7.5, and a reaction time of 3 to 5 h (Hájek et al. 2012). Maximum biodiesel was obtained by the transesterification of sewage sludge with a KOH catalyst amount of 1.5% (Wu et al. 2017). An optimum biodiesel yield of 95% was achieved from transesterification of cottonseed oil with an NaOH amount of 0.75 wt.% and a 7:1 alcohol to oil molar ratio at 110°C (Shankar, Pentapati, and Prasad 2017).

Rapeseed oil transesterification with an NaOH catalyst concentration 1.34 wt.% and an oil to methanol molar ratio of 1:6.9 gives 98% of biodiesel yield (Gu et al.

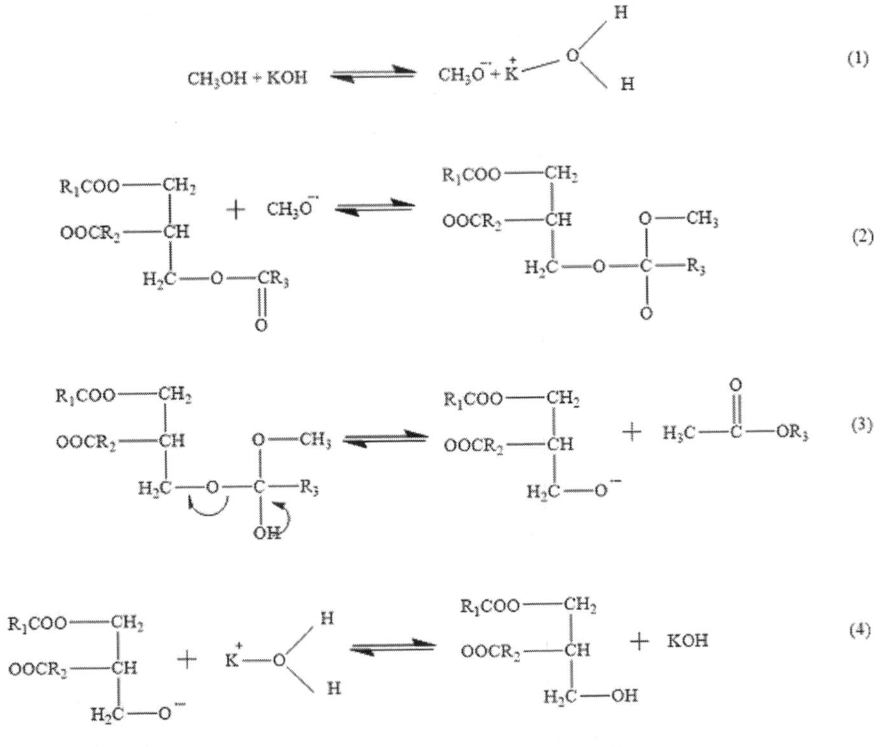

**FIGURE 2.2** The reaction mechanism of alkali homogenous catalyzed transesterification.

2015). Alkaline sodium methoxide homogeneous catalyst has a high number of active site and gives a higher yield of biodiesel 98% by methanolysis of *Jatropha curcas* oil with a 1 wt.% catalyst, an oil to methanol ratio of 1:9, a stirring speed of 1000 rpm, and 30 min of reaction time at 50°C (Silitonga et al. 2013). The transesterification of *Sesamum indicum* L. seed oil gives 87.8% of biodiesel yield with 0.75 wt.% of sodium methoxide catalyst concentration and an oil to methanol molar ratio of 1:6 (Dawodu, Ayodele, and Bolanle-Ojo 2014). Using potassium methoxide homogeneous alkaline catalyst obtained 98.46% yield of methyl esters from transesterification of sunflower oil with 1 wt.% catalyst concentration, 65°C reaction temperature, and a 1:6 oil to methanol molar ratio (Rashid, Anwar, Moser, and Ashraf 2008). The highest biodiesel yield, 98.9%, was obtained from soybean oil by using a potassium methoxide alkaline catalyst concentration of 1 wt.% and a 1:6 molar ratio of oil-methanol at 60°C (Saydut et al. 2016).

Saponification of triglycerides or esters is less with a metal methoxide catalyst, such as potassium methoxide ($CH_3OK$), and sodium methoxide ($CH_3ONa$), than it is with a metal hydroxide (KOH, NaOH) catalyst. During the reaction, the metal alkoxide catalyst is converted into $CH3O^-$ and $K^+$ or $Na^+$, thereby giving no chance for water formation (Atadashi et al. 2013). Vicente et al. studied a comparative analysis of four different homogeneous alkaline catalysts, namely KOH, NaOH, $CH_3OK$, and $CH_3ONa$. The highest biodiesel yields, of 99.33% and 98.46%, were obtained for $CH_3OK$ and $CH_3ONa$ catalyst (Vicente, Martínez, and Aracil 2004). Homogeneous alkaline catalyst has many advantages like a high reaction rate, shorter period of conversion, and lower reaction temperature required for transesterification (Lam, Lee, and Mohamed 2010).

However, homogenous catalysts have many drawbacks over heterogeneous catalysts. Homogeneous catalysts do not work with oils with higher FFA content. Use of a homogeneous alkali catalyst for transesterification of high-FFA oil leads to soap formation. The soap formation is illustrated in Figure 2.3. Separation of homogenous catalysts is difficult, and the process generates a large amount of wastewater. Hence, the production process is costly and ultimately difficult (Mansir et al. 2018). In order to overcome the issues related to the homogeneous catalyst, there is a need for heterogeneous catalysts. Heterogeneous catalysts have some advantages over homogeneous catalysts like reusability, less toxicity, a decrease in the soaping problem, ease

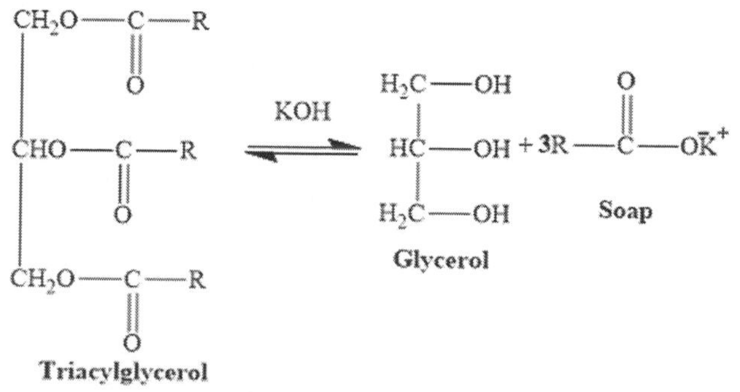

**FIGURE 2.3**   Formation of soap during transesterification process.

of separation, and a decrease in water usage (Yan et al. 2010; Bournay et al. 2005; DaSilveira Neto et al. 2007; López Granados et al. 2010; Boffito et al. 2013).

### 2.2.2 Heterogeneous Catalyst

The heterogeneous catalyst, which usually exists in solid form, acts in the liquid reaction mixture in another phase that is in contrast to the homogeneous catalyst phase. The homogeneous catalysts are widely used in industrial-scale biodiesel production due to high abundance, shorter processing time, and higher reaction rate (Borges and Díaz 2012). Moreover, there are some issues with the use of homogeneous catalysts during the esterification and transesterification processes. The homogeneous acid–catalyzed ($H_2SO_4$) esterification results in a significant amount of sulfur in the biodiesel product, since as per the standard, the maximum limit for sulfur is about 10 ppm. Further product separation, removal of catalyst from biodiesel, and washing of the product with an acid neutralization step that can generate large amounts of residual water are required for homogeneous alkaline catalysts. An unfavorable aqueous emulsion is formed while washing the products, which makes the separation of the glycerol ester process complicated. The wastewater generated as a by-product often needs further treatment, adding an extra cost of production. When using a heterogeneous catalyst, processing costs are reduced because the catalyst can be separated from the liquid phase quickly and reused for a higher number of cycles.

Besides, the side saponification reaction can be prevented, so no further separation process is required (Meher et al. 2013). The general biodiesel production process conditions are also less severe, with minimum energy consumption and less corrosivity and toxicity, since it requires a longer time to complete (Leung, Wu, and Leung 2010). In recent days, scientists have successfully synthesized the acid and base heterogeneous catalysts to produce biodiesel. The primary requirements of a heterogeneous catalyst are (i) high porosity, (ii) huge surface area, (iii) high concentration of basic/acidic sites, (iv) low leaching and deactivation tendencies (hydrophobic), (v) low cost, (vi) environmental friendliness, and (vii) abundance/ready availability.

#### 2.2.2.1 Acid Catalyst

The heterogeneous acid catalysts are a promising alternative substitute for the esterification of oils with higher FFA content. Moreover, these catalysts require mild conditions to reduce the FFA content as compared to the homogeneous acid catalyst. The heterogeneous acid catalyst has beneficial effects on the environment and the biodiesel produced meets the specifications required by ASTM (Meher et al. 2013). Moreover, the heterogeneous acid catalysts have weak catalytic activity that needs elevated reaction temperatures and more time to achieve higher biodiesel performance (Mardhiah et al. 2017).

Heterogeneous base catalyst produces a higher yield of biodiesel when oil has less FFA content. FFA and water content in oil are responsible for soap formation in heterogeneous base-catalyzed transesterification because heterogeneous base catalysts are highly sensitive towards FFA and water contained in oil (Liu et al. 2008). The heterogeneous base catalyst synthesis process is costly, complicated, time-consuming, and difficult (Leung, Wu, and Leung 2010). Purification, separation, and cleaning of heterogeneous base catalysts and biodiesel require a large amount of water. Leaching of catalyst takes place more in heterogeneous base-catalyzed transesterification than in heterogeneous acid-catalyzed transesterification (Kawashima, Matsubara, and Honda 2009b). Biodiesel yield does not depend on the water and FFA contains heterogeneous acid-catalyzed transesterification because the heterogeneous acid catalyst is not sensitive toward FFA and water content in oil (Tariq, Ali, and Khalid 2012). Heterogeneous acid catalysts are non-corrosive to system parts. By using heterogeneous acid catalyst, *in situ* transesterification of oils are possible, which reduces the process time of biodiesel production (Helwani et al. 2009). In biodiesel production, the most-used heterogeneous acid catalysts are zirconium oxide ($ZrO_2$), zinc oxide ($ZnO_2$), titanium oxide (TiO), sulfonic modified mesostructured silica, ion-exchange resin, sulfonated carbon-based catalyst, HPA, and zeolites (Abdullah et al. 2017).

Transesterification of palm oil with zirconium oxide catalyst with a condition reaction temperature of 200°C, a 350 rpm stirring sped, a 6:1 methanol to oil molar ratio, and a 1 h reaction time gives a 64.2% conversion yield of methyl ester. The transesterification of coconut oil in the same condition with the same catalyst gives a 49.3% conversion yield of methyl ester. The catalytic activity of zirconium oxide increases by sulfonating, and sulfonated zirconium oxide catalyst gives higher yield, 90.3% and 86.3% for palm oil and coconut oil, respectively, in the same reaction condition (Jitputti et al. 2006). Soybean oil transesterification gives more than 66% yield using sulfonated zirconium oxide catalyst, which was calcined at 675°C for 1.5 h in the following reaction condition: reaction time of 4 h, 4 g of catalyst loading, 1:40 oil to methanol molar ratio, and reaction temperature of 175°C. Sulfonated zirconium oxide is more acidic than 100% concentrated $H_2SO_4$ (Furuta, Matsuhashi, and Arata 2004). The production of biodiesel from sunflower oil with a $ZrO_2$-supported $La_2O_3$ catalyst gives 84.9% biodiesel yield in the following reaction conditions: 3:1 methanol to oil molar ratio, 5 h reaction time, 60°C reaction temperature, and 2 wt.% of oil catalyst loading (Antunes, Veloso, and Henriques 2008). Soybean oil transesterification gives 98.6% biodiesel yield by using S–$ZrO_2$ catalyst with 20:1 methanol to oil molar ratio, reaction temperature 120°C, 1 h reaction time, and 5 wt.% of oil catalyst loading (Yang and Xie 2007). Transesterification of purified palm oil with sulfonated $ZrO_2$ catalyst gives 90% biodiesel yield in the following reaction condition: time 10 minutes, 1:25 oil to methanol molar ratio, reaction temperature 250°C, and 0.5 wt.% of oil catalyst loading (Petchmala et al. 2010). Transesterification of sunflower oil with $SO_4^{2-}$/$ZrO_2$/$SiO_2$ catalyst gives 91.5% biodiesel yield in following reaction condition: time 6 h, 1:12 oil to methanol molar ratio, reaction temperature 200°C, and 14. 6 wt.% of oil catalyst loading (Jiménez-Morales et al. 2011).

Soybean oil transesterification gives 82% biodiesel yield by using the ZnO catalyst in the following reaction condition: a 55:1 methanol to oil molar ratio and a 7 h processing time at 100°C (Antunes, Veloso, and Henriques 2008).Transesterification of palm oil with zinc oxide (ZnO) catalyst gives 86.1% conversion yield of methyl ester in the following condition: reaction temperature 300°C, 6:1 methanol to oil molar ratio, reaction time 1 h, and stirring speed of 350 rpm. Transesterification of coconut oil in the same condition with the same catalyst gives 77.5% conversion yield of methyl ester (Jitputti et al. 2006). Cottonseed oil transesterification gives more than 85% yield using sulfonated zinc oxide ($ZnO/SO_4^-$) catalyst in the following reaction condition: reaction time 8 h and 1:12 oil to methanol molar ratio. Sulfonated ZnO gives a better yield of conversion than pure ZnO (Vicente, Martínez, and Aracil 2007).

### 2.2.2.2 Base Catalyst

Low reaction conditions are enough for the solid base heterogeneous catalyst transesterification reaction, in contrast to the solid acid heterogeneous catalyzed reaction. An economical solid base heterogeneous catalyst can be prepared from waste material available from industries and their surroundings. Awareness of research into waste materials may help develop solid base heterogeneous catalysts for biodiesel production in sustainable and environmentally friendly ways (Marwaha et al. 2018). The recent trend shows the use of natural biological sources such as calcium and carbon for the preparation of solid catalyst supports and basic catalysts to produce biodiesel. According to Smith and Notheisz (1999), the performance of the catalyst depends on many things, such as high surface area, pore volume, pore size, and active site distribution. Metal oxide structure consists of positive metal ions (cations) that exhibit Lewis acidity, i.e., these act as electron acceptors, and as negative oxygen ions (anions) that act as proton acceptors. Brønsted bases also have adsorption effects. In oil methanolysis, metal oxide structure provides excellent adsorptive sites for methanol where the (O-H) bonds break into methoxide anions and hydrogen cations readily. The anions of methoxide then react with triglyceride molecules to create methyl esters (Smith and Notheisz 1999).

Calcium oxide is the most frequently used metal oxide catalyst for biodiesel synthesis, possibly because of its low price, low toxicity, and ease of access (Taufiq-Yap et al. 2011). It can be synthesized from inexpensive sources such as calcium carbonate and hydrated lime: Zabeti et al. reported that because of its low solubility in methanol and its simpler handling compared to KOH, calcium oxide has relatively high basic strength and less environmental effects (Zabeti, Daud, and Aroua 2010). Kawashima et al. (2008) examined 13 different kinds of metal oxides that contain calcium, barium, magnesium, or lanthanum for biodiesel production. The findings show that the catalysts in the Ca sequence have greater catalytic activity for the biodiesel synthesis (Kawashima, Matsubara, and Honda 2008). Overall it can be concluded that CaO is likely to produce as strong efficiency as NaOH does. The catalysts prepared from natural sources are eco-friendly and economical because they are non-corrosive, non-toxic, abundantly available, low-cost and have no need for water purification (Abdullah et al. 2017). Hence, the following reviews highlight the preparation of activated carbon- and calcium-based catalyst from waste sources.

*CaO from natural and waste sources*: The basic heterogeneous catalysts obtained from waste sources can be categorized into four categories: shells, ashes, rocks, and natural clays. Catalysts produced from renewable sources have a high potential for being applied as catalysts for commercial biodiesel production. The advantages of using these catalysts are availability, low cost, a large number of resources, suitable catalytic activity, and renewability.

*Waste shells*: The massive consumption of eggs and shellfish produces a large number of shells, which is a solid waste disposal problem. The calcium oxide can be obtained by calcination of these shells at 600–1000°C because shells are mostly comprised of calcium carbonate. Bird eggshells have been shown to be reliable sources of CaO. The most common bird eggshell is the hen eggshell (Tan, Abdullah, and Nolasco-Hipolito 2015). Wei et al. synthesized the first chicken eggshell-based CaO for biodiesel production. A maximum biodiesel yield of 95% was obtained with calcined eggshell at 1000°C. The ostrich egg is the biggest bird egg in the world, with a length of about 15 cm, a width of 13 cm, and a weight of 1.4 to 2.0 kg or more. The calcined ostrich eggshell at 1000°C showed higher activity compared to calcined hen eggshell (96% vs. 94%) due to smaller particle size (Tan et al. 2015).

*Dolomitic rock*: Dolomites are a low-price and high-alkalinity source of natural calcium. The main constituents of dolomitic rocks are $MgCO_3$ and $CaCO_3$. Calcium oxide derived from natural dolomite rocks produced 99.9% biodiesel yield in 3 h reaction time with 15:1 molar ratio, 10 wt.% catalyst amount. The catalyst produced below 20% of biodiesel yield after the fifth reuse due to a reduction in catalyst activity (Ngamcharussrivichai, Wiwatnimit, and Wangnoi 2007). Very similar research was performed by Siyada Jaiyen using dolomite and waste shells as the catalyst: five different types of seashells and dolomitic rock were provided by a dolomite company, calcination was done at around the same temperature of 600–800°C, and the particles were maintained less than 2 μm using a sieve. Dolomite calcined at 600 and 700°C did not produce any yield, whereas at 800°C the right amount of yield was formed due to the formation of CaO, which is highly basic in nature; similar to the previous case the maximum yield occurred at 10 wt.%. compared to dolomitic rocks. CaO is more easily extracted from waste shells; the catalyst can be reused using centrifugation. Calcined dolomite can be used up to ten times without considerable loss in yield, whereas calcined shells can be reused to only 6 or 7 times (Jaiyen, Naree, and Ngamcharussrivichai 2015).

*Pure CaO heterogeneous catalyst*: A solid base heterogeneous catalyst is prepared from waste shells to produce biodiesel. *Anadara granosa* waste cockle shell was used to prepare the calcium-rich catalyst to synthesize methyl ester from palm olein oil. The shell was calcined at 900°C for 2 h. The calcined catalyst produced 97.5% of fatty acid methyl ester at 3 h of reaction time (Boey et al. 2011). Suryaputra et al. (2013) studied the effect of calcination temperature of the *Turbonilla striatula* waste shell for the transesterification of mustard oil. A maximum biodiesel yield of 93.3% was achieved with catalyst calcined at 700°C. The leaching of active sites was confirmed from the reusability test. The catalyst was reactivated by calcinated at 900°C for 3 h (Suryaputra et al. 2013). *Amusium cristatum* Capiz shell was calcined at 900°C for 2 h in a furnace to prepare CaO. A higher biodiesel yield of 93% was achieved with 3 wt.% of catalyst amount in a 6 h reaction time. The activity of the catalyst was suddenly decreased to 50% after the third cycle due to absorbance of $CO_2$ and water on its surface. Besides, leaching of CaO by solvent and side reactions with FFA reduce the number of active sites (Suryaputra et al. 2013).

The eggshell-derived CaO catalyst was successfully synthesized and used to produce soybean methyl ester. A higher biodiesel yield of 97% was achieved with 5.8 wt.% of catalyst amount and a 1:6 ratio of oil to methanol (Piker et al. 2016). Sinha and Murugavelh (2016) produced biodiesel from waste cotton cooking oil using eggshell-derived catalyst and achieved 92% biodiesel yield with 3 wt.% catalyst loading at 60°C (Sinha and Murugavelh 2016). In another work Roschat et al. (2017) synthesized CaO catalyst from eggshell and waste coral for the production of biodiesel rubber seed oil. Eggshell catalyst yields 97% methyl ester with 9 wt.% of catalyst loading and a 1:12 oil to methanol molar ratio in 180 min at 65°C, whereas calcined waste coral catalyst required a longer reaction time of 210 min and more amount of catalyst (Roschat et al. 2017).

CaO doped with organic compound and alkali was shown to increase the basic site activity of the catalyst. Alkali-doped CaO catalyst using the incipient wetness method with calcination shows catalytic activity the following order: $LiNO_3$, $NaNO_3$, and $KNO_3$. Transesterification of rapeseed oil using CaO catalyst doped with the above compound gives a 99% conversion yield in 3 h. Without calcination 1.25% doping of alkali metal, Li on CaO gives higher catalytic activity than 1.25% doping of alkali metal Na and K. As the alkali metal Li doping percentage increases, the catalytic activity of catalyst increases at the start of reaction. At different percentages of alkali metal doping, Li without calcination gives different yields (and a maximum yield of 94.39% for 1.23% doping). If the FFA percentage of the oil increases, the conversion yield decreases; for example, when FFA content increases from 0.48% to 5.75%, the yield decreases by 4.6 % from 94.9% to 90.3% (MacLeod et al. 2008). In the transesterification of canola oil, Li/CaO-doped catalyst shows higher catalytic activity compared with other alkali metal-doped catalysts and gives 70.7% conversion yield (D'Cruz et al. 2007). The calcination temperature plays a vital role in catalyst performance, calcination temperature between 500°C and 700°C gives a higher catalytic activity than other temperatures, and catalytic activity does not significantly change below 500°C (Alonso et al. 2009).

Transesterification of high FFA contains non-edible oil and waste oil without pre-treatments done with alkali metal K- and Zn-doped CaO catalysts which were synthesized with the wet impregnation method followed by calcination (Kumar and Ali 2012). A 100% conversion of oil was achieved using K- and Zn-doped CaO catalysts when 1.5% Zn or 3.5% K was doped on CaO. Moisture and FFA containment variations that do not significantly affect the conversion yield of transesterification for K/CaO and Zn/CaO catalysts (Kumar et al. 2013). For different metal-doped calcium oxide catalysts, the conversion yields of biodiesel increase in the following order: CO/CaO, pure CaO, Cu/CaO, Mn/CaO, Fe/CaO, Cd/CaO, Ni/CaO, Zn/CaO. In the above catalyst from CO/CaO to Fe/CaO, catalyst yield was below 20%, Cd/CaO and Ni/CaO have a better yield at 80%, and Zn /CaO gives an even higher yield at 98% (Kumar and Ali 2012).

The 25% KF-doped CaO catalyst prepared by the impregnation method gives 96% conversion yield for transesterification of raw tallow seed oil, which contains FFA in the range of 40 to 70% in mild reaction conditions for 2.5 h. The catalyst was shown to have 16-fold higher catalytic activity and the yield was reduced only by 5% at 16 cycles, which indicates that the catalyst has more reusability (Wen et al. 2010).

***Loaded CaO heterogeneous catalyst:*** Nowadays carriers are used for increasing the catalytic activity of CaO. CaO was loaded on the different carrier compounds by using impregnation, precipitation, or co-precipitation of precursor salts of CaO and activate catalyst by calcination. The most widely used carriers for CaO loading are alumina, metal oxide, or silica. Loaded CaO catalyst shows higher catalytic activity than pure CaO due to the effect of carrier structural and chemical properties on the dispersion of CaO particles. Strong bonding in particles increases the stability of loaded catalyst over pure catalyst against the effects of water content, catalyst leaching, and FFA content (Tang et al. 2013).

The production of palm methyl ester using $CaO/Al_2O_3$ catalyst synthesized by an impregnation process preceded through calcination gives 98% of biodiesel with a 12:1 methanol to oil molar ratio, a 5 h reaction time, a 1000 rpm stirring speed, and catalyst loading 6 wt.% at 75°C (Zabeti, Daud, and Aroua 2009). Soybean oil transesterification using $CaO/Al_2O_3$ catalyst gives different yields according to the catalyst's neutral, acidic, or basic nature: 90%, 72%, or 82%, respectively, in the following reaction condition: 9:1 methanol to oil molar ratio, 500 rpm stirring speed, 3 wt.% of oil catalyst loading, and 6 h reaction time at 150°C (Pasupulety et al. 2013). Sunflower oil-catalyzed transesterification using a CaO/SBA-15 catalyst concentration of 1 wt.% of oil with a reaction condition of 5 h reaction time, a 12:1 methanol to oil molar ratio, 1024 rpm stirring speed at 60°C gives 95% of biodiesel conversion yield (Albuquerque et al. 2008). A different catalyst prepared by an impregnation method followed by calcination gives different yields: 92% for CaO/MgO, 60% for $CaO/SiO_2$, and 36% for $CaO/Al_2O_3$ for refined rapeseed oil with 6 h reaction time, 63°C reaction temperature, 2 wt.% of oil catalyst loading, and 950 rpm stirring speed (Yan, Lu, and Liang 2008).

*Calcium oxide mixed heterogeneous catalyst:* Pure oxide catalyst has a lower basic nature than mixed oxide catalyst. Nowadays, mixed oxide catalysts are being used to produce biodiesel from edible and non-edible oil because mixed oxide has higher alkalinity strength. Mostly, metal oxides like lanthanide group oxides, cerium oxide, lanthanum oxide and transition metal oxides like magnesium oxide, zirconium oxide, zinc oxide, etc. are mixed with calcium oxide to increase catalytic activity, and stability against water content and FFA content (Taufiq-Yap et al. 2014). The synthesis of soybean methyl ester using $CaO–La_2O_3$ mixed catalyst via a co-precipitation method gives 94.3% biodiesel yield with a 3:1 molar ratio of Ca: La followed by calcination under the following reaction conditions: time 1 h, 1:7 oil to methanol molar ratio, 450 rpm stirring speed, reaction temperature 58°C, and 5 wt.% of oil catalyst loading [66].

*Jatropha curcas* oil transesterification gives 86.51% biodiesel yield by using $CaO–La_2O_3$ mixed catalyst prepared by co-precipitation method followed by calcination under the following reaction conditions: 6 h reaction time, reaction temperature 65°C, 24:1 methanol to oil molar ratio, and 4 wt.% of oil catalyst loading (Taufiq-Yap et al. 2014). Transesterification of waste cooking oil with CaO-ZrO mixed catalyst prepared by co-precipitation method gives 92.1% biodiesel yield in a 1:2 molar ratio of Ca: Zr followed by calcination under the following reaction condition: time 2 h, 1:30 oil to methanol molar ratio, 500 rpm stirring speed, reaction temperature 65°C, and 10 wt.% of oil catalyst loading (Molaei Dehkordi and Ghasemi 2012).

Palm oil transesterification gives 95% biodiesel yield by using $CaO-CeO_2$ mixed catalyst prepared by a co-precipitation method followed by calcination under the following reaction conditions: 3 h process time, reaction temperature 85°C, oil to methanol molar ratio 1:12, 500 rpm stirring speed, and 5 wt.% of oil catalyst loading (Thitsartarn and Kawi 2011). Refined oils of sunflower, soybean, and palm kernel gave more than 94% biodiesel yield when CaO–ZnO catalyzed transesterification was done with 500 rpm stirring speed, 30:1 methanol to oil molar ratio, 3 h process time, and 10 wt.% of oil catalyst loading at a reaction temperature of 60°C. CaO–ZnO prepared by the co-precipitation method was followed by calcination at 800°C for 2 h (Ngamcharussrivichai, Totarat, and Bunyakiat 2008).

Preparation of CaMgO and CaZnO catalyst by a co-precipitation method was followed by calcination at 800°C and 900°C respectively, for 6 h. For preparing the catalyst, the CaMgO constituent was used in 0.3 molar ratio of Ca: Mg and CaZnO constituent in 0.22 molar of Ca: Zn. CaMgO and CaZnO mixed oxide catalyst was used for the transesterification of *Jatropha curcas* oil with a 15:1 methanol to oil molar ratio, a reaction temperature of 65°C, 6 h processing time, 4 wt.% of oil catalyst loading, and vigorous stirring, which gave 83% and 81% biodiesel yield, respectively (Taufiq-Yap et al. 2011). Transesterification of rapeseed oil with $Ca_{12}Al_{14}O_{33}$ catalyst prepared by a chemical synthesis method in a 3:2 molar ratio of Ca: Al followed by calcination gives 94.34% biodiesel yield under the following reaction condition: time 3 h, 1:15 oil to methanol molar ratio, 270 rpm stirring speed, reaction temperature 65°C, and 6 wt.% of oil catalyst loading (Wang et al. 2013). Transesterification of sunflower oil with $Ca_{12}Al_{14}O_{33}$ catalyst prepared by chemical synthesis method in 2:1 molar ratio of Ca: Al followed by calcination gives 98% biodiesel yield in the following reaction condition: time 3 h, 1:12 oil to methanol molar ratio, 1000 rpm stirring speed, reaction temperature 60°C, and 1 wt.% of oil catalyst loading (Campos-Molina et al. 2010).

It can be concluded from all the studies cited above that surface area and amount of calcium content play a vital role in biodiesel production (Laca, Laca, and Díaz 2017). Also, the optimum temperature for the complete conversion of calcium carbonate into calcium oxide was found to be 800–900°C (Yin et al. 2016). However, the main difficulty of using calcium oxide as a catalyst is that it quickly converts to calcium carbonate and calcium hydroxide because of the chemisorption of carbon dioxide and water on the surface-active sites. This poisoning effect reduces the catalytic activity and stability of the catalyst. In order to overcome these problems, support materials such as activated carbon and ash can be used or converted into calcium methoxide (Jain, Khatri, and Rani 2010).

Much research has been done on using activated carbon as the support material for CaO catalyst. However, few works have been carried out on the synthesis of CaO from biomass or waste material. Preparation of CaO from biomass or waste material can be an efficient, economical substitute for commercial CaO catalysts (Chen et al. 2015; Chakraborty, Bepari, and Banerjee 2010). Rice husk and eggshell were successfully converted into catalyst support and CaO for transesterification of palm oil. In this, the optimum calcination to get maximum basicity was 800°C. Transesterification of palm oil using rice husk-derived ash supported on CaO catalyst produced 91.5% biodiesel at optimum conditions of 1:9 oil to methanol molar ratio, catalyst loading of 7 wt.%, and reaction time of 4 h at 65°C (Chen et al. 2015).

Buasri et al. (2012) prepared activated carbon from *Jatropha curcas* shells and used it as supporting material for the KOH catalyst. They found that the highest biodiesel yield of 87% was obtained with 31.3% of catalyst loading, 20:1 methanol to oil molar ratio, and 2 h reaction time at 60°C. Alike, Dhawane et al. (2016) produced rubber seed biodiesel with 89.85% yield using KOH supported on activated carbon prepared from *Flamboyant* pods catalyst. They found the optimum conditions to be a 1:15 oil to methanol molar ratio, catalyst loading of 3.5 wt.%, and 60 min reaction time (Dhawane, Kumar, and Halder 2016). This is because of the different types of biomass used in those works. Hence, it reveals the significance in the selection of proper biomass for the preparation of activated carbon. Biodiesel production cost can be minimized by utilizing this waste material for the preparation of the heterogeneous green catalyst.

### 2.2.2.3 Enzymatic Catalyst

The use of an enzymatic catalyst in the transesterification process overcomes the issues associated with a heterogeneous catalyst such as complex synthesis process, higher energy consumption, and production cost. The enzyme used as a catalyst can be prepared by immobilization with a material that allows resistance to changes in pH, humidity, and temperature. However, the enzymatic catalysts are sensitive to polar solvents such as methanol, water, and phospholipids. This is alleviated by the gradual addition of alcohol (Mardhiah et al. 2017). The performance of enzymatic catalysts, namely *Pseudomonas cepacian*, *Candida rugosa*, *Mucor javanicus*, *Pseudomonas cepacian*, and *Pseudomonas fluorescens*, was assessed. Maximum biodiesel of 98% was obtained with celite-immobilized enzyme catalyst with 4–5 wt.% water level and 8 h at 50°C compared to commercially prepared lipase (Shah and Gupta 2007).

In another study, Abdulla and Ravindra synthesized an immobilized *Burkholderia cepacia* enzymatic catalyst with glutaraldehyde for the transesterification of *Jatropha* oil (Abdulla and Ravindra 2013). The prepared catalyst showed 100% fatty acid methyl ester conversion with 24 h reaction time and a 10:1 methanol to oil molar ratio at 35°C. Similarly, *Rhizopus oryzae* lipase (ROL) was immobilized on the microporous resin (MI-ROL) and anion exchange resin (AI-ROL) to convert *Pistacia chinensis* bge seed oil to methyl ester without alcohol. The MI-ROL and AI-ROL catalyst produced maximum biodiesel yields of 92% and 94%, respectively. Hence, the AI-ROL catalyst showed better catalytic activity without loss of biodiesel yield in repeated cycles (Li et al. 2012). However, the enzymatic catalyst has a few drawbacks such as sensitivity to polar solvent, glycerol adsorption, and complex preparation procedures (Mardhiah et al. 2017).

## 2.3 CO-SOLVENT TRANSESTERIFICATION

Methanol and vegetable oil are non-miscible and occur in two different phases. It results in less mass transfer, since the bulk transition happens only at the interface of two phases. In this regard, numerous approaches for overcoming this limitation have been studied and mentioned in the literature. One alternative to this issue is to merge the two phases with a proper stimulus design. An alternative would be to raise the temperature of the reaction, which results in increased energy consumption in the system. The boiling point of methanol even limits temperature increase. The safest way to handle the lower solubility of oil and methanol is by using a co-solvent that can directly influence the reaction rate (Encinar, Pardal, and Sánchez 2016). The addition of co-solvents such as diethyl ether, acetone, ethanol, etc. have both polar and non-polar sites, which increases the miscibility and molecule-molecule interaction and thereby decreases the reaction time (Singh, Yadav, and Sharma 2017).

Diethyl ether was used as a co-solvent with calcium oxide catalyst in the transesterification of linseed oil. Diethyl ether could successfully overcome the obstacle to mass transfer and increased the biodiesel from 75.83% to 98.08% (Hashemzadeh Gargari and Sadrameli 2018). Similarly, Sahani et al. investigated the effect of co-solvents, namely, n-hexane, di-isopropyl ether, acetone, and toluene on *Schleichera oleosa* biodiesel yield. The di-isopropyl ether showed better miscibility and mass transfer compared to other solvents. A maximum biodiesel yield of 96% was obtained with a 1:1 ratio of cosolvent to oil (Sahani, Banerjee, and Sharma 2018).

In another study, the effect of acetone co-solvent on biodiesel production from waste vegetable oil using calcium aluminium oxide as a heterogeneous catalyst was assessed. The highest methyl ester conversion of 97.89% was achieved with 20 wt.% of acetone and 25 min of reaction time at 55°C. The amount of acetone increase reduced the viscosity of mixture at the same time that the density between biodiesel and glycerol increased, thereby reducing the time required for phase separation (Singh, Yadav, and Sharma 2017). Likewise, the mutual effect of nano-catalyst and co-solvent, namely acetone and tetrahydrofuran, on biodiesel production was studied. The addition of co-solvent increased the reaction rate and reduced the methanol amount required for typical reactions (Ambat et al. 2020).

## 2.4   MICROWAVE-ASSISTED TRANSESTERIFICATION

In the electromagnetic spectrum, 0.3 to 300 GHz frequency ranges correspond to microwaves, which are located between infrared and radio wave frequencies. Microwaves generate heat through polarization of polar molecules such as alcohol and water. The polar molecules oscillate to rearrange their position against intermolecular force through a continually changing electric field. The molecules' absorbed energy is released as heat by the collision of polar molecules. Nevertheless, the oscillation is expeditious at high frequency, resulting in no heat generation. This is due to the stronger intermolecular force between molecules that stops the motion. Hence, household and laboratory microwave ovens are operated at 2.45 GHz. Microwave-assisted (MA) techniques are extensively used in many applications, including chemical synthesis and catalyst preparation. The primary benefits of MA techniques are higher yield in shorter reaction time, rapid heat generation, and eco-friendliness compared to conventional techniques. Moreover, in the MA technique heat is generated from inside out, which results in higher local temperatures and reduces the activation energy required for the transesterification reaction (Motasemi and Ani 2012).

Calcium oxide is a low-cost catalyst widely used to produce biodiesel using microwave-assisted techniques. Kawashima et al. investigated the effect of calcium-based catalysts such as CaO, $Ca(OH)_2$, and $CaCO_3$ in transesterification of soybean oil and obtained 93%, 12%, and 0% of biodiesel yield respectively in 1 h reaction time (Kawashima, Matsubara, and Honda 2009a). Lue et al. studied microwave-assisted transesterification of soybean oil using calcium methoxide catalyst. A maximum biodiesel yield of 98% was obtained in 2 h reaction time. Lewis's basic theory and experimental results revealed that calcium methoxide ($Ca(CH_3O)_2$) catalyst has stronger basicity than CaO (Dai, Chen, and Chen 2014). Ye et al. compared the kinetic study of CaO-catalyzed transesterification of palm oil using microwave-assisted technique and conventional heating. The study confirmed that the third-order reaction in MA techniques is faster than the first-order reaction in conventional heating. The optimum process levels are 5 wt.% catalyst loading, 450 rpm stirring speed, 1:9 ratio of oil to methanol, and 150 W microwave power in 60 min (Ye et al. 2016).

Besides, microwave-absorbing materials such as carbon/activated carbon and silicon carbide can be used as supporting materials to enhance the biodiesel conversion in MA transesterification process. Nevertheless, only limited studies were carried out using a carbon-supported heterogeneous base catalyst in MA biodiesel production. Wang et al. synthesized a carbon-supported solid base $K_2SiO_3$ catalyst via the impregnation method and used it for microwave-assisted transesterification of soybean oil. They reported that carbon support absorbed more microwave energy and increased the localized temperature, which reduced the activation energy and thereby enhanced the methyl ester conversion (Wang, Chen, and Chen 2011). A heterogeneous carbon-supported nanocomposite ($Ca_{12}Al_{14}O_{33}$) catalyst was prepared by the microwave combustion method. The study reported that potassium hydroxide-supported nanocomposite produced less biodiesel yield compared to carbon-supported catalysts. This is due to higher surface area and pore size, which results in the absorbance of more microwave energy (Nayebzadeh et al. 2018).

Biodiesel production involves mixing catalysts with alcohol, chemical transformation, and separation of catalyst and glycerol from biodiesel. Existing reactors using conventional and assisted techniques such as ultrasonic and microwave techniques are costly because of their larger size, and reactions consume more time (Aghel et al. 2014). Excessive alcohol is required to move a reaction in a forward manner. Residence time is high, and also extra energy is provided for maintaining the chemical equilibrium. Microreactor-assisted intensification techniques can overcome those drawbacks for biodiesel production. In process intensification, continuous processing from the batch is done by the use of microreactor technologies, where uniform mixing and heat transfer rates occur compared with the conventional process. It enhances the improvement of solution mixing and the heat/mass transfer rate, which significantly influences the rate of the reaction and reduces the residence time (Rahimi et al. 2014). Microreactors can accomplish reaction rates by using a higher surface area to the volume ratio, which shortens the rate of diffusion and improves the efficiency of mass transfer (Santana et al. 2017).

## 2.5 ULTRASOUND-ASSISTED TRANSESTERIFICATION

Ultrasonic sound wave frequency is higher than the level detectable by humans: the 20 kHz to 10 MHz range. Nowadays, ultrasound waves are commonly used in chemical synthesis, oil extraction, and wastewater treatment (Tan et al. 2019). The usage of ultrasound technology is expanding, and its use in the production of biodiesel is timely and justified through current research. The ultrasonic wave enhances the mass transfer rate between two miscible liquid phases like alcohol and oil, which in turn increases the reaction rate and thereby reduces reaction time (Ho, Ng, and Gan

2016). Costa et al. compared the mixing performance of ultrasound and mechanical steering in biodiesel production. They found that ultrasound is more effective in mixing, which produced higher biodiesel yield (Costa-Felix and Ferreira 2015). An ultrasound wave generates continuous expansion and contraction in the liquid interface during irradiation.

This effect causes the formation of micro-bubbles, which expand to a larger extent and collapse violently, thus raising local temperature and pressure in a shorter lifetime. The emulsification of alcohol and oil is enhanced by the radial motion of cavitation bubbles and free radicals formed when bubbles collapse. All these effects increase the interfacial surface area between alcohol and oil thereby increasing the conversion rate of methyl ester (Veljković, Avramović, and Stamenković 2012). Many researchers have used the ultrasound-assisted technique (UA) to produce biodiesel from the various feedstocks. These UA techniques have been shown to increase the reaction rate of the transesterification of vegetable oils. Moreover, the UA technique overcomes the problems related to conventional batch type with mechanical stirring such as longer reaction time, slow reaction rate, and difficulties in phase separation.

Jogi et al. (2016) achieved 98% biodiesel yield using UA techniques from degummed *Jatropha curcas* oil; the optimum reaction conditions were 1 wt.% catalyst amount of tri-basic potassium phosphate, a 12:1 methanol to oil ratio, 600 W ultrasonic power, and a 45 min reaction time at 55°C (Jogi et al. 2016). The transesterification of *Karabi* oil was carried out in a conventional mechanical stirring and UA method with CaO catalyst. It was found that UA produced a maximum biodiesel yield of 94% using 5 wt.% of CaO, 1:12 ratio of oil to methanol in 120 min at 65°C (Yadav et al. 2016). Ali et al. synthesized a mixture of metal oxide catalysts such as strontium oxide and calcium oxide for the production of biodiesel from *Jatropha* oil. A maximum yield of 98% methyl ester conversion was obtained in 30 min using 6 wt.% of SrO–CaO, 210 W ultrasonic power at 65°C (Ali et al. 2017). The majority of research work demonstrated that ultrasound-assisted techniques could be a promising method to substitute for mechanical stirring in the production of biodiesel. It is capable of providing a localized temperature rise during the ultrasonic irradiation, but that is not enough to attain the required process temperature (Tan et al. 2019). Hence, there is a need for an external heating sources in UA biodiesel production.

## 2.6   NON-CATALYTIC SUPERCRITICAL METHANOL TRANSESTERIFICATION

Saka and Kusdiana (2001) initially suggested producing biodiesel using alcohol under pressure and in a catalyst-free condition (Saka and Kusdiana 2001). This catalyst-free process works well with oils having water and higher FFA content since the hydrolysis, esterification, and transesterification reactions take place simultaneously. In addition, the raw material used may be of low quality. The biodiesel production process is completed in minutes and mass transfer limitations are minimized as there is greater solubility between the oil and alcohol phases compared to other approaches, which improves the reaction rates (Silva et al. 2007). The key downside of this method is higher energy utilization (2.4 MW per 10,000 metric tons/year) due to the high temperatures and pressures used in the process, raising the final costs of producing biodiesel (Tan and Lee 2011).

Tan et al. produced biodiesel from waste frying oil (WFO) under supercritical methanol conditions without a catalyst. They achieved 80% biodiesel yield under optimal conditions (Tan, Lee, and Mohamed 2011). The effect of methanol and ethanol on supercritical methanol transesterification of WFO without a catalyst was studied. The ethanol showed a better conversion of 80% esters compared to methanol. In another work, Ghoreishi and Moein, 95.27% of esters were obtained under a 1:33.8 methanol molar ratio, 23.1 MPa pressure, and 20.4 min at 271.1°C (Ghoreishi and Moein 2013).

## 2.7  PURIFICATION OF BIODIESEL

Several standards need to be respected in order to maintain the quality of the final biodiesel. Biodiesel should be free of impurities, because this can damage the engine components as they settle in the nozzles and cause corrosive incrustations. Such residues originate from the unsaponifiable substances in the raw oil itself, namely residues of catalysts, water, glycerol, and excess alcohol from the reaction. Hence, these unwanted materials should be removed in order to enhance the biodiesel quality. A few different methods are available to remove the impurities from biodiesel. Among these, wet washing and chemical cleaning are mostly used at the industrial scale and also have been studied in recent years (Fonseca et al. 2019).

*Wet washing*: A traditional way of eliminating biodiesel impurities. Even though the effectiveness of the wet washing method has been demonstrated, this method has drawbacks like wastewater generation, significant product losses, and emulsion formation when biodiesel is made from high-FFA oils (Berrios and Skelton 2008). Wet washing involves applying a specified quantity of water with soft stirring to prevent emulsion formation. The water washing cycle goes on until the water becomes colorless.

*Chemical cleaning*: The phosphoric acid-water solution, distilled water, and mixture of organic solvent and water can be used for water washing (Fonseca et al. 2019). The addition of acid in the wash solution is advantageous because it can neutralize the excess amount of catalyst. Finally, the biodiesel is fully treated with water to eliminate the remaining impurities (Atadashi et al. 2011). Predojevic (2008) and Hingu et al. (2010) used 5% of phosphoric acid to remove impurities from the biodiesel, which increased the methyl ester content from 89.5% to 93.5% (Hingu, Gogate, and Rathod 2010; Predojević 2008).

## 2.8  CONCLUSION

The production of alternative fuel from renewable sources presents many challenges. The exhaustion of fossil fuel resources, the unstable price of oil, and environmental considerations are the primary reasons for searching alternative and eco-friendly fuels. Biodiesel is a renewable and promising alternative fuel that scientists are focusing on and striving to produce at a lower cost and with excellent fuel characteristics. Transesterification is one of the best methods to reduce the viscosity of the vegetable oil and to produce biodiesel. Homogeneous (acid and alkali), heterogeneous (acid, alkali, and enzymatic), and non-catalytic transesterification are various methods that have been used to produce biodiesel.

Each method has its own pros and cons. Alkali-catalyzed transesterification is a heavily used method for biodiesel production, but it has few drawbacks such as sensitivity to water and FFA. Acid catalyst efficiently converts high-FFA oil into esters, but it requires a longer reaction time. Enzymatic catalysts have shown better activity in transesterification, but they require complex preparation procedures. Non-catalytic supercritical methanol biodiesel production without a catalyst is not economical due to its higher production cost. Hence, researchers are keen to utilize advanced assisted techniques such as ultrasonic and microwave transesterification using a heterogeneous catalyst to produce biodiesel.

## REFERENCES

Abdulla, Rahmath, and Pogaku Ravindra. 2013. "Immobilized Burkholderia Cepacia Lipase for Biodiesel Production from Crude Jatropha Curcas L. Oil." *Biomass and Bioenergy*. doi:10.1016/j.biombioe.2013.04.010.

Abdullah, Sharifah Hanis Yasmin Sayid, Nur Hanis Mohamad Hanapi, Azman Azid, Roslan Umar, Hafizan Juahir, Helena Khatoon, and Azizah Endut. 2017. "A Review of Biomass-Derived Heterogeneous Catalyst for a Sustainable Biodiesel Production." *Renewable and Sustainable Energy Reviews*. doi:10.1016/j.rser.2016.12.008.

Agarwal, Madhu, Garima Chauhan, S. P. Chaurasia, and Kailash Singh. 2012. "Study of Catalytic Behavior of KOH as Homogeneous and Heterogeneous Catalyst for Biodiesel Production." *Journal of the Taiwan Institute of Chemical Engineers*. doi:10.1016/j.jtice.2011.06.003.

Aghel, Babak, Masoud Rahimi, Arash Sepahvand, Mohammad Alitabar, and Hamid Reza Ghasempour. 2014. "Using a Wire Coil Insert for Biodiesel Production Enhancement in a Microreactor." *Energy Conversion and Management* 84. Elsevier Ltd: 541–49. doi:10.1016/j.enconman.2014.05.009.

Ahmad, A. L., N. H. Mat Yasin, C. J. C. Derek, and J. K. Lim. 2011. "Microalgae as a Sustainable Energy Source for Biodiesel Production: A Review." *Renewable and Sustainable Energy Reviews* 15 (1). Elsevier Ltd: 584–93. doi:10.1016/j.rser.2010.09.018.

Albuquerque, Mônica C. G., Inmaculada Jiménez-Urbistondo, José Santamaría-González, Josefa M. Mérida-Robles, Ramón Moreno-Tost, Enrique Rodríguez-Castellón, Antonio Jiménez-López, Diana C. S. Azevedo, Célio L. Cavalcante, and Pedro Maireles-Torres. 2008. "CaO Supported on Mesoporous Silicas as Basic Catalysts for Transesterification Reactions." *Applied Catalysis A: General*. doi:10.1016/j.apcata.2007.09.028.

Ali, Syed Danish, Isma Noreen Javed, Usman Ali Rana, Muhammad Faizan Nazar, Waqas Ahmed, Asifa Junaid, Mahmood Pasha, Rumana Nazir, and Rizwana Nazir. 2017. "Novel SrO-CaO Mixed Metal Oxides Catalyst for Ultrasonic-Assisted Transesterification of Jatropha Oil into Biodiesel." *Australian Journal of Chemistry* 70 (3). CSIRO PUBLISHING: 258. doi:10.1071/CH16236.

Alonso, D. Martín, R. Mariscal, M. López Granados, and P. Maireles-Torres. 2009. "Biodiesel Preparation Using Li/CaO Catalysts: Activation Process and Homogeneous Contribution." *Catalysis Today* 143 (1–2): 167–71. doi:10.1016/j.cattod.2008.09.021.

Ambat, Indu, Varsha Srivastava, Sidra Iftekhar, Esa Haapaniemi, and Mika Sillanpää. 2020. "Effect of Different Co-Solvents on Biodiesel Production from Various Low-Cost Feedstocks Using Sr–Al Double Oxides." *Renewable Energy*. doi:10.1016/j.renene.2019.08.061.

Antunes, Wallace Magalhães, Cláudia de Oliveira Veloso, and Cristiane Assumpção Henriques. 2008. "Transesterification of Soybean Oil with Methanol Catalyzed by Basic Solids." *Catalysis Today.* doi:10.1016/j.cattod.2007.12.055.

Armas, O., R. Ballesteros, F. J. Martos, and J. R. Agudelo. 2005. "Characterization of Light Duty Diesel Engine Pollutant Emissions Using Water-Emulsified Fuel." *Fuel* 84 (7–8): 1011–18. doi:10.1016/j.fuel.2004.11.015.

Atabani, A. E., and Aldara Da Silva César. 2014. "Calophyllum Inophyllum L. – A Prospective Non-Edible Biodiesel Feedstock. Study of Biodiesel Production, Properties, Fatty Acid Composition, Blending and Engine Performance." *Renewable and Sustainable Energy Reviews* 37. Elsevier: 644–55. doi:10.1016/j.rser.2014.05.037.

Atabani, A. E., A. S. Silitonga, Irfan Anjum Badruddin, T. M. I. Mahlia, H. H. Masjuki, and S. Mekhilef. 2012. "A Comprehensive Review on Biodiesel as an Alternative Energy Resource and Its Characteristics." *Renewable and Sustainable Energy Reviews* 16 (4). Elsevier Ltd: 2070–93. doi:10.1016/j.rser.2012.01.003.

Atabani, A. E., T. M. I. Mahlia, H. H. Masjuki, Irfan Anjum Badruddin, Hafizuddin Wan Yussof, W. T. Chong, and Keat Teong Lee. 2013. "A Comparative Evaluation of Physical and Chemical Properties of Biodiesel Synthesized from Edible and Non-Edible Oils and Study on the Effect of Biodiesel Blending." *Energy* 58: 296–304. doi:10.1016/j. energy.2013.05.040.

Atadashi, I. M., M. K. Aroua, and A. Abdul Aziz. 2010. "High Quality Biodiesel and Its Diesel Engine Application: A Review." *Renewable and Sustainable Energy Reviews* 14 (7). Elsevier Ltd: 1999–2008. doi:10.1016/j.rser.2010.03.020.

Atadashi, I. M., M. K. Aroua, A. R. Abdul Aziz, and N. M. N. Sulaiman. 2011. "Refining Technologies for the Purification of Crude Biodiesel." *Applied Energy.* doi:10.1016/j. apenergy.2011.05.029.

Atadashi, I. M., M. K. Aroua, A. R. Abdul Aziz, and N. M. N. Sulaiman. 2013. "The Effects of Catalysts in Biodiesel Production: A Review." *Journal of Industrial and Engineering Chemistry* 19 (1). The Korean Society of Industrial and Engineering Chemistry: 14–26. doi:10.1016/j.jiec.2012.07.009.

Balat, Mustafa, and Havva Balat. 2010. "Progress in Biodiesel Processing." *Applied Energy* 87 (6). Elsevier Ltd: 1815–35. doi:10.1016/j.apenergy.2010.01.012.

Berrios, M., and R. L. Skelton. 2008. "Comparison of Purification Methods for Biodiesel." *Chemical Engineering Journal.* doi:10.1016/j.cej.2008.07.019.

Bhuiya, M. M. K., M. G. Rasul, M. M. K. Khan, N. Ashwath, A. K. Azad, and M. A. Hazrat. 2016. "Prospects of 2nd Generation Biodiesel as a Sustainable Fuel – Part 2: Properties, Performance and Emission Characteristics." *Renewable and Sustainable Energy Reviews* 55. Elsevier: 1129–46. doi:10.1016/j.rser.2015.09.086.

Boey, Peng Lim, Gaanty Pragas Maniam, Shafida Abd Hamid, and Dafaalla Mohamed Hag Ali. 2011. "Utilization of Waste Cockle Shell (Anadara Granosa) in Biodiesel Production from Palm Olein: Optimization Using Response Surface Methodology." *Fuel.* doi:10.1016/j.fuel.2011.03.002.

Boffito, D. C., V. Crocellà, C. Pirola, B. Neppolian, G. Cerrato, M. Ashokkumar, and C. L. Bianchi. 2013. "Ultrasonic Enhancement of the Acidity, Surface Area and Free Fatty Acids Esterification Catalytic Activity of Sulphated ZrO2-TiO2 Systems." *Journal of Catalysis* 297. Elsevier Inc.: 17–26. doi:10.1016/j.jcat.2012.09.013.

Borges, M. E., and L. Díaz. 2012. "Recent Developments on Heterogeneous Catalysts for Biodiesel Production by Oil Esterification and Transesterification Reactions: A Review." *Renewable and Sustainable Energy Reviews* 16 (5). Elsevier Ltd: 2839–49. doi:10.1016/j.rser.2012.01.071.

Bournay, L., D. Casanave, B. Delfort, G. Hillion, and J. A. Chodorge. 2005. "New Heterogeneous Process for Biodiesel Production: A Way to Improve the Quality and the Value of the Crude Glycerin Produced by Biodiesel Plants." *Catalysis Today* 106 (1–4): 190–92. doi:10.1016/j.cattod.2005.07.181.

Buasri, Achanai, Nattawut Chaiyut, Vorrada Loryuenyong, Chao Rodklum, Techit Chaikwan, Nanthakrit Kumphan, Kritsanapong Jadee, Pathravut Klinklom, and WittayaWittayarounayut. 2012. "Transesterification of Waste Frying Oil for Synthesizing Biodiesel by KOH Supported on Coconut Shell Activated Carbon in Packed Bed Reactor." *ScienceAsia* 38 (3): 283–88. doi:10.2306/scienceasia1513-1874.2012.38.283.

Campos-Molina, María Jośe, Jośe Santamaría-Gonźalez, Josefa Mosefaaría-Go, Rasef Moreno-Tost, Monica C. G. Albuquerque, Sebastían Bruque-Ǵamez, Enrique Rodríguez-Castelĺon, Antonio Jitonioez-Cas, and Pedro Maireles-Torres. 2010. "Base Catalysts Derived from Hydrocalumite for the Transesterification of Sunflower Oil." *Energy and Fuels.* doi:10.1021/ef9009394.

Canakci, M., and H. Sanli. 2008. "Biodiesel Production from Various Feedstocks and Their Effects on the Fuel Properties." *Journal of Industrial Microbiology and Biotechnology* 35 (5): 431–41. doi:10.1007/s10295-008-0337-6.

Cao, Hechun, Zhiling Zhang, Xuwen Wu, and Xiaoling Miao. 2013. "Direct Biodiesel Production from Wet Microalgae Biomass of Chlorella Pyrenoidosa through in Situ Transesterification." *BioMed Research International.* doi:10.1155/2013/930686.

Chakraborty, R., S. Bepari, and A. Banerjee. 2010. "Transesterification of Soybean Oil Catalyzed by Fly Ash and Egg Shell Derived Solid Catalysts." *Chemical Engineering Journal.* doi:10.1016/j.cej.2010.10.019.

Chen, Guan Yi, Rui Shan, Jia Fu Shi, and Bei Bei Yan. 2015. "Transesterification of Palm Oil to Biodiesel Using Rice Husk Ash-Based Catalysts." *Fuel Processing Technology* 133. Elsevier B. V.: 8–13. doi:10.1016/j.fuproc.2015.01.005.

Costa-Felix, Rodrigo P. B., and Jerusa R. L. Ferreira. 2015. "Comparing Ultrasound and Mechanical Steering in a Biodiesel Production Process." *Physics Procedia.* doi:10.1016/j.phpro.2015.08.227.

D'Cruz, Amanda, Mangesh G. Kulkarni, Lekha Charan Meher, and Ajay K. Dalai. 2007. "Synthesis of Biodiesel from Canola Oil Using Heterogeneous Base Catalyst." *JAOCS, Journal of the American Oil Chemists' Society.* doi:10.1007/s11746-007-1121-x.

Dai, Yong Ming, Kung Tung Chen, and Chiing Chang Chen. 2014. "Study of the Microwave Lipid Extraction from Microalgae for Biodiesel Production." *Chemical Engineering Journal.* doi:10.1016/j.cej.2014.04.031.

DaSilveira Neto, Brenno Amaro, Melquizedeque B. Alves, Alexandre A. M. Lapis, Fabiane M. Nachtigall, Marcos N. Eberlin, Jaïrton Dupont, and Paulo A. Z. Suarez. 2007. "1-n-Butyl-3-Methylimidazolium Tetrachloro-Indate (BMI ŝ InCl4) as a Media for the Synthesis of Biodiesel from Vegetable Oils." *Journal of Catalysis* 249 (2): 154–61. doi:10.1016/j.jcat.2007.04.015.

Dawodu, F. A., O. O. Ayodele, and T. Bolanle-Ojo. 2014. "Biodiesel Production from Sesamum Indicum L. Seed Oil: An Optimization Study." *Egyptian Journal of Petroleum.* doi:10.1016/j.ejpe.2014.05.006.

DeMello, Jared A., Catherine A. Carmichael, Emily E. Peacock, Robert K. Nelson, J. Samuel Arey, and Christopher M. Reddy. 2007. "Biodegradation and Environmental Behavior of Biodiesel Mixtures in the Sea: An Initial Study." *Marine Pollution Bulletin* 54 (7): 894–904. doi:10.1016/j.marpolbul.2007.02.016.

Devi, Anuchaya, Vijay K. Das, and Dhanapati Deka. 2017. "Ginger Extract as a Nature Based Robust Additive and Its Influence on the Oxidation Stability of Biodiesel Synthesized from Non-Edible Oil." *Fuel* 187. doi:10.1016/j.fuel.2016.09.063.

Dhawane, Sumit H., Tarkeshwar Kumar, and Gopinath Halder. 2016. "Biodiesel Synthesis from Hevea Brasiliensis Oil Employing Carbon Supported Heterogeneous Catalyst: Optimization by Taguchi Method." *Renewable Energy* 89. Elsevier Ltd: 506–14. doi:10.1016/j.renene.2015.12.027.

Dias, Joana M., Maria C. M. Alvim-Ferraz, and Manuel F. Almeida. 2008. "Comparison of the Performance of Different Homogeneous Alkali Catalysts during Transesterification of Waste and Virgin Oils and Evaluation of Biodiesel Quality." *Fuel*. doi:10.1016/j.fuel.2008.06.014.

Ehiri, R. C., I. I. Ikelle, and O. F. Ozoaku. 2014. "Acid-Catalyzed Transesterification Reaction of Beef Tallow For Biodiesel Production By Factor Variation." *American Journal of Engineering Research* 3 (7): 174–177.

Encinar, José M., Ana Pardal, and Nuria Sánchez. 2016. "An Improvement to the Transesterification Process by the Use of Co-Solvents to Produce Biodiesel." *Fuel*. doi:10.1016/j.fuel.2015.10.110.

Fabiano, Demian Patrick, Berna Hamad, Dilson Cardoso, and Nadine Essayem. 2010. "On the Understanding of the Remarkable Activity of Template-Containing Mesoporous Molecular Sieves in the Transesterification of Rapeseed Oil with Ethanol." *Journal of Catalysis* 276 (1). Elsevier Inc.: 190–96. doi:10.1016/j.jcat.2010.09.015.

Fernández-Álvarez, P., J. Vila, J. M. Garrido, M. Grifoll, G. Feijoo, and J. M. Lema. 2007. "Evaluation of Biodiesel as Bioremediation Agent for the Treatment of the Shore Affected by the Heavy Oil Spill of the Prestige." *Journal of Hazardous Materials* 147 (3): 914–22. doi:10.1016/j.jhazmat.2007.01.135.

Fonseca, Jhessica Marchini, Joel Gustavo Teleken, Vitor de Cinque Almeida, and Camila da Silva. 2019. "Biodiesel from Waste Frying Oils: Methods of Production and Purification." *Energy Conversion and Management* 184 (December 2018): 205–18. doi:10.1016/j.enconman.2019.01.061.

Furuta, Satoshi, Hiromi Matsuhashi, and Kazushi Arata. 2004. "Biodiesel Fuel Production with Solid Superacid Catalysis in Fixed Bed Reactor under Atmospheric Pressure." *Catalysis Communications*. doi:10.1016/j.catcom.2004.09.001.

Ghoreishi, S. M., and P. Moein. 2013. "Biodiesel Synthesis from Waste Vegetable Oil via Transesterification Reaction in Supercritical Methanol." *Journal of Supercritical Fluids*. doi:10.1016/j.supflu.2013.01.011.

Gu, Ling, Wei Huang, Shaokun Tang, Songjiang Tian, and Xiangwen Zhang. 2015. "A Novel Deep Eutectic Solvent for Biodiesel Preparation Using a Homogeneous Base Catalyst." *Chemical Engineering Journal*. doi:10.1016/j.cej.2014.08.026.

Hájek, Martin, Franti Šek Skopal, Libor Čapek, Michal Černoch, and Petr Kutálek. 2012. "Ethanolysis of Rapeseed Oil by KOH as Homogeneous and as Heterogeneous Catalyst Supported on Alumina and CaO." *Energy*. doi:10.1016/j.energy.2012.06.052.

Hashemzadeh, Gargari M., and S. M. Sadrameli. 2018. "Investigating Continuous Biodiesel Production from Linseed Oil in the Presence of a Co-Solvent and a Heterogeneous Based Catalyst in a Packed Bed Reactor." *Energy*. doi:10.1016/j.energy.2018.01.105.

Helwani, Z., M. R. Othman, N. Aziz, J. Kim, and W. J. N. Fernando. 2009. "Solid Heterogeneous Catalysts for Transesterification of Triglycerides with Methanol: A Review." *Applied Catalysis A: General*. doi:10.1016/j.apcata.2009.05.021.

Hingu, Shishir M., Parag R. Gogate, and Virendra K. Rathod. 2010. "Ultrasonics Sonochemistry Synthesis of Biodiesel from Waste Cooking Oil Using Sonochemical Reactors." *Ultrasonics – Sonochemistry* 17 (5). Elsevier B. V.: 827–32. doi:10.1016/j.ultsonch.2010.02.010.

Ho, Wilson Wei Sheng, Hoon Kiat Ng, and Suyin Gan. 2016. "Advances in Ultrasound-Assisted Transesterification for Biodiesel Production." *Applied Thermal Engineering* 100. Elsevier Ltd: 553–63. doi:10.1016/j.applthermaleng.2016.02.058.

Jain, Deepti, Chitralekha Khatri, and Ashu Rani. 2010. "Fly Ash Supported Calcium Oxide as Recyclable Solid Base Catalyst for Knoevenagel Condensation Reaction." *Fuel Processing Technology*. doi:10.1016/j.fuproc.2010.02.021.

Jain, Mohit, Usha Chandrakant, Valérie Orsat, and Vijaya Raghavan. 2018. "A Review on Assessment of Biodiesel Production Methodologies from Calophyllum Inophyllum Seed Oil." *Industrial Crops and Products* 114 (September 2017). Elsevier: 28–44. doi:10.1016/j.indcrop.2018.01.051.

Jain, Siddharth, and M. P. Sharma. 2011. "Oxidation Stability of Blends of Jatropha Biodiesel with Diesel." *Fuel* 90 (10). Elsevier Ltd: 3014–20. doi:10.1016/j.fuel.2011.05.003.

Jaiyen, Siyada, Thikumporn Naree, and Chawalit Ngamcharussrivichai. 2015. "Comparative Study of Natural Dolomitic Rock and Waste Mixed Seashells as Heterogeneous Catalysts for the Methanolysis of Palm Oil to Biodiesel." *Renewable Energy* 74. Elsevier Ltd: 433–40. doi:10.1016/j.renene.2014.08.050.

Janaun, Jidon, and Naoko Ellis. 2010. "Perspectives on Biodiesel as a Sustainable Fuel." *Renewable and Sustainable Energy Reviews* 14 (4). Elsevier Ltd: 1312–20. doi:10.1016/j.rser.2009.12.011.

Jiménez-Morales, I., J. Santamaría-González, P. Maireles-Torres, and A. Jiménez-López. 2011. "Calcined Zirconium Sulfate Supported on MCM-41 Silica as Acid Catalyst for Ethanolysis of Sunflower Oil." *Applied Catalysis B: Environmental.* doi:10.1016/j.apcatb.2011.01.014.

Jitputti, Jaturong, Boonyarach Kitiyanan, Pramoch Rangsunvigit, Kunchana Bunyakiat, Lalita Attanatho, and Peesamai Jenvanitpanjakul. 2006. "Transesterification of Crude Palm Kernel Oil and Crude Coconut Oil by Different Solid Catalysts." *Chemical Engineering Journal.* doi:10.1016/j.cej.2005.09.025.

Jogi, Ramakrishna, Y. V. V. Satyanarayana Murthy, M. R. S. Satyanarayana, T. Nagarjuna Rao, and Syed Javed. 2016. "Biodiesel Production from Degummed *Jatropha Curcas* Oil Using Constant-Temperature Ultrasonic Water Bath." *Energy Sources, Part A: Recovery, Utilization, and Environmental Effects* 38 (17). Taylor & Francis: 2610–16. doi:10.1080/15567036.2015.1093044.

Kafuku, G., and M. Mbarawa. 2010. "Biodiesel Production from Croton Megalocarpus Oil and Its Process Optimization." *Fuel* 89 (9). Elsevier Ltd: 2556–60. doi:10.1016/j.fuel.2010.03.039.

Karmakar, Aninidita, Subrata Karmakar, and Souti Mukherjee. 2010. "Properties of Various Plants and Animals Feedstocks for Biodiesel Production." *Bioresource Technology* 101 (19). Elsevier Ltd: 7201–10. doi:10.1016/j.biortech.2010.04.079.

Kawashima, Ayato, Koh Matsubara, and Katsuhisa Honda. 2008. "Development of Heterogeneous Base Catalysts for Biodiesel Production." *Bioresource Technology.* doi:10.1016/j.biortech.2007.08.009.

———. 2009a. "Acceleration of Catalytic Activity of Calcium Oxide for Biodiesel Production." *Bioresource Technology.* doi:10.1016/j.biortech.2008.06.049.

———. 2009b. "Bioresource Technology Acceleration of Catalytic Activity of Calcium Oxide for Biodiesel Production." *Bioresource Technology* 100 (2). Elsevier Ltd: 696–700. doi:10.1016/j.biortech.2008.06.049.

Knothe, Gerhard, and Luis F. Razon. 2017. "Biodiesel Fuels." *Progress in Energy and Combustion Science* 58. Elsevier Ltd: 36–59. doi:10.1016/j.pecs.2016.08.001.

Kojima, Yoshihiro, Hiroki Imazu, and Keiichi Nishida. 2014. "Physical and Chemical Characteristics of Ultrasonically-Prepared Water-in-Diesel Fuel: Effects of Ultrasonic Horn Position and Water Content." *Ultrasonics Sonochemistry* 21 (2). Elsevier B. V.: 722–28. doi:10.1016/j.ultsonch.2013.09.019.

Kumar, Dinesh, and Amjad Ali. 2012. "Nanocrystalline K-CaO for the Transesterification of a Variety of Feedstocks: Structure, Kinetics and Catalytic Properties." *Biomass and Bioenergy.* doi:10.1016/j.biombioe.2012.06.040.

———. 2013. "Transesterification of Low-Quality Triglycerides over a Zn/CaO Heterogeneous Catalyst: Kinetics and Reusability Studies." *Energy and Fuels.* doi:10.1021/ef400594t.

Kumar, Shiv, T. Ram, S. Mukesh, M. Ali, and S. Arya. 2013. "Sustainability of Biodiesel Production as Vehicular Fuel in Indian Perspective." *Renewable and Sustainable Energy Reviews* 25. Elsevier: 251–59. doi:10.1016/j.rser.2013.04.024.

Laca, Amanda, Adriana Laca, and Mario Díaz. 2017. "Eggshell Waste as Catalyst: A Review." *Journal of Environmental Management* 197. Elsevier Ltd: 351–59. doi:10.1016/j.jenvman.2017.03.088.

Lam, Man Kee, Keat Teong Lee, and Abdul Rahman Mohamed. 2010. "Homogeneous, Heterogeneous and Enzymatic Catalysis for Transesterification of High Free Fatty Acid Oil (Waste Cooking Oil) to Biodiesel: A Review." *Biotechnology Advances.* doi:10.1016/j.biotechadv.2010.03.002.

Leung, Dennis Y. C., Xuan Wu, and M. K. H. Leung. 2010. "A Review on Biodiesel Production Using Catalyzed Transesterification." *Applied Energy.* doi:10.1016/j.apenergy.2009.10.006.

Li, Xun, Xiao Yun He, Zhi Lin Li, You Dong Wang, Chun Yu Wang, Hao Shi, and Fei Wang. 2012. "Enzymatic Production of Biodiesel from Pistacia Chinensis Bge Seed Oil Using Immobilized Lipase." *Fuel.* doi:10.1016/j.fuel.2011.06.048.

Lim, Steven, and Lee Keat Teong. 2010. "Recent Trends, Opportunities and Challenges of Biodiesel in Malaysia: An Overview." *Renewable and Sustainable Energy Reviews* 14 (3): 938–54. doi:10.1016/j.rser.2009.10.027.

Lin, Lin, Zhou Cunshan, Saritporn Vittayapadung, Shen Xiangqian, and Dong Mingdong. 2011. "Opportunities and Challenges for Biodiesel Fuel." *Applied Energy* 88 (4). Elsevier Ltd: 1020–31. doi:10.1016/j.apenergy.2010.09.029.

Liu, Xuejun, Huayang He, Yujun Wang, Shenlin Zhu, and Xianglan Piao. 2008. "Transesterification of Soybean Oil to Biodiesel Using CaO as a Solid Base Catalyst." *Fuel.* doi:10.1016/j.fuel.2007.04.013.

López Granados, M., A. C. Alba-Rubio, F. Vila, D. Martín Alonso, and R. Mariscal. 2010. "Surface Chemical Promotion of Ca Oxide Catalysts in Biodiesel Production Reaction by the Addition of Monoglycerides, Diglycerides and Glycerol." *Journal of Catalysis* 276 (2): 229–36. doi:10.1016/j.jcat.2010.09.016.

Lucena, Izabelly L., Giovanilton F. Silva, and Fabiano A. N. Fernandes. 2008. "Biodiesel Production by Esterification of Oleic Acid with Methanol Using a Water Adsorption Apparatus." *Industrial and Engineering Chemistry Research.* doi:10.1021/ie800547h.

MacLeod, Claire S., Adam P. Harvey, Adam F. Lee, and Karen Wilson. 2008. "Evaluation of the Activity and Stability of Alkali-Doped Metal Oxide Catalysts for Application to an Intensified Method of Biodiesel Production." *Chemical Engineering Journal.* doi:10.1016/j.cej.2007.04.014.

Maiboom, Alain, and Xavier Tauzia. 2011. "NOx and PM Emissions Reduction on an Automotive HSDI Diesel Engine with Water-in-Diesel Emulsion and EGR: An Experimental Study." *Fuel* 90 (11). Elsevier Ltd: 3179–92. doi:10.1016/j.fuel.2011.06.014.

Mansir, Nasar, Siow Hwa Teo, Umer Rashid, and Yun Hin Taufiq-Yap. 2018. "Efficient Waste Gallus Domesticus Shell Derived Calcium-Based Catalyst for Biodiesel Production." *Fuel* 211 (May 2017): 67–75. doi:10.1016/j.fuel.2017.09.014.

Mardhiah, H. Haziratul, Hwai Chyuan Ong, H. H. Masjuki, Steven Lim, and H. V. Lee. 2017. "A Review on Latest Developments and Future Prospects of Heterogeneous Catalyst in Biodiesel Production from Non-Edible Oils." *Renewable and Sustainable Energy Reviews* 67. Elsevier: 1225–36. doi:10.1016/j.rser.2016.09.036.

Marwaha, Akshey, Pali Rosha, Saroj Kumar Mohapatra, Sunil Kumar Mahla, and Amit Dhir. 2018. "Waste Materials as Potential Catalysts for Biodiesel Production: Current State and Future Scope." *Fuel Processing Technology.* doi:10.1016/j.fuproc.2018.09.011.

Mat Yasin, Mohd Hafizil, Rizalman Mamat, G. Najafi, Obed Majeed Ali, Ahmad Fitri Yusop, and Mohd Hafiz Ali. 2017. "Potentials of Palm Oil as New Feedstock Oil for a Global Alternative Fuel: A Review." *Renewable and Sustainable Energy Reviews* 79 (February): 1034–49. doi:10.1016/j.rser.2017.05.186.

Meher, L. C., C. P. Churamani, Md Arif, Z. Ahmed, and S. N. Naik. 2013. "Jatropha Curcas as a Renewable Source for Bio-Fuels – A Review." *Renewable and Sustainable Energy Reviews* 26. Elsevier: 397–407. doi:10.1016/j.rser.2013.05.065.

Millo, Federico, Biplab Kumar Debnath, Theodoros Vlachos, Claudio Ciaravino, Lucio Postrioti, and Giacomo Buitoni. 2015. "Effects of Different Biofuels Blends on Performance and Emissions of an Automotive Diesel Engine." *Fuel* 159 (x). Elsevier Ltd: 614–27. doi:10.1016/j.fuel.2015.06.096.

Mofijur, M., A. E. Atabani, H. H. Masjuki, M. A. Kalam, and B. M. Masum. 2013. "A Study on the Effects of Promising Edible and Non-Edible Biodiesel Feedstocks on Engine Performance and Emissions Production: A Comparative Evaluation." *Renewable and Sustainable Energy Reviews* 23: 391–404. doi:10.1016/j.rser.2013.03.009.

Molaei Dehkordi, Asghar, and Mohammad Ghasemi. 2012. "Transesterification of Waste Cooking Oil to Biodiesel Using Ca and Zr Mixed Oxides as Heterogeneous Base Catalysts." *Fuel Processing Technology*. doi:10.1016/j.fuproc.2012.01.010.

Motasemi, F., and F. N. Ani. 2012. "A Review on Microwave-Assisted Production of Biodiesel." *Renewable and Sustainable Energy Reviews* 16 (7). Elsevier: 4719–33. doi:10.1016/j.rser.2012.03.069.

Mudge, S. M. 1999. "Shoreline Treatment of Spilled Vegetable Oils." *Spill Science & Technology Bulletin* 5 (5): 303–04. doi:10.1016/S1353-2561(00)00065-7.

Mudge, Stephen M., and Gloria Pereira. 1999. "Stimulating the Biodegradation of Crude Oil with Biodiesel Preliminary Results." *Spill Science and Technology Bulletin* 5 (5–6): 353–55. doi:10.1016/S1353-2561(99)00075-4.

Nayebzadeh, Hamed, Naser Saghatoleslami, Mohammad Haghighi, Mohammad Tabasizadeh, and Ehsan Binaeian. 2018. "Comparative Assessment of the Ability of a Microwave Absorber Nanocatalyst in the Microwave-Assisted Biodiesel Production Process." *Comptes Rendus Chimie* 21 (7). Elsevier Masson: 676–83. doi:10.1016/J.CRCI.2018.04.003.

Ng, Yan Fei, Liya Ge, Wen Kiat Chan, Swee Ngin Tan, Jean Wan Hong Yong, and Timothy Thatt Yang Tan. 2015. "An Environmentally Friendly Approach to Treat Oil Spill: Investigating the Biodegradation of Petrodiesel in the Presence of Different Biodiesels." *Fuel* 139. Elsevier Ltd: 523–28. doi:10.1016/j.fuel.2014.08.073.

Ngamcharussrivichai, Chawalit, Prangsinan Totarat, and Kunchana Bunyakiat. 2008. "Ca and Zn Mixed Oxide as a Heterogeneous Base Catalyst for Transesterification of Palm Kernel Oil." *Applied Catalysis A: General*. doi:10.1016/j.apcata.2008.02.020.

Ngamcharussrivichai, Chawalit, Wipawee Wiwatnimit, and Sarinyarak Wangnoi. 2007. "Modified Dolomites as Catalysts for Palm Kernel Oil Transesterification." *Journal of Molecular Catalysis A: Chemical* 276 (1–2): 24–33. doi:10.1016/j.molcata.2007.06.015.

Pasupulety, Nagaraju, Kamalakar Gunda, Yuanqing Liu, Garry L. Rempel, and Flora T. T. Ng. 2013. "Production of Biodiesel from Soybean Oil on CaO/Al2O3 Solid Base Catalysts." *Applied Catalysis A: General* 452. Elsevier B. V.: 189–202. doi:10.1016/j.apcata.2012.10.006.

Petchmala, Akaraphol, Navadol Laosiripojana, Bunjerd Jongsomjit, Motonobu Goto, Joongjai Panpranot, Okorn Mekasuwandumrong, and Artiwan Shotipruk. 2010. "Transesterification of Palm Oil and Esterification of Palm Fatty Acid in Near- and Super-Critical Methanol with SO4-ZrO2 Catalysts." *Fuel*. doi:10.1016/j.fuel.2010.04.010.

Piker, Alla, Betina Tabah, Nina Perkas, and Aharon Gedanken. 2016. "A Green and Low-Cost Room Temperature Biodiesel Production Method from Waste Oil Using Egg Shells as Catalyst." *Fuel*. doi:10.1016/j.fuel.2016.05.078.

Predojević, Zlatica J. 2008. "The Production of Biodiesel from Waste Frying Oils: A Comparison of Different Purification Steps." *Fuel.* doi:10.1016/j.fuel.2008.07.003.

Prince, Roger C., Christine Haitmanek, and Catherine Coyle Lee. 2008. "The Primary Aerobic Biodegradation of Biodiesel B20." *Chemosphere* 71 (8): 1446–51. doi:10.1016/j.chemosphere.2007.12.010.

Rahimi, Masoud, Babak Aghel, Mohammad Alitabar, Arash Sepahvand, and Hamid Reza Ghasempour. 2014. "Optimization of Biodiesel Production from Soybean Oil in a Microreactor." *Energy Conversion and Management* 79. Elsevier Ltd: 599–605. doi:10.1016/j.enconman.2013.12.065.

Rashid, Umer, Farooq Anwar, Bryan R. Moser, and Gerhard Knothe. 2008. "Moringa Oleifera Oil: A Possible Source of Biodiesel." *Bioresource Technology* 99 (17): 8175–79. doi:10.1016/j.biortech.2008.03.066.

Rashid, Umer, Farooq Anwar, Bryan R. Moser, and Samia Ashraf. 2008. "Production of Sunflower Oil Methyl Esters by Optimized Alkali-Catalyzed Methanolysis." *Biomass and Bioenergy.* doi:10.1016/j.biombioe.2008.03.001.

Roschat, Wuttichai, Theeranun Siritanon, Boonyawan Yoosuk, Taweesak Sudyoadsuk, and Vinich Promarak. 2017. "Rubber Seed Oil as Potential Non-Edible Feedstock for Biodiesel Production Using Heterogeneous Catalyst in Thailand." *Renewable Energy* 101. Elsevier Ltd: 937–44. doi:10.1016/j.renene.2016.09.057.

Sadeghinezhad, E., S. N. Kazi, A. Badarudin, C. S. Oon, M. N. M. Zubir, and Mohammad Mehrali. 2013. "A Comprehensive Review of Bio-Diesel as Alternative Fuel for Compression Ignition Engines." *Renewable and Sustainable Energy Reviews* 28. Elsevier: 410–24. doi:10.1016/j.rser.2013.08.003.

Sagiroglu, Ayten, Isbilir Selen, Mevlut Ozcan, Hatice Paluzar, and Neslihan Toprakkiran. 2011. "Comparison of Biodiesel Productivities of Different Vegetable Oils by Acidic Catalysis." *Chemical Industry and Chemical Engineering Quarterly.* doi:10.2298/ciceq100114054s.

Sahani, Shalini, Sushmita Banerjee, and Yogesh C. Sharma. 2018. "Study of 'Co-Solvent Effect' on Production of Biodiesel from Schleichera Oleosa Oil Using a Mixed Metal Oxide as a Potential Catalyst." *Journal of the Taiwan Institute of Chemical Engineers.* doi:10.1016/j.jtice.2018.01.029.

Saka, S., and D. Kusdiana. 2001. "Biodiesel Fuel from Rapeseed Oil as Prepared in Supercritical Methanol." *Fuel.* doi:10.1016/S0016-2361(00)00083-1.

Santana, Harrson S., Deborah S. Tortola, João L. Silva, and Osvaldir P. Taranto. 2017. "Biodiesel Synthesis in Micromixer with Static Elements." *Energy Conversion and Management* 141. Elsevier Ltd: 28–39. doi:10.1016/j.enconman.2016.03.089.

Sarin, Rakesh, Meeta Sharma, S. Sinharay, and R. K. Malhotra. 2007. "Jatropha-Palm Biodiesel Blends: An Optimum Mix for Asia." *Fuel.* doi:10.1016/j.fuel.2006.11.040.

Saydut, A., A. B. Kafadar, F. Aydin, S. Erdogan, C. Kaya, and C. Hamamci. 2016. "Effect of Homogeneous Alkaline Catalyst Type on Biodiesel Production from Soybean [Glycine Max (L.) Merrill] Oil." *Indian Journal of Biotechnology* 15 (4): 596–600.

Shah, Shweta, and Munishwar N. Gupta. 2007. "Lipase Catalyzed Preparation of Biodiesel from Jatropha Oil in a Solvent Free System." *Process Biochemistry.* doi:10.1016/j.procbio.2006.09.024.

Shahabuddin, M., M. A. Kalam, H. H. Masjuki, M. M. K. Bhuiya, and M. Mofijur. 2012. "An Experimental Investigation into Biodiesel Stability by Means of Oxidation and Property Determination." *Energy* 44 (1): 616–22. doi:10.1016/j.energy.2012.05.032.

Shankar, A. Arun, Prudhvi Raj Pentapati, and R. Krishna Prasad. 2017. "Biodiesel Synthesis from Cottonseed Oil Using Homogeneous Alkali Catalyst and Using Heterogeneous Multi Walled Carbon Nanotubes: Characterization and Blending Studies." *Egyptian Journal of Petroleum.* doi:10.1016/j.ejpe.2016.04.001.

Silitonga, A. S., H. H. Masjuki, T. M. I. Mahlia, H. C. Ong, A. E. Atabani, and W. T. Chong. 2013. "A Global Comparative Review of Biodiesel Production from Jatropha Curcas Using Different Homogeneous Acid and Alkaline Catalysts: Study of Physical and Chemical Properties." *Renewable and Sustainable Energy Reviews*. doi:10.1016/j.rser.2013.03.044.

Silva, C., T. A. Weschenfelder, S. Rovani, F. C. Corazza, M. L. Corazza, C. Dariva, and J. Vladimir Oliveira. 2007. "Continuous Production of Fatty Acid Ethyl Esters from Soybean Oil in Compressed Ethanol." *Industrial and Engineering Chemistry Research*. doi:10.1021/ie070310r.

Singh, Veena, Meena Yadav, and Yogesh Chandra Sharma. 2017. "Effect of Co-Solvent on Biodiesel Production Using Calcium Aluminium Oxide as a Reusable Catalyst and Waste Vegetable Oil." *Fuel*. doi:10.1016/j.fuel.2017.04.111.

Sinha, Duple, and S. Murugavelh. 2016. "Comparative Studies on Biodiesel Production from Waste Cotton Cooking Oil Using Alkaline, Calcined Eggshell and Pistachio Shell Catalyst." In *2016 International Conference on Energy Efficient Technologies for Sustainability, ICEETS 2016*. doi:10.1109/ICEETS.2016.7582912.

Smith, Gerard V., and Ferenc Notheisz. 1999. "Immobilized Homogeneous Catalysts." In *Heterogeneous Catalysis in Organic Chemistry*. doi:10.1016/b978-012651645-6/50007-x.

Suryaputra, Wijaya, Indra Winata, Nani Indraswati, and Suryadi Ismadji. 2013. "Waste Capiz (Amusium Cristatum) Shell as a New Heterogeneous Catalyst for Biodiesel Production." *Renewable Energy*. doi:10.1016/j.renene.2012.08.060.

Tan, K. T., K. T. Lee, and A. R. Mohamed. 2011. "Potential of Waste Palm Cooking Oil for Catalyst-Free Biodiesel Production." *Energy*. doi:10.1016/j.energy.2010.05.003.

Tan, Kok Tat, and Keat Teong Lee. 2011. "A Review on Supercritical Fluids (SCF) Technology in Sustainable Biodiesel Production: Potential and Challenges." *Renewable and Sustainable Energy Reviews*. doi:10.1016/j.rser.2011.02.012.

Tan, Shiou Xuan, Steven Lim, Hwai Chyuan Ong, and Yean Ling Pang. 2019. "State of the Art Review on Development of Ultrasound-Assisted Catalytic Transesterification Process for Biodiesel Production." *Fuel* 235 (March 2018). Elsevier: 886–907. doi:10.1016/j.fuel.2018.08.021.

Tan, Yie Hua, Mohammad Omar Abdullah, and Cirilo Nolasco-Hipolito. 2015. "The Potential of Waste Cooking Oil-Based Biodiesel Using Heterogeneous Catalyst Derived from Various Calcined Eggshells Coupled with an Emulsification Technique: A Review on the Emission Reduction and Engine Performance." *Renewable and Sustainable Energy Reviews*. doi:10.1016/j.rser.2015.03.048.

Tan, Yie Hua, Mohammad Omar Abdullah, Cirilo Nolasco-Hipolito, and Yun Hin Taufiq-Yap. 2015. "Waste Ostrich- and Chicken-Eggshells as Heterogeneous Base Catalyst for Biodiesel Production from Used Cooking Oil: Catalyst Characterization and Biodiesel Yield Performance." *Applied Energy* 160. Elsevier Ltd: 58–70. doi:10.1016/j.apenergy.2015.09.023.

Tang, Ying, Jingfang Xu, Jie Zhang, and Yong Lu. 2013. "Biodiesel Production from Vegetable Oil by Using Modified CaO as Solid Basic Catalysts." *Journal of Cleaner Production*. doi:10.1016/j.jclepro.2012.11.001.

Tarabet, L., K. Loubar, M. S. Lounici, K. Khiari, T. Belmrabet, and M. Tazerout. 2014. "Experimental Investigation of Di Diesel Engine Operating with Eucalyptus Biodiesel/ Natural Gas under Dual Fuel Mode." *Fuel* 133. Elsevier Ltd: 129–38. doi:10.1016/j.fuel.2014.05.008.

Tariq, Muhammad, Saqib Ali, and Nasir Khalid. 2012. "Activity of Homogeneous and Heterogeneous Catalysts, Spectroscopic and Chromatographic Characterization of Biodiesel: A Review." *Renewable and Sustainable Energy Reviews* 16: 6303–16.

Taufiq-Yap, Y. H., H. V. Lee, M. Z. Hussein, and R. Yunus. 2011. "Calcium-Based Mixed Oxide Catalysts for Methanolysis of Jatropha Curcas Oil to Biodiesel." *Biomass and Bioenergy*. doi:10.1016/j.biombioe.2010.11.011.

Taufiq-Yap, Yun Hin, Siow Hwa Teo, Umer Rashid, Aminul Islam, Mohd Zobir Hussien, and Keat Teong Lee. 2014. "Transesterification of Jatropha Curcas Crude Oil to Biodiesel on Calcium Lanthanum Mixed Oxide Catalyst: Effect of Stoichiometric Composition." *Energy Conversion and Management.* doi:10.1016/j.enconman.2013.12.075.

Thanh, Le Tu, Kenji Okitsu, Luu Van Boi, and Yasuaki Maeda. 2012. "Catalytic Technologies for Biodiesel Fuel Production and Utilization of Glycerol: A Review." *Catalysts* 2: 191–222. doi:10.3390/catal2010191.

Thanh, Le Tu, Kenji Okitsu, Yasuaki Maeda, and Hiroshi Bandow. 2014. "Ultrasound Assisted Production of Fatty Acid Methyl Esters from Transesterification of Triglycerides with Methanol in the Presence of KOH Catalyst: Optimization, Mechanism and Kinetics." *Ultrasonics Sonochemistry.* doi:10.1016/j.ultsonch.2013.09.015.

Thirugnanasambandham, K., K. Shine, A. Agatheeshwaren, and V. Sivakumar. 2016. "Biodiesel Production from Castor Oil Using Potassium Hydroxide as a Catalyst: Simulation and Validation." *Energy Sources, Part A: Recovery, Utilization and Environmental Effects.* doi:10.1080/15567036.2016.1179363.

Thitsartarn, W., and S. Kawi. 2011. "An Active and Stable CaO-CeO 2 Catalyst for Transesterification of Oil to Biodiesel." *Green Chemistry.* doi:10.1039/c1gc15596b.

Utlu, Zafer, and Mevlüt Süreyya Koçak. 2008. "The Effect of Biodiesel Fuel Obtained from Waste Frying Oil on Direct Injection Diesel Engine Performance and Exhaust Emissions." *Renewable Energy* 33 (8): 1936–41. doi:10.1016/j.renene.2007.10.006.

Varatharajan, K., M. Cheralathan, and R. Velraj. 2011. "Mitigation of NOx Emissions from a Jatropha Biodiesel Fuelled Di Diesel Engine Using Antioxidant Additives." *Fuel* 90 (8). Elsevier Ltd: 2721–25. doi:10.1016/j.fuel.2011.03.047.

Veljković, Vlada B., Jelena M. Avramović, and Olivera S. Stamenković. 2012. "Biodiesel Production by Ultrasound-Assisted Transesterification: State of the Art and the Perspectives." *Renewable and Sustainable Energy Reviews* 16 (2): 1193–209. doi:10.1016/j.rser.2011.11.022.

Vicente, Gemma, Mercedes Martínez, and José Aracil. 2004. "Integrated Biodiesel Production: A Comparison of Different Homogeneous Catalysts Systems." *Bioresource Technology* 92 (3): 297–305. doi:10.1016/j.biortech.2003.08.014.

———. 2007. "Optimisation of Integrated Biodiesel Production. Part I. A Study of the Biodiesel Purity and Yield." *Bioresource Technology.* doi:10.1016/j.biortech.2006.07.024.

Wang, Boyang, Shufen Li, Songjiang Tian, Rihua Feng, and Yonglu Meng. 2013. "A New Solid Base Catalyst for the Transesterification of Rapeseed Oil to Biodiesel with Methanol." *Fuel.* doi:10.1016/j.fuel.2012.08.034.

Wang, Jianxun, Kungtung Chen, and Chiingchang Chen. 2011. "Biodiesel Production from Soybean Oil Catalyzed by K2SiO3/C." *Chinese Journal of Catalysis* 32 (9–10). Elsevier: 1592–96. doi:10.1016/S1872-2067(10)60265-3.

Wen, Libai, Yun Wang, Donglian Lu, Shengyang Hu, and Heyou Han. 2010. "Preparation of KF/CaO Nanocatalyst and Its Application in Biodiesel Production from Chinese Tallow Seed Oil." *Fuel* 89 (9). Elsevier Ltd: 2267–71. doi:10.1016/j.fuel.2010.01.028.

Wu, Xuemin, Fenfen Zhu, Juanjuan Qi, Luyao Zhao, Fawei Yan, and Chenghui Li. 2017. "Challenge of Biodiesel Production from Sewage Sludge Catalyzed by KOH, KOH/ Activated Carbon, and KOH/CaO." *Frontiers of Environmental Science and Engineering.* doi:10.1007/s11783-017-0913-y.

Yadav, Ashok Kumar, Mohd Emran Khan, Amit Pal, and Balbir Singh. 2016. "Ultrasonic-Assisted Optimization of Biodiesel Production from Karabi Oil Using Heterogeneous Catalyst." *Biofuels* 9 (1). Taylor & Francis: 101–12. doi:10.1080/17597269.2016.125 9522.

Yan, Shuli, Craig Dimaggio, Siddharth Mohan, Manhoe Kim, Steven O. Salley, and K. Y. Simon Ng. 2010. "Advancements in Heterogeneous Catalysis for Biodiesel Synthesis." *Topics in Catalysis* 53 (11–12): 721–36. doi:10.1007/s11244-010-9460-5.

Yan, Shuli, Houfang Lu, and Bin Liang. 2008. "Supported CaO Catalysts Used in the Transesterification of Rapeseed Oil for the Purpose of Biodiesel Production." *Energy and Fuels*. doi:10.1021/ef070105o.

Yang, Zhenqiang, and Wenlei Xie. 2007. "Soybean Oil Transesterification over Zinc Oxide Modified with Alkali Earth Metals." *Fuel Processing Technology*. doi:10.1016/j.fuproc.2007.02.006.

Ye, Wei, Yujie Gao, Hui Ding, Mingchao Liu, Shejiang Liu, Xu Han, and Jinlong Qi. 2016. "Kinetics of Transesterification of Palm Oil under Conventional Heating and Microwave Irradiation, Using CaO as Heterogeneous Catalyst." *Fuel* 180 (September). Elsevier: 574–79. doi:10.1016/J.FUEL.2016.04.084.

Yin, Xiulian, Xiuli Duan, Qinghong You, Chunhua Dai, Zhongbiao Tan, and Xiaoyan Zhu. 2016. "Biodiesel Production from Soybean Oil Deodorizer Distillate Usingcalcined Duck Eggshell as Catalyst." *Energy Conversion and Management*. doi:10.1016/j.enconman.2016.01.026.

Zabeti, Masoud, Wan Mohd Ashri Wan Daud, and Mohamed Kheireddine Aroua. 2009. "Optimization of the Activity of CaO/Al2O3 Catalyst for Biodiesel Production Using Response Surface Methodology." *Applied Catalysis A: General* 366 (1): 154–59. doi:10.1016/j.apcata.2009.06.047.

———. 2010. "Biodiesel Production Using Alumina-Supported Calcium Oxide: An Optimization Study." *Fuel Processing Technology*. doi:10.1016/j.fuproc.2009.10.004.

# 3 Physicochemical and Thermal Properties of Biodiesel

## 3.1 INTRODUCTION

Fossil fuel usage has increased from 6 Gtoe to 15 Gtoe of energy consumption during the period from 1970 to 2015 (Singh et al. 2020). As the world population increases day by day, the need for petroleum products and the depletion of petroleum reserves also increase rapidly. Demand for energy has made a way for research into alternative fuels (Saravanan et al. 2020). Biodiesel is one alternative fuel similar to petrol diesel which can be derived from biomass. The chemical properties of biodiesel are similar to those of diesel fuel, and biodiesel can help in the substitution of the petroleum fuels. The merits of biodiesel are domestic origin, biodegradability, nontoxicity, and inherent lubricity (Fauzan et al. 2020).

Biodiesel is a combination of methyl esters of fatty acid with specific standards. Biodiesels are mainly produced from vegetable oils, animal fats, algae, and waste oils (Yesilyurt et al. 2020). Biodiesel can be used in diesel engines with blending on diesel without any engine modifications. Research into biodiesel has shown that it produces less toxic emissions compared to diesel (Madheshiya and Vedrtnam 2018). The usage of edible vegetable oils can create an energy imbalance and an availability problem. So to overcome this, non-edible sources have been identified for biodiesel production. All over the world, around 360 species of feedstocks are identified as a potential for biodiesel. Usage of non-edible sources can improve the economy and it will have less impact on the environment (Kumar and Sharma 2016). Normally vegetable oils have greater viscosity, which makes them difficult in direct usage in diesel engines. Biodiesels are prepared using various techniques that are involved in reducing the viscosity of feedstocks. Pyrolysis, dilution, micro-emulsion, and transesterification are the methods. Among these, transesterification is the most commonly used technique for biodiesel production (Atabani, Silitonga, et al. 2013).

Transesterification is a chemical process between vegetable oils and alcohols with the presence of catalyst to form alkyl esters. Alcohol usage does not determine the yield; cost and preparation determines the selection of alcohol. The most frequently consumed alcohols are ethanol and methanol. Transesterification is a three-stage process in which triglycerides of oil are converted to diglycerides following conversion of monoglycerides. During the reaction, methyl ester is formed with glycerol as a by-product, and glycerol has commercial value (Ogunkunle and Ahmed 2019).

The properties of biodiesel depend upon the various pretreatment processes, alcohol used, production techniques, and post-treatment processes. The physicochemical properties of the biodiesel changes from one feedstock to another because the composition of fatty acids varies. To ensure quality biodiesel, international standards EN14214 and ASTM D6751 were established, and these standards give information about the quality requirements and test procedures for determining the properties associated with the biodiesel such as its physicochemical properties and its various blends (Caldeira et al. 2017). During the transesterification reaction, the composition of fatty acids does not change. Strongly dependent on the fatty acid characteristics of feedstocks like the number of bonds and chain length are the physicochemical properties of biodiesel, such as density, viscosity, iodine value (IV), lubricity, calorific value, cetane number, cold flow properties, and oxidation stability (Yaşar 2020).

Biodiesel has a high flash point, enhanced biodegradability, and less toxicity (Knothe 2006). The oxidation stability of the biodiesel is very low and it also has poor cold flow properties. The decrement in the stability of oxidation of biodiesel is because of the presence of polysaturated fatty acid methyl esters (FAME), as it can further react to form to aldehydes, acids, ketones, and peroxides, which can change properties of biodiesel. Due to oxidation reaction, biodiesel can have saponification at an undesirable rate, which can lead to an increase in kinematic viscosity and acid number (Tang et al. 2008). The oxidation stability is a major factor involved in storage of FAME for longer periods (Adu-Mensah et al. 2019). Atomization and vaporization characteristics of the fuel are highly dependent on the properties of kinematic viscosity and density of the biodiesel (Silitonga et al. 2013). Cetane number improves the performance of the engine and characteristics of the combustion; also, the cetane number of the fuel is a link between density of the fuel and fuel's calorific value. Biodiesel has a high cetane number because of its rich oxygen content (Miraboutalebi, Kazemi, and Bahrami 2016).

Engine fuel not only provides energy; it should also provide lubricity to the fuel flow system. The tribological properties of engine parts are influenced by the various properties of biodiesel (Fazal, Haseeb, and Masjuki 2013). The lubricity of biodiesel is superior compared to that of ordinary diesel. The lubricity properties of various biodiesels vary depending upon their chemical properties. Before the usage of biodiesel in a conventional diesel engine, its physicochemical properties and tribological behavior of the biodiesel should be found out for better results (Wakil et al. 2015). This chapter is about the determination and influence of the physicochemical and thermal properties of biodiesel for various feedstocks.

## 3.2   ACID VALUE

The feedstocks used for the biodiesel production should have free fatty acid (FFA) content below 1% by mass (Zhou et al. 2017). If the FFA values are above 2% of its mass, then it requires a reaction of esterification consisting of two step transesterification in order to reduce the acidic value present in biodiesel. Esterification process is the first in which the reaction utilizes methanol and catalyst for producing FFA esters. The transesterification process is the second step, followed by esterification, in which base catalyst is used with the esterification products, and this yields larger amounts of methyl esters of the fatty acid and glycerol. The drawbacks of this two-step transesterification are usage of more alcohol content and increased reaction time and heat energy required for reaction (Sahabdheen and Arivarasu 2019).The high percentage of tri-glycerides content increases the acid value, and the saponification value of the biodiesel because the acid value and saponification value are proportional to the triglycerides content. Similarly, for lower acid value, number of fatty acids will be less in number. It was found that the value of acid of Aegle Marmelos Correa (AMC) oil was 2.6 mg KOH/g of oil. After the two-step transesterification process, the AMC's acid value was found to be 0.2 mg KOH/g of oil (Thangarasu, Siddharth, and Ramanathan 2020). The acid value of Jatropha oil was 5–10 mg KOH/g and after a two-step transesterification, the acid value of oil was 0.154 with a nano-based catalyst (Deng et al. 2011).

The presence of the amount of free fatty acids in the fat was determined by the acid value. With the standard EN14104 titration method, the acid value of the biodiesel and any oil can be determined. A beaker was taken for titration process with a diethyl ether and ethanol mixture of 50 mL, a 1-g sample to be tested is taken, and it is dissolved. The mixture was added with 2–3 drops of phenolphthalein indicator. The titration was performed with a burette solution of an aqueous solution of KOH also known as potassium hydroxide with a normality of 0.1 N. The following equation was used for calculating the acid value of the tested sample.

$$\text{Acid value} = \frac{MW \times N \times V}{w} \tag{3.1}$$

where $MW$, KOH molecular weight, g/mol; $V$, volume of KOH in mL; $N$, KOH normality; $w$, weight of the sample.

## 3.3 SAPONIFICATION VALUE

The saponification value is the required quantity of alkali to saponify a definite amount of oil or fat. The titration method of ASTM D1962 standard was used for calculation of the saponification value of biodiesel and oil samples. A unit gram of sample fat was used in the beaker for testing, which was then dissolved with ethanol solvent of 10 mL. In addition, the combination of the oil and the solvent was quantitatively discharged into ethanolic solution of 25 mL and KOH of 0.5 normality, and the result was labeled as the test sample. Another plain sample was prepared with the same procedure without any addition of sample fat. Both the test and plain samples are taken and fitted to the reflux condenser, thereby heating them for 30 min at the temperature of boiling water. After this method, the samples were cooled to reach room temperature. Eventually, a measure of 2–3 decreases of phenolphthalein was mixed to both samples and titration was done against 0.5 N hydrochloric acid (Thangarasu and Anand 2019). The following equation was used to estimate the saponification value.

$$\text{Saponification value} = \frac{MW \times N \times \left(V_{\text{plain}} - V_{\text{test}}\right)}{w} \tag{3.2}$$

where $MW$, KOH molecular weight, g/mol; $V_{plain}$, HCl volume for plain sample, mL; $V_{test}$, HCl volume for test sample, mL; $N$, KOH normality, mol/mL; $w$, sample weight, g.

The saponification value shows the saponifiable unit qualities (acyl groups) per unit weight of the given oil. A higher value of saponification suggests a greater portion of the presence of the higher-molecular-weight fatty acids which are low in the oil or vice versa (Diwakar et al. 2010). The saponification value calculates the oil's average molecular weight and it is measured in potassium hydroxide milligrams (mg KOH g⁻¹ oil) (Ismail and Ali 2015). It calculates the total conversion potential of the oil through conventional transesterification methods based on biodiesel production. If the saponification value is higher, it can lead to greater corrosion problems in diesel engine parts (Belagur and Chitimi 2013). Saponification values of different feedstocks were reported previously in the ranges of 169–312 mg KOH g⁻¹ oil (Azam, Waris, and Nahar 2005), 173–200 mg KOH g⁻¹ oil (Toscano et al. 2012), and 168–202 mg KOH g⁻¹ oil (Belagur and Chitimi 2013). Oil of higher saponification values can be used for soap and shampoo products. The soap-making ability of the oil depends on its saponification value (Odoom and Edusei 2015). The saponification value is 183.9 mg KOH g⁻¹ for Jatropha biodiesel and 200.2 mg KOH g⁻¹ for *Calophyllum inophyllum* biodiesel. A higher saponification value of 212–268 mg KOH g⁻¹ was seen in coconut oil (Marutani, Soria, and Martinez 2018). The saponification value of Aegle Marmelos Correa biodiesel is 224 mg KOH g⁻¹ (Thangarasu, Balaji, and Ramanathan 2019). Karanja biodiesel has a saponification value of 189 mg KOH g⁻¹ (Bart, Palmeri, and Cavallaro 2010).

## 3.4 IODINE VALUE

Iodine value can be measured to find the average number of double bonds in an oil or fat. Double bonds are created by the addition of halogen, with consumption of 1 mol of halogen for every double bond. The amount of iodine consumed will give the average range of the double bonds existing in it. Based on the standard of EN14111, titration was done to determine the value of iodine of biodiesel and oil. The titration was started with 1 g of oil sample in a beaker; 10 mL of chloroform solvent was dissolved, and the resulting mixture was named as the "test sample." Afterwards, it was mixed with iodine monochloride reagent of 20 mL thoroughly. Similarly, a blank sample was prepared by the above procedure without the addition of a sample with fat. The two samples were placed in a black room for a 30-min incubation. Then, sample was added to potassium iodine solution of 10 mL and the beaker side was rinsed with distilled water of 50 mL. The titration was done with both samples against the aqueous solution of sodium thiosulfate ($Na_2S_2O_3$), which has normality of 0.1; the titration was done until the color changed to pale straw. The solution was then given 1 mL of starch indicator, and the color of the solution switched to violet. The solution was titrated until it became colorless. The following equation was used for the calculation of the value of iodine in the test sample.

$$\text{Iodine value} = \frac{MW \times N \times \left(V_{\text{plain}} - V_{\text{test}}\right) \times 100 \times 10^{-3}}{w} \qquad (3.3)$$

where $MW$, molecular weight $Na_2S_2O_3$, g/mol; $V_{plain}$, $Na_2S_2O_3$ volume for blank sample, mL; $V_{test}$, $Na_2S_2O_3$ volume for test sample, mL; $N$, normality of $Na_2S_2O_3$, mol/mL; $w$, sample weight, g.

Classically, the iodine value criterion was used to calculate the total degree of unsaturation in vegetable oils and fats. Measurement of the IV in biodiesel is useful as a reference to avoid problems in engines such as polymerization and deposition on piston rings, engine nozzles, and piston ring grooves. Greater IV amounts are also associated with the development of degradation of various materials, which can adversely affect engine operability and reduction in lubrication property (Bouaid, Martinez, and Aracil 2007). The IV is a chemical parameter that is based on the alkyl double bond reactivity. This is measured by calculating the amount of mono-halide iodine contained in a sample. Regional and international standards organizations provide for certain protocols for evaluating IV in biodiesel. The iodine value is related to oxidative stability and represents the oil or fat's propensity to oxidize, as well as to polymerize and forming deposits in engines. However, several profiles of fatty acids and different fatty acid structures can produce the same IV. In addition, there was no clear observation of links between IV and oxidative stability in an investigation of the biodiesel with an extensive IV range.

Furthermore, since engine efficiency tests with a combination of oils from vegetables of various IV values lead to failure in the production of the required outcomes, which leads to justification for a low IV, it is not completely shocking that IV is excluded from all national biodiesel requirements (Bart, Palmeri, and Cavallaro 2010). Jatropha biodiesel has 82.9 mg iodine $g^{-1}$ and *Calophyllum inophyllum* biodiesel has 84.9 mg iodine $g^{-1}$ (Marutani, Soria, and Martinez 2018). Balakrishnan, Parthasarathy, and Gollahalli (2016) reported that the iodine values of those biodiesels in the 59–185 range, which emits lower NO emissions than diesel. The IV of AMC biodiesel was 56, which is within allowable limits according to the standards (Thangarasu, Balaji, and Ramanathan 2019). Karanja biodiesel has an iodine value of 86.5 and it has greater carbon–carbon bonds due to the greater iodine value (Das et al. 2009).

## 3.5 CETANE NUMBER

Cetane number is an indicator of any fuel's ignition characteristics. Better ignition properties will be observed for fuels of higher cetane number. Vegetable oils have lower cetane number, and this corresponds to knocking at low engine speeds. Similarity, delay in ignition is high for vegetable oils, which also have slow rates of combustion compared to diesel. As said above, cetane number has an inverse correlation with the delay time of the ignition of fuel. The delay in the ignition of a diesel engine is related to the air-fuel ratio, or the amount of the fuel consumed per unit of air. The same relationship can be used for calculating the cetane number. In Cooperative Fuel Research (CFR) engines, the rate of consumption of air is variable and the rate of consumption of fuel is constant, so this can help in determining cetane number by measuring the fuel consumption rate and co-relating it with the relevant figure (consumption rate of the cetane-air) which has been given by the providers of the CFR engine.

Before starting the experiment, the engine is calibrated with a known normal cetane base fuel. After this, the fuel sample whose cetane number is to be calculated is discharged into the engine. Here in the engine, the rate of consumption of fuel remains constant and for determining the delay in the ignition, the only variable in the rate of consumption of air should be known after obtaining a constant delay time of ignition. After the experiment, the fuel's cetane number is obtained with the help of an air consumption measurement gauge and a standard diagram provided by the manufacturers.

The level of saturation of the fatty acids was strongly affected by the cetane amount, but further properties also show an active part. CN, combustion heat, melting point, and viscosity of neat fat compounds typically escalate with rise in the chain length (i.e., uninterrupted $CH_2$ molecules) and fall of the increased unsaturation. Saturated fatty acids have a beneficial effect on the number of cetanes and the resistance of oxidation. Saturated fatty acid esters, such as palmitic (C16:0) and stearic (C18:0) acids, are observed with high CNs. In addition to escalation in the double bonds (and their chain position), chain branching often lowers CNs. Cetane numbers of 49.5, 42.9, and 58, respectively, were found for Jatropha, Karanja, and AMC biodiesel (Thangarasu and Anand 2019).

## 3.6  CLOUD AND POUR POINT

With the help of subzero equipment, the cloud of diesel and biodiesel and pour point of diesel and biodiesel can be determined as per the ASTM D2500 standard. Tubes of height of 12 m and diameter of 3 cm were taken for the subzero equipment. The tubes are kept in a refrigerated chamber, where they are surrounded by a copper vessel. A 50 mL sample of fuel is filled into a glass tube, which is fitted with a rubber cork. The sample-filled fuel tubes are kept inside the refrigerated chamber. At every temperature interval of 1°C, the glass tubes were taken out to check the cloud and pour point from the copper vessel. The temperature at which cloud formed was called cloud point temperature. Primarily after the cloud point temperature check, the sample was checked for motionlessness by tilting the glass tube to a horizontal position for 5 s. The point where it is motionless is known as pour point temperature.

The cloud point is the fuel temperature where observation of the small solid crystals can be done for cooling down as the fuel. The cold filter plug point (CFPP) is the temperature at which the crystallized fuel components result in fuel filter clogging. The pour point is the temperature at which there is absolutely no fuel part movement when the container is turned. AMC biodiesel, Karanja biodiesel, palm biodiesel, *Calophyllum inophyllum*, and Jatropha biodiesel cloud scales are found to be around 1°C, 4°C, 6°C, 4°C, and 6°C, respectively. AMC biodiesel, Karanja biodiesel, palm biodiesel, *Calophyllum inophyllum*, and Jatropha biodiesel pour points are found to be −4°C, −2°C, 6°C, 2°C, and −7°C, respectively (Silitonga et al. 2016). Major problems linked with using biodiesel include its low-temperature properties, which are very poor; these are seen in cold climates with stearic and palmitic acid methyl esters from palm biodiesel. As long as the carbon chains in biodiesel remained unsaturated, the low-temperature properties remain poor. Biodiesels derived from almond, olive, maize, rapeseed, castor seed, and high oleic sunflower oils have the best inclusive properties, as they have higher mono-unsaturated content (Ramos et al. 2009). At present, combining the fatty acid esters along with fossil diesel is the most used technique for enhancing biodiesel's cold properties. Fossil diesel present in the blended fuel behaves as a solvent of low-temperature precipitated crystals, waxes, or gels (Mejía, Salgado, and Orrego 2013).

## 3.7 COLD FILTER PLUGGING POINT

CFPP is defined as the lowest temperature at which a standard volume of diesel-like fuel can pass through a standard filter device in a given time when it is cooled under certain operating conditions and it is expressed in 1°C increments. The CFPP gives an idea about the lowest temperature that the fuel can flow without giving any trouble in the fuel system (Mejía, Salgado, and Orrego 2013). For countries which have very cold temperatures, a high CFPP can block engines easily.

CFPP is assessed according to ASTM D6371 standards, in an automatic pour point tester. A flat-bottomed test jar of cylindrical clear glass with a diameter of 32 mm and a height of about 120 mm can be used. The jar should have graduated markings up to the 50 mL level. A thermometer was selected for measuring temperatures of –38°C to 50°C. The bath temperature can range up to –34°C. The type of cooling bath is optional; the bath has a fitted covering with one or more holes to support a ring. About 50 mL of the sample at room temperature should be filled into the jar, and in case of room temperature below 15°C dry filter paper should be used. Once the sample fuel reaches the 45 mL mark, the jar should be closed with the stopper carrying the pipet with the filter unit and the graduated thermometer. Prepare the cooling bath and jacket; after that place the jar in the cooling jacket, which is kept at –34°C and connected with the vacuum source. It is best to immediately start the testing once the test jar is assembled into the jacket. If there are known cloud points, then the sample is allowed to pause until the sample has fallen to a temperature of not less than 5°C above the cloud point of the sample.

Under specified conditions, cooling of the sample is done at breaks of 1°C; the specimen is drawn through a standardized wire mesh filter into a pipet under a controlled vacuum. For cooling the specimen, the procedure is repeated for every 1°C below the first test temperature. Continuation if the testing is done unless the quantity of wax crystal isolated from the solution is enough to reduce the flow such that the pipet fill time exceeds 60 s or the fuel fails to return fully to the test jar; then, the fuel is cooled a further 1°C. The last filtration of the indicated temperature commenced was recorded as the CFPP.

Diesel, palm methyl ester, and *Calophyllum inophyllum* methyl ester have CFPP values of –8°C, 4°C, and 0°C respectively. Jatropha methyl ester has a CFPP of 0°C (Echim, Maes, and Greyt 2012). Cloud point, pour point, and cold filter plug point can all be found to be related to the crystallization of the diesel. A high cloud point, pour point, and cold filter plug point are generally undesirable because the presence of wax crystals affects the fuel's fluidity, which in effect obstructs fuel injectors and filters and disturbs engine operation, particularly in cold weather (Atabani, Mahlia, et al. 2013).

## 3.8 KINEMATIC VISCOSITY

The kinematic viscosities of the fuel samples can be measured with a Redwood viscometer. As per the ASTM D445 standard, the experiment should be maintained at 40°C. In a Redwood viscometer, the biodiesel was filled into a heat chamber and it was heated to 40°C. The viscometer stopper was removed to drain out the heated biodiesel, and the drained biodiesel was collected with a graduated beaker placed below the stopper. When 50 mL of biodiesel was bought into the beaker, the flow of the biodiesel was stopped using the stopper. The time period for the collection of 50 mL of biodiesel was noted. For calculating the viscosity of the biodiesel, the following equation is used (Karmakar, Kundu, and Rajor 2018).

$$\text{Kinematic viscosity} = \left(A \times \text{time}\right) - \frac{B}{\text{time}} \qquad (3.4)$$

where $A$ and $B$ are constants for the specific Redwood viscometer, $A = 0.26$, $B = 179$ (for times less than 100 s) or $A = 0.24$, $B = 50$ (for times more than 100 s).

Kinematic viscosity affects the ease of engine start and the spray pattern of the fuel; as the viscosity becomes higher, the bigger will the fuel droplets be, which results in excessive mixing of air and fuel, resulting in low combustion and high PM emissions (Nabi et al. 2015). On the other hand, when the norm is bigger than the viscosity, the heat release rate of the peak is decreased, and there is a decrease in the degree of fuel impingement, and the air-fuel mixing rate is also reduced (Hellier, Ladommatos, and Yusaf 2015). According to ASTM D6751, the range of the biodiesel viscosity is from 1.9 to 6.0 mm²/s with biodiesel dropping from 1.9 and 6.0 mm²/s; if there is very low viscosity, the outflow would result in a loss in power of the engine, but if it happens to be high, there are lower possibilities that the injection pump will be able to provide sufficient fuel to seal the pumping chamber, resulting in power loss once again (Sani et al. 2018). It is found that the viscosities of AMC biodiesel, Karanja biodiesel, and Jatropha biodiesel are 3.6, 5.60, and 4.84 centistokes, respectively. The *Calophyllum inophyllum* methyl ester and palm methyl ester kinematic viscosities are found to be 5.21 and 4.18 centistokes, respectively (Silitonga et al. 2016). It is seen that there is a factor of two between the biodiesels' viscosities, which is often greater than petroleum diesel; as the biodiesel level decreases, the viscosity decreases.

## 3.9 DENSITY

As per the ASTM D1298 and IS 1448: Part 32: 1992 standard at 15°C, determination of the density of oil, diesel, and biodiesel was done. A 60 mL empty vessel was taken and it was taken and weighed, and the fuel sample to be tested was transferred into the vessel up to the graduated marks and weighed. The testing fuel was kept at 15°C by placing it in the defreezer chamber. By subtracting the empty vessel weight of the taken fuel sample from the filled one, the weight of the sample fuel was calculated. The following equation was used for finding the density of the fuel sample.

$$\text{Density} = \frac{\text{weight of fuel 15°C}}{\text{volume of fuel at 15°C}} \qquad (3.5)$$

The density or specific gravity data are critical for the various operations of chemical engineering units. In industries related to oleochemistry, lipid density data are required for preparing the reactor's design for fatty acid spilling or converting fatty acids into their derivatives, for distillation units for fatty acid separation, for storage tanks, and for process piping. Biodiesel density data are required for modeling the combustion processes and other applications as a function of the temperature. The injection systems (pump and injectors) of the engine are designed for the production of a specified fuel volume, while the main parameter in the combustion chamber is mass air-fuel ratio (Veny et al. 2009). The density of the Jatropha biodiesel was estimated to be 884.2 kg/m$^3$ (Pramanik 2003) and the density of the AMC oil was found to be 880 kg/m$^3$ (Thangarasu, Siddharth, and Ramanathan 2020).

## 3.10 CARBON RESIDUE

As per the ASTM D4530 method, the carbon residue experiment was done to measure the presence of the quantity of carbon residue in the fuel sample. Presence of the quantity of carbon residue in the fuel sample after pyrolysis method was measured. In the Conradson carbon residue experiment, a moisture-free sample of 5 g was placed in an iron crucible. The crucible was then kept in the middle of a Skidmore crucible apparatus. After placement of the crucible, the crucible was closed with a lid with an exit portal for allowing formed vapors to escape. The oven was heated electrically. The oven temperature was raised to 500°C slowly at a 10°C /min rate, and it was kept for 15 min for pyrolyzing the fuel sample. With a rate of flow of 600 mL/min, the nitrogen gas was purged for the pyrolysis process. After the pyrolysis of fuel sample, the power supply for the oven was shut down the nitrogen flow was continued until the temperature reached 150°C. After the oven temperature reached 150°C, the crucible was shifted out and the sample temperature was reduced to 30°C by placing it in a desiccator. After this, carbon residue was weighed using a precision weighing balance, and the mass percentage of carbon residue was calculated according to the following equation.

$$\% \text{ carbon residue} = \frac{CR \times 100}{w} \qquad (3.6)$$

where $CR$, carbon residue, g; $w$, weight of sample, g.

One of the important indicators for calculating the propensity is the carbon residue which is used for the formation of the deposits of carbonaceous materials in engines, causing several problems associated with operation like nozzle blockage, corrosion, and part cracking. In the case of biodiesel, the indication of the carbon residue is not only the quantity of material remaining next after the process of vaporization and pyrolysis but also the amount of glycerides which are free glycerol, partially reacted, or unreacted glycerides along with added residues which remain in the biodiesel product (free fatty acids and catalyst residues) (Phan and Phan 2008).

Increasing a volume of unconverted/partially converted glycerides increased the amount of carbon residue according to Fernando et al. Hence, the development of a large amount of residue may be attributed to the polymerization of unsaturated alkyl chains (about 10 wt.%) and the degradation of glycerides and free fatty acid remaining at a high temperature in the biodiesel (Fernando et al. 2007). Additionally, glyceride degradation at high temperatures can also serve as a catalyst for the polymerization of unsaturated fatty acids (Nas and Berktay 2007). The percentages of carbon residue for the Jatropha biodiesel and diesel are 0.22% and 0.10%, respectively (Pranab 2011). The Karanja biodiesel percentage of carbon residue was estimated to be 0.07% (Dhar and Agarwal 2014). The AMC biodiesel was found to be 0.025%. The degree of unsaturation in the case of Jatropha biodiesel was found to be 77.2%. Biodiesel for Karanja was found to be 73.93%. The degree of saturation for AMC biodiesel was found to be 66.6%.

## 3.11  COPPER STRIP CORROSION

Corrosion of metals happens due to the presence of water molecules in it. Fuel flows through various metallic and non-metallic parts of the engine system. Copper strip corrosion is a qualitative method used to determine the level of corrosion in fuel samples such as gasoline, diesel fuel, and other hydrocarbons. It helps in the determination of various aromatic compounds and harmful corrosive substances present in fuels.

The test was done using a copper corrosion apparatus as the per ASTM D130-12 standard, the instrument is used for determination of corrosiveness to copper for fuel samples. A test tube of 25 mm in diameter and 150 mm in length was taken and the test tube was filled with 30 mL of fuel sample. A polished copper strip of 12.5 mm in width, 1.5 to 3.5 mm in thickness, and 75 mm in length was made to slide inside the sample test tube. The glass tube was closed with a vented stopper and it was placed inside a bath which is maintained at 50°C. After 2 h, the test tubes are withdrawn from the bath and the strips examined as per chart of the ASTM copper strip corrosion standards (ASTM method D 130 / IP 154).

Meenakshi et al. studied Karanja biodiesel and found that the degree of tarnish at the corroded strip corresponds with the fuel sample's overall corrosiveness. ASTM certifies a maximum corrosion value of 3 on copper strips for biodiesel. The corrosion property of the copper strip of the investigated biodiesel was found to be 2, well within ASTM D6751 requirements (Parameswaran, Anand, and Krishnamurthy 2013). The Jatropha biodiesel was found to be within the limits, having less corrosive effect on the engine components (Kywe and Oo 2009).

## 3.12  FLASH AND FIRE POINT

As per the standard ASTM D93, the flash and fire point of fuel samples are measured by the Open Cup Cleveland apparatus. The fuel sample was taken in the test cup and filled up to a particular limit. Heat is supplied to the cup electrically and the rise in temperature was measured with the help of a thermometer. For every 1°C temperature rise, a flame was introduced above the surface of the fuel sample using a match stick. The temperature at which flash is observed over the fuel surface when the flame is introduced is recorded as the flash point. If the fire caught over the surface of the fuel sample when the flame source was introduced and the fire continued for a minimum of 5 min even after removal of the flame source, the temperature was recorded as the fire point temperature.

The flash point is defined as the lowest temperature at which a liquid generates sufficient vapors to ignite in the presence of an ignition source and is a major property used in industrial processes for flammable liquids while assessing process safety (Lazzús 2010). Likewise, the fire point is described as the lowest temperature at which the material's vapors can easily be attracted to fire and proceed to burn also after removal of the source of ignition. The flash point is lower than the fire point, as the creation of the vapors at the flash point is not sufficient for the ignition of the fuel. The volatility of the biodiesel highly affects the flash and fire points. Volatility is the material's tendency to vaporize, and is directly related to the pressure of the vapor of the respective biodiesel at that specific temperature. The higher the vapor pressure exhibited by a biodiesel at a specified temperature, the more volatile it is compared to those which exhibit low vapor pressure at the same temperature. The flash point and fire point of the biodiesel derived from palm oil were found to be 180°C and 194°C, respectively (Gorey et al. 2017). AMC biodiesel was found to have flash and fire points of 164°C and 175°C, respectively. The Karanja biodiesel flash point was found to be at 97.8°C. The Jatropha biodiesel flash point was found to be at 192°C.

## 3.13 SULFUR CONTENT

The sulfur content in biodiesel is very much less than that in fossil fuel. Due to the diversity of the feedstocks, there is variance in their sulfur contents. The sulfur content in biodiesel should be below 15 ppm. As per the ASTM D5373 standard, elemental analysis of carbon, nitrogen, hydrogen, sulfur, and oxygen can be done using CHNS analysis. The Flash Smart Elemental Analyzer can be used in conjunction with the Flame Photometric Detector (Thermoscientific) for sulfur determination in the fuel samples. The fuel samples are weighed in a hard tin container. The combustion process produces gases which were driven by the helium flow to a copper-filled layer, which is then sent through a water trap column, a short gas chromatography (GC) column. Finally, the determination of sulfur is made by the Flame Photometric Detector. The running time for sulfur determination is 5 minutes. The sulfur content in biodiesel is less than that in diesel. The amount of sulfur in diesel is 449.65 mg/kg, whereas the amounts of sulfur in palm methyl ester and *Calophyllum inophyllum* methyl ester are 3.82 mg/kg and 9.76 mg/kg, respectively. Jatropha biodiesel has 20 mg/kg sulfur content, which is the allowable limit according to standards (Liu et al. 2019).

## 3.14 METAL CONTENT

Feedstocks gain their metal contents from soil and water during their living period. Similarly, during the production and storage of biodiesel, there is a chance of metals getting into the composition of the biodiesel (de Oliveira et al. 2009). The presence of metallic elements like Na, K, Ca, and Mg are monitored in biodiesel because the presence of these elements even at low concentrations can damage engine components, decrease performance, reduce the oxidation stability of biodiesel, and causes corrosion problems, and lead to contamination of fuel over storage or during usage (De Souza, Leocádio, and Da Silveira 2008).

The presence of sodium and potassium in biodiesel should be monitored because the catalysts used are their peroxides. These elements can exist as abrasive solids and insoluble soaps, leading to destruction and deterioration of engine parts (Chaves et al. 2008). Elements of calcium and magnesium may get into the biodiesel during the washing process, if hard water was employed or through contamination from the adsorbents used in cleaning process. These elements can lead to soap formation of different undesirable compounds in engines (Bondioli 2007). Atomic absorption spectrometry (AAS) is the most commonly used technique to find metal contents in biodiesel. Recently, flame AAS has been a widely adopted technique because it is capable of measuring the most important organometallic compounds in fuel (Lyra et al. 2010).

The PerkinElmer – PinAAcle 900T instrument for the study of atomic absorption spectrometry can be used for the determination of metal content. A 2-g sample was taken in the beaker, dissolved with a mixture of 5 mL $H_2SO_4$ (96% w/v concentration) and 5 mL of $HNO_3$ (60% v/v concentration), and heated for the elimination of nitrous vapor. After completing this, a new mixture of 5 mL of $HNO_3$ and 5 mL of $H_2SO_4$ was added to the digested product. The final solution was diluted in distilled water in a 50 mL volumetric flask. The sample was heated for 4 h daily at 80°C for 9 days to ensure complete digestion of the products. In AAS, analysis was done by aspiration (air/acetylene flame) for finding out the metal content in the product. The metal contents are determined with wavelengths of 248.3 nm, 357.87 nm, 279.48 nm, 213 nm, 217 nm, and 327.40 nm for Fe, Cr, Mn, Zn, Pb, and Cu, respectively (Santhoshkumar and Ramanathan 2020).

The presence of alkaline and alkaline earth ions (Na, K, Ca, and Mg) can result in residual deposits which obstruct the fuel injection system (Almeida et al. 2014). In this report, both the combined amounts of Na + K and Ca + Mg contained in Jatropha biodiesel complied with the requirements provided in ASTM D 6751 and EN14214 (almost 5 mg/kg). It was found that the combined amount of Na and K in biodiesel derived from Jatropha was 1.9 mg/kg, which suggested adequate washing out of the biodiesel trace catalyst of KOH. The combined amount of Ca and Mg in biodiesel derived from Jatropha, on the other hand, was 3.9 mg/kg. The total volume of Ca and Mg was greater than that of Na and K, which may be attributed to the characteristics in the raw materials like the natural presence of Ca and Mg and them remaining in the finished product (Tan et al. 2019).

## 3.15   METHANOL CONTENT

The residual methanol content present in the biodiesel should be monitored because even a small amount of methanol can decrease the flash point of biodiesel. The presence of residual methanol can affect engine parts like seals, fuel pumps, and elastomers, which can cause poor combustion characteristics. As per the EN 14110 standard, determination of residue methanol content was done using GC method. The GC system uses a megapore capillary column and a flame-ionizing detector with integrated computer software. Nitrogen gas was sent in at a flow of 3 mL/min. The detector temperature and injector port temperature were maintained at 210°C and 280°C, respectively. A splitless injection system was used for injecting the fuel and to avoid contamination between the samples (before every experiment of this type, the syringes should be cleaned, heated, and rinsed with distilled water). The oven was initially maintained at 38°C for 3 min and was raised to 250°C at the rate of 50°C/min; then it was maintained for 1 min with the help of a computer program. Silitonga et al. found that excess methanol present in the methyl ester of palm oil and *Calophyllum inophyllum* was at levels of 0.24% and 7.18% in the yield which includes glycerin and other impurities (Silitonga et al. 2016).

## 3.16 PHOSPHORUS CONTENT

Crude plant oils have various impurities like acylglycerols, phospholipids, sugar, steroid, free fatty acids, and trace metals (Lin, Rhee, and Koseoglu 1997; AOCS 2009; Fan, Burton, and Austic 2010). Crude oil may have phosphorus content above 100 mg/kg [1] (Korn et al. 2007). According to the standard EN14214, the presence of phosphorus in biodiesel is regulated to less than 10 ppm. The presence of phosphorus is due to the oil-refining quality. Phospholipids contain glycerol with two fatty acids; this glycerol is attached to a phosphoric molecule (Mendow et al. 2011). In biodiesel, the presence of phosphorus content can damage the catalytic convertor in the exhaust line in diesel engines, which in turn can generate several pollutants such as particulate matter, carbon monoxide, and sulfur dioxide (Mittelbach 1996; Munack 2005).

The phosphorus content in the biodiesel is primarily removed by water degumming, and Emanuel et al. (2018) reported that usage of phosphoric acid can help to achieve better fuel properties than – raw *Crambe abyssinica* seed oil (Costa et al. 2018). According to the EN14214 standards, the phosphorus content in diesel can be determined by an atomic absorption spectrometer method (Dos Santos et al. 2007; Lira et al. 2011). The amount of phosphorus content present in crude soybean oil is 226 ppm. The coconut has phosphorus content of approximately 500 ppm, and phosphorus content of 7.1 ppm was obtained for biodiesel, which can be further reduced to 4.1 ppm with further treatment (Mendow et al. 2011). The phosphorous content in Karanja biodiesel was found to be 5 ppm, which is within permissible limits.

## 3.17 FREE GLYCEROL AND TOTAL GLYCERIN

Glycerol is a by-product obtained during the transesterification in the presence of a chemically catalyzed reaction mixture of fatty acids. It is hygroscopic and immiscible in nature with FAME, it settles as a layer in biodiesel, and decantation is done when the reaction is completed. Glycerin in trace amounts is obtained as a valuable by-product in production biodiesel with some traces of mono-, di-, and triglycerides formed as intermediates. The residuals of catalyst, methanol, glycerol, and mono-, di-, and triglycerides are removed by water washing of biodiesel. Water of thrice the volume of biodiesel was used for the removal of the glycerol. Glycerol and glycerin are common words of almost the same meaning, but with subtle differences. Glycerol is a pure compound, but glycerin refers to commercial scales, irrespective of their purity. Glycerin is a co-product obtained during the production of soap from oils, fats, fatty esters, and fatty acids. Glycerin has very low solubility in methyl esters, so it is easily removable with the help of water from biodiesel because of its higher density. As per the EN 1405/ ASTM D6584 standard, the GC method is used for the determination of free glycerol and total glycerin.

Excessive free glycerol fuel may lead to material incompatibility, engine deposits, and engine combustion problems. A biodiesel's free glycerol content should not exceed 0.020 wt.% and the overall total glycerol content should be 0.240 wt.% according to the ASTM D6751 standard (Meenakshi and Shyamala 2015). The free glycerin and total glycerin of Karanja biodiesel were within the limit of 0.0064 wt.% and 0.082 wt.% (Harreh et al. 2018). The Jatropha methyl ester has free glycerol and total glycerol of 0.01 and 0.012, respectively (Jain and Sharma 2014).

## 3.18   MONO-, DI-, AND TRIGLYCERIDES

During the transesterification, three reversible reactions happen, in which 1 mole of fatty methyl esters is released in every step. During the first reaction, the triglyceride is reacted to form diglyceride; in the second step, the diglyceride gets converted to monoglyceride; and in the final step, it is converted to glycerol. Monoglycerides are fatty acid esters of the mono type of glycerol. They are formed through chemical processes, and during the degradation they are formed as intermediates between triglyceride and diglyceride. Diglyceride consists of two fatty acids which are esterified to the trihydric alcohol glycerol. To find the amount of mono-, di-, and triglycerides present in the biodiesel sample, a gas chromatograph with a flame ionization detector is used as per the EN 14105 standard. In a temperature-controlled oven, a high-resolution silica capillary house is installed. The maximum capacity of the oven is 22.6 L. Electronically pressure-controlled systems are used to control the gases in the gas chromatography. There are separate injectors and detectors for liquid and gaseous samples. A flame-ionizing detector is used for liquid sample analysis and a thermal conductivity detector is used for gaseous samples. The temperature range of the setup is 30°C to 500°C with a 1°C set point resolution of accuracy. The heating rate varies from 1 to 50°C/min, and it has 1 to 7 segments of temperature profile. Using a TR-FAME capillary column, the biodiesel analysis was made. Capillary tubes will be 10 m in length and of 0.22 mm inner diameter, and the thickness of the film will be 0.25 μm.

The Karanja biodiesel was has monoglyceride, diglyceride, and triglyceride contents of 2.63 wt.%, 0.78 wt.%, and 0.06 wt.%, respectively. The Karanja biodiesel has glyceride contents within the standard limits of EN 14214, and the maximum allowable contents of monoglyceride, diglyceride, triglyceride are <0.8, <0.2, and <0.2, respectively (Harreh et al. 2018). The Jatropha biodiesel has monoglyceride, diglyceride, and triglyceride contents of 0.1 wt.%, 0.0 wt.%, and 0.0 wt.% respectively. The relationship between triglyceride, diglyceride, monoglyceride, and free fatty acid weights and the test sample mass was used to find the amount of the glyceride fractions in Jatropha oil (Amalia Kartika et al. 2013).

## 3.19   ESTER CONTENT

Ester content present in the biodiesel helps in determining the purity of biodiesel, where oil conversion into biodiesel has happened fully. Esterification is a reversible reaction of acid with an alcohol in the presence of catalyst to form alcoholic esters. In the biodiesel preparation, esterification and transesterification are done to reduce the acid value of the fatty substances or oil. A nuclear magnetic resonance (NMR) spectrometer is used to determine the formation of methyl esters. As per the ASTM D6751, an AVANCE II 300 MHz instrument was selected for the analysis; this instrument is capable of doing any one-dimensional or two-dimensional experiments with a frequency resolution of 0.005 Hz on a 2-way channel NMR spectrometer. It has a 5 min BBO probe with a differential temperature option. The required time for frequency setting, amplitude, and phase is 25 ns. The selected BBO has the capability of autotuning to process with various dielectric solvents. It is a chemical approach which helps in the determination of molecular structure and immaculateness of the component. The compositions of various mixtures of sample are analyzed quantitatively.

The ester content of Karanja biodiesel was 98.6 wt.%, which satisfies the ASTM and European standards for biodiesel. According to the European EN14214 standard, a minimum of 96.5 wt.% of ester content should be present in the biodiesel (Meher et al. 2006). The palm oil and *Calophyllum inophyllum* had yields of 99.76% and 92.82%, which are comparatively suitable for the production of biodiesel. The maximum yield of Jatropha oil was found to be 99% with the use of an ultrasonic technique for the conversion process (Tan et al. 2019).

## 3.20 LUBRICITY

The measure for the reduction and/or wear by a lubricant is known as lubricity. Lubricity is measured by the amount of wear made to a surface due to friction, leading to changes in the integrity of the object over a particular standard period condition and time. If two fluids have identical viscosities, if either of the fluids produces less wear, then that fluid has added lubricity. The lubricity of a fluid can also be called its anti-wear property. The lubricity property of biodiesel fuel is more significant for the injection of fuel in the fuel injection pumps. Some parts of engines are not lubricated by the engine lubricant oil; these will be lubricated by the fuel flow. For example, the fuel injector is lubricated by the fuel flow. Lubricity helps in the indication of the quantity of wear and tear occurring between any two constituents, both when they come in contact with each other and also when they are fully enclosed within the biodiesel. Biodiesel with high lubricity as will lead to better component life and minimized scarring. If the lubricity of the biodiesel is low, then the scarring will be high and its usage will be critical.

The characteristics of the friction and wear of the biodiesel and fuel samples were determined using a four-ball wear tester (DUCOM, TR-30L-IAS). As per the ASTM D2266 standard, tribological tests were done. Lubricants or lubricity of any new grease was commonly tested with the help of four-ball tribotesting. The setup contains four balls, where three balls are held in the lubricant cup rigidly against each other with the help of a clamping ring and the leftmost ball is bonded with a rotating chuck kept above these balls. Before starting the experimental setup, the lubricant cup and steel balls were cleaned with toluene and then dried. About 10 mL of the fuel sample to be tested was poured in the lubricant cup, and it should be ensured that the sample has covered the three balls at the bottom and to the height of 3 mm over the balls. The test was run for 60 min with a chuck speed of 1200 rpm and a 40 kg load, both of which were held constant. After the test run, three stationary balls were bought together for measurement wear scar diameter using scanning electron microscopy analysis. A friction-measuring device uses a calibrated spring to measure the frictional torque. The following equation was used for the calculation of the coefficient of friction.

$$\text{Coefficient of friction } (\mu) = \frac{T\sqrt{6}}{3Wr} \tag{3.7}$$

where $T$ is the frictional torque in Nm, $W$ is the load applied in N, $r$ is the distance from center of the contact surface of the bottom three balls to the rotation axis in mm, and $\mu$ is the coefficient of friction.

Flash temperature, a parameter which helps in determination of stable temperature of the lubricating film, is also determined. The following relation was used to calculate the flash temperature parameter with the load applied and mean diameter of the wear scar.

$$\text{Flash temperature} = \frac{W}{d^{1.4}} \tag{3.8}$$

where $d$ is the mean wear scar diameter of the three-bottom ball in mm and $W$ is the applied load in kg.

The steel balls were taken for wear analysis to find the wear scar diameter. The experiment was done as per the ASTM D4712 standards, using an optical microscope with a high optical resolution of 0.01 mm to analyze wear. The steel balls were dried and cleaned with toluene before they were examined. If a scar was found, the balls were thoroughly checked over their entire surface for any wear scars; all scars were viewed using microscope at proper magnification and focus to measure them the wear scar diameter. If a scar was identified, a picture was taken and analyzed using software to determine the wear scar diameter of the ball. The same steps were repeated for all balls after their experimental run.

The wear track diameter and the mean friction coefficient for Jatropha biodiesel were 170.21 μm and 0.1, respectively. The diameter of the wear track was also slightly smaller than the European standard value EN590 (460 μm). A higher reading of the oil film meant that the metal surface was in a separate state, creating a strong protective chemical layer. Jatropha biodiesel's oil film formulation content was 90.38%, further demonstrating the related good properties of anti-wear and anti-friction (Liu et al. 2019).

## 3.21  OXIDATION STABILITY

Oxidation stability means the tendency of fuels to react with oxygen near to their ambient temperature. It is an important biodiesel parameter that gives information about the various parameters that affects storage for longer periods. Normally, the unsaturated compounds present in the biodiesel will tend to be unstable and deteriorate upon oxidation; this process makes fuel usage undesirable and darkens the fuel. The reactions are much slower than reactions occurring at higher temperatures.

Storing the biodiesel for longer periods and using it in engines over a prolonged period may lead to aging of the biodiesel, leading in turn to the formation of hydroperoxides, acids, aldehydes, and polymers. In some advanced cases, oxidation causes acidity and causes fuel system corrosion. If all the corrosive acids get deposited in the engine, this can lead to wear in fuel engine pumps and injectors. If there are water molecules inside the fuel, this can lead to formation of rust and corrosiveness exacerbated by the presence of acids. Due to the polymerization reaction, there can be an impact of sediment and gum formation on the fuel flow system, which can lead to fuel filter plugging. It can also clog fuel lines and pumps. Thus, a high degree of oxidation can lead to fuel pump and injector problems.

Oxidation stability is studied to find the susceptibility of biodiesel to oxidize due to unsaturated ester content; testing was done as per EN 14112 standard. The oxidation stability experiment is done using the Rancimat method, which requires an 873 Rancimat instrument, manufactured by Metrohm. This method is similar to the Oil Stability Index (OSI) method, which is the method prescribed by the American Oil Chemists Society (AOCS). The method involves passing air at a constant flow rate of 10 L/h through a sample of 3 g, which is held in a heated reaction vessel at 110°C. After this, air from the heated vessel is passed into a measuring vessel which contains an absorption solution of distilled water and an air sample carrying volatile secondary oxidation products of water-soluble short-chain carboxylic acids. In the measuring vessel, the electrical conductivity is monitored continuously by cell electrodes. Elevation in the cell conductivity indicates the gathering of volatile acids inside the water, because of the oxidation of the sample. Earlier acids which are formed are formic acids. The oxidation curve will initially have a slow rate and gradually climb steeply, indicating the conductivity. Here the automated Rancimat software is used in the calculation of the induction period, which evaluates the extreme 2nd derivate of conductivity with respect to time.

Many metals like nickel, copper, iron, tin, and brass will substantially lead to decrements in the OSI of the biodiesel. Saturated or unsaturated long-chain fatty acids directly affect the OSI of the biodiesel. Linolenic < linoleic < oleic was found to be the trend; this trend corresponds to a higher presence of unsaturated long-chain fatty acids, which have high oxidation stability. In AMC biodiesel, the percentage of oleic acid is 26.69%, the percentage of linoleic acid is 24.48%, and the percentage of linolenic acid is 15.43%. In biodiesel derived from Jatropha, the percentage of oleic acid is 40%, the percentage of linoleic acid is 36.9%, and the percentage of linolenic acid is 0.2%. In Karanja biodiesel, the percentage of oleic acid is 51.59%, the percentage of linoleic acid is 16.64%, and the percentage of the linolenic acid is 5.7%. Hence, it was found that AMC biodiesel's oxidation stability was low compared to Karanja and Jatropha.

## 3.22 THERMAL BEHAVIOR

The thermal behavior of biodiesel was evaluated using thermogravimetric analysis (TG)/derivate thermogravimetric analysis (DTG). The mass amount of biodiesel and the reduction rate of the mass were evaluated using TG/DTG, which uses temperature and time as the function under restrained atmospheres. The results gained can be used for correlation of the composition of the substance. A sample of 100 mg was taken for the DTG/TG analysis with an operational temperature range of 25–1000°C, and the curves were obtained with 0.1 µg of resolution and a reading of 0.1% lower than uncertainty. This was done under the condition of air at 60 mL/min and nitrogen at 50 mL/min. Results of the DTG/TG curves included a high resolution slope of 10–600°C, and the sample was maintained at a continuously controlled heating rate based on alterations in the sample decomposition rate (Ávila and Sodré 2012).

Dantas et al. (2007) experimented with corn oil biodiesel; corn oil biodiesel was stable up to 225°C. Methanol biodiesel was stable up to 139°C and ethanol biodiesel was stable up to 159°C in air. In a nitrogen gas atmosphere, stability of the corn oil was up to 336°C, methanol biodiesel was up to 145°C, and ethanol biodiesel was up to 169°C. Ethyl ester biodiesel was observed to be more stable than methanol biodiesel and other biodiesels produced in the experiment at atmospheric condition. The thermal stability of the Jatropha biodiesel was found to be 113°C for dry air and that under a nitrogen atmosphere was 128°C (Jain and Sharma 2012). According to Salaheldeen et al. (2015), Moringa peregrine biodiesel was stable up to 125°C.

## 3.23  CONCLUSION

The study tells about the various properties of biodiesel and their influences on engines. The selection of the feedstocks is based on the physicochemical and thermal properties of the biodiesels derived from them, along with their availability. This chapter has discussed the various property determination methods and the presence of various properties in the different types of biodiesel. Alternative fuels are progressing, and thus it is necessary to understand biodiesel properties.

## REFERENCES

Adu-Mensah, Derick, Deqing Mei, Lei Zuo, Qi Zhang, and Junfeng Wang. 2019. "A Review on Partial Hydrogenation of Biodiesel and Its Influence on Fuel Properties." *Fuel*. doi:10.1016/j.fuel.2019.04.036.
Almeida, Joseany M. S., Rafael M. Dornellas, Sakae Yotsumoto-Neto, Mirela Ghisi, Jethânia G. C. Furtado, Edmar P. Marques, Ricardo Q. Aucélio, and Aldaléa L. B. Marques. 2014. "A Simple Electroanalytical Procedure for the Determination of Calcium in Biodiesel." *Fuel*. doi:10.1016/j.fuel.2013.07.088.
Amalia Kartika, I., M. Yani, D. Ariono, Ph Evon, and L. Rigal. 2013. "Biodiesel Production from Jatropha Seeds: Solvent Extraction and in Situ Transesterification in a Single Step." *Fuel*. doi:10.1016/j.fuel.2013.01.021.
AOCS. 2009. *Official Methods and Recommended Practices of the AOCS*. American Oil Chemist's Society.
Atabani, A. E., T. M. I. Mahlia, H. H. Masjuki, Irfan Anjum Badruddin, Hafizuddin Wan Yussof, W. T. Chong, and Keat Teong Lee. 2013. "A Comparative Evaluation of Physical and Chemical Properties of Biodiesel Synthesized from Edible and Non-Edible Oils and Study on the Effect of Biodiesel Blending." *Energy*. doi:10.1016/j.energy.2013.05.040.
Atabani, A. E., A. S. Silitonga, H. C. Ong, T. M. I. Mahlia, H. H. Masjuki, Irfan Anjum Badruddin, and H. Fayaz. 2013. "Non-Edible Vegetable Oils: A Critical Evaluation of Oil Extraction, Fatty Acid Compositions, Biodiesel Production, Characteristics, Engine Performance and Emissions Production." *Renewable and Sustainable Energy Reviews*. doi:10.1016/j.rser.2012.10.013.

Ávila, Ronaldo Nunes de Andrade, and José Ricardo Sodré. 2012. "Physical-Chemical Properties and Thermal Behavior of Fodder Radish Crude Oil and Biodiesel." *Industrial Crops and Products.* doi:10.1016/j.indcrop.2012.01.007.

Azam, M. Mohibbe, Amtul Waris, and N. M. Nahar. 2005. "Prospects and Potential of Fatty Acid Methyl Esters of Some Non-Traditional Seed Oils for Use as Biodiesel in India." *Biomass and Bioenergy.* doi:10.1016/j.biombioe.2005.05.001.

Balakrishnan, A., R. N. Parthasarathy, and S. R. Gollahalli. 2016. "Effects of Degree of Fuel Unsaturation on NOx Emission from Petroleum and Biofuel Flames." *Fuel.* doi:10.1016/j.fuel.2016.06.052.

Bart, Jan C. J., Natale Palmeri, and Stefano Cavallaro. 2010. "Emerging New Energy Crops for Biodiesel Production." In *Biodiesel Science and Technology.* doi:10.1533/9781845697761.226.

Belagur, Venkanna K., and Venkataramana Reddy Chitimi. 2013. "Few Physical, Chemical and Fuel Related Properties of Calophyllum Inophyllum Linn (Honne) Oil and Its Blends with Diesel Fuel for Their Use in Diesel Engine." *Fuel.* doi:10.1016/j.fuel.2013.02.015.

Bondioli, Paolo. 2007. "Book Reviews: The Biodiesel Handbook. Edited by G. Knothe, J. Krahl, J. Van Gerpen." *CLEAN – Soil, Air, Water.* doi:10.1002/clen.200790016.

Bouaid, Abderrahim, Mercedes Martinez, and José Aracil. 2007. "Long Storage Stability of Biodiesel from Vegetable and Used Frying Oils." *Fuel.* doi:10.1016/j.fucl.2007.02.014.

Caldeira, Carla, Fausto Freire, Elsa A. Olivetti, and Randolph Kirchain. 2017. "Fatty Acid Based Prediction Models for Biodiesel Properties Incorporating Compositional Uncertainty." *Fuel.* doi:10.1016/j.fuel.2017.01.074.

Chaves, Eduardo Sidinei, Tatiana Dillenburg Saint'Pierre, Eder José Dos Santos, Luciano Tormen, Vera Lúcia Azzolin Frescura Bascunan, and Adilson José Curtius. 2008. "Determination of Na and K in Biodiesel by Flame Atomic Emission Spectrometry and Microemulsion Sample Preparation." *Journal of the Brazilian Chemical Society.* doi:10.1590/S0103-50532008000500008.

Costa, Emanuel, Manuel Fonseca Almeida, Maria da Conceição Alvim-Ferraz, and Joana Maia Dias. 2018. "Effect of Crambe Abyssinica Oil Degumming in Phosphorus Concentration of Refined Oil and Derived Biodiesel." *Renewable Energy.* doi:10.1016/j.renene.2017.08.089.

Dantas, M. B., Marta M. Conceição, V. J. Fernandes, Nataly A. Santos, R. Rosenhaim, Aldalea L. B. Marques, Iêda M. G. Santos, and A. G. Souza. 2007. "Thermal and Kinetic Study of Corn Biodiesel Obtained by the Methanol and Ethanol Routes." *Journal of Thermal Analysis and Calorimetry.* doi:10.1007/s10973-006-7780-2.

Das, L. M., Dilip Kumar Bora, Subhalaxmi Pradhan, Malaya K. Naik, and S. N. Naik. 2009. "Long-Term Storage Stability of Biodiesel Produced from Karanja Oil." *Fuel.* doi:10.1016/j.fuel.2009.05.005.

Deng, Xin, Zhen Fang, Yun hu Liu, and Chang Liu Yu. 2011. "Production of Biodiesel from Jatropha Oil Catalyzed by Nanosized Solid Basic Catalyst." *Energy.* doi:10.1016/j.energy.2010.12.043.

Dhar, Atul, and Avinash Kumar Agarwal. 2014. "Experimental Investigations of Effect of Karanja Biodiesel on Tribological Properties of Lubricating Oil in a Compression Ignition Engine." *Fuel.* doi:10.1016/j.fuel.2014.03.066.

Diwakar, Bastihalli Tukaram, Pinto Kumar Dutta, Belur Ramaswamy Lokesh, and Kamatham Akhilender Naidu. 2010. "Physicochemical Properties of Garden Cress (Lepidium Sativum l.) Seed Oil." *JAOCS, Journal of the American Oil Chemists' Society.* doi:10.1007/s11746-009-1523-z.

Echim, Camelia, Jeroen Maes, and Wim De Greyt. 2012. "Improvement of Cold Filter Plugging Point of Biodiesel from Alternative Feedstocks." *Fuel.* doi:10.1016/j.fuel.2011.11.036.

Fan, Xiaohu, Rachel Burton, and Greg Austic. 2010. "Conversion of Degummed Soybean Oil to Biodiesel: Optimization of Degumming Methods and Evaluation of Fuel Properties." *International Journal of Green Energy.* doi:10.1080/15435075.2010.529403.

Fauzan, Nur Afiqah, Tan Ee Sann, Pua Fei Ling, and Gopinathan Muthaiyah. 2020. "Physiochemical Properties Evaluation of Calophyllum Inophyllum Biodiesel for Gas Turbine Application." *South African Journal of Chemical Engineering* 32: 56–61.

Fazal, M. A., A. S. M. A. Haseeb, and H. H. Masjuki. 2013. "Investigation of Friction and Wear Characteristics of Palm Biodiesel." *Energy Conversion and Management.* doi:10.1016/j.enconman.2012.12.002.

Fernando, Sandun, Prashanth Karra, Rafael Hernandez, and Saroj Kumar Jha. 2007. "Effect of Incompletely Converted Soybean Oil on Biodiesel Quality." *Energy.* doi:10.1016/j.energy.2006.06.019.

Gorey, Neeraj, Shankha Ghosh, Priyank Srivastava, and Vivek Kumar. 2017. "Characterization of Palm Oil as Biodiesel." *IOP Conference Series: Materials Science and Engineering* 225 (1). doi:10.1088/1757-899X/225/1/012220.

Harreh, Dewi, A. A. Saleh, A. N. R. Reddy, and S. Hamdan. 2018. "An Experimental Investigation of Karanja Biodiesel Production in Sarawak, Malaysia." *Journal of Engineering (United Kingdom).* doi:10.1155/2018/4174205.

Hellier, Paul, Nicos Ladommatos, and Talal Yusaf. 2015. "The Influence of Straight Vegetable Oil Fatty Acid Composition on Compression Ignition Combustion and Emissions." *Fuel.* doi:10.1016/j.fuel.2014.11.021.

Ismail, Samir Abd Elmonem A., and Rehab Farouk M. Ali. 2015. "Physico-Chemical Properties of Biodiesel Manufactured from Waste Frying Oil Using Domestic Adsorbents." *Science and Technology of Advanced Materials.* doi:10.1088/1468-6996/16/3/034602.

Jain, Siddharth, and M. P. Sharma. 2012. "Application of Thermogravimetric Analysis for Thermal Stability of Jatropha Curcas Biodiesel." *Fuel.* doi:10.1016/j.fuel.2011.09.002.

Jain, Siddharth, and M. P. Sharma. 2014. "Effect of Metal Contents on Oxidation Stability of Biodiesel/Diesel Blends." *Fuel* 116. Elsevier Ltd: 14–18. doi:10.1016/j.fuel.2013.07.104.

Karmakar, Rachan, Krishnendu Kundu, and Anita Rajor. 2018. "Fuel Properties and Emission Characteristics of Biodiesel Produced from Unused Algae Grown in India." *Petroleum Science.* doi:10.1007/s12182-017-0209-7.

Knothe, Gerhard. 2006. "Analysis of Oxidized Biodiesel by 1H-NMR and Effect of Contact Area with Air." *European Journal of Lipid Science and Technology.* doi:10.1002/ejlt.200500345.

Korn, Maria das Graças Andrade, Denilson Santana Sodré dos Santos, Bernhard Welz, Maria Goreti Rodrigues Vale, Alete Paixão Teixeira, Daniel de Castro Lima, and Sérgio Luis Costa Ferreira. 2007. "Atomic Spectrometric Methods for the Determination of Metals and Metalloids in Automotive Fuels – A Review." *Talanta.* doi:10.1016/j.talanta.2007.03.036.

Kumar, Mukesh, and Mahendra Pal Sharma. 2016. "Selection of Potential Oils for Biodiesel Production." *Renewable and Sustainable Energy Reviews.* doi:10.1016/j.rser.2015.12.032.

Kywe, Tint Tint, and Mya Mya Oo. 2009. "Production of Biodiesel from Jatropha Oil (Jatropha Curcas) in Pilot Plant." *World Academy of Science, Engineering and Technology* 50: 477–483.

Lazzús, Juan A. 2010. "Prediction of Flash Point Temperature of Organic Compounds Using a Hybrid Method of Group Contribution + Neural Network + Particle Swarm Optimization." *Chinese Journal of Chemical Engineering.* doi:10.1016/S1004-9541(09)60133-6.

Lin, L., K. C. Rhee, and S. S. Koseoglu. 1997. "Bench-Scale Membrane Degumming of Crude Vegetable Oil: Process Optimization." *Journal of Membrane Science.* doi:10.1016/S0376-7388(97)00098-7.

Lira, Liliana Fátima Bezerra, Daniele C. M. B. Dos Santos, Mauro A. B. Guida, Luiz Stragevitch, Maria Das Graas A. Korn, Maria Fernanda Pimentel, and Ana Paula Silveira Paim. 2011. "Determination of Phosphorus in Biodiesel Using FIA with Spectrophotometric Detection." *Fuel.* doi:10.1016/j.fuel.2011.06.022.

Liu, Zuowen, Fashe Li, Jiaxu Shen, and hua Wang. 2019. "Effect of Oxidation of Jatropha Curcas-Derived Biodiesel on Its Lubricating Properties." *Energy for Sustainable Development.* doi:10.1016/j.esd.2019.06.003.

Lyra, Fernanda Henrique, Maria Tereza Weitzel Dias Carneiro, Geisamanda Pedrini Brandão, Helen Moura Pessoa, and Eustáquio Vinícius de Castro. 2010. "Determination of Na, K, Ca and Mg in Biodiesel Samples by Flame Atomic Absorption Spectrometry (F AAS) Using Microemulsion as Sample Preparation." *Microchemical Journal.* doi:10.1016/j.microc.2010.03.005.

Madheshiya, Arvind Kumar, and Ajitanshu Vedrtnam. 2018. "Energy-Exergy Analysis of Biodiesel Fuels Produced from Waste Cooking Oil and Mustard Oil." *Fuel.* doi:10.1016/j.fuel.2017.11.060.

Marutani, Mari, Juan Andres Soria, and Mario A. Martinez. 2018. "Characterization of Crude and Biodiesel Oils of Jatropha Curcas and Calophyllum Inophyllum in Guam *." *Micronesica* 2018 (01):1–15.

Meenakshi, H. N., and R. Shyamala. 2015. "Effect of Flow and Dissolved Oxygen on the Compatibility of Pongamia Pinnata Biodiesel with Common Construction Materials Used in Storage and Transportation." *International Journal of Chemical Engineering.* doi:10.1155/2015/463064.

Meher, Lekha Charan, Mangesh G. Kulkarni, Ajay K. Dalai, and Satya Narayan Naik. 2006. "Transesterification of Karanja (Pongamia Pinnata) Oil by Solid Basic Catalysts." *European Journal of Lipid Science and Technology.* doi:10.1002/ejlt.200500307.

Mejía, J. D., N. Salgado, and C. E. Orrego. 2013. "Effect of Blends of Diesel and Palm-Castor Biodiesels on Viscosity, Cloud Point and Flash Point." *Industrial Crops and Products.* doi:10.1016/j.indcrop.2012.08.026.

Mendow, G., F. C. Monella, M. L. Pisarello, and C. A. Querini. 2011. "Biodiesel Production from Non-Degummed Vegetable Oils: Phosphorus Balance throughout the Process." *Fuel Processing Technology.* doi:10.1016/j.fuproc.2010.11.029.

Miraboutalebi, Seyed Mohammadreza, Pezhman Kazemi, and Peyman Bahrami. 2016. "Fatty Acid Methyl Ester (FAME) Composition Used for Estimation of Biodiesel Cetane Number Employing Random Forest and Artificial Neural Networks: A New Approach." *Fuel.* doi:10.1016/j.fuel.2015.10.118.

Mittelbach, Martin. 1996. "Diesel Fuel Derived from Vegetable Oils, VI: Specifications and Quality Control of Biodiesel." *Bioresource Technology.* doi:10.1016/0960-8524(95)00172-7.

Munack, Axel. 2005. "Book Review: Biodiesel – A Comprehensive Handbook Edited by Martin Mittelbach and Claudia Remschmidt." *European Journal of Lipid Science and Technology.* doi:10.1002/ejlt.200590065.

Nabi, Md Nurun, Md Mostafizur Rahman, Muhammad Aminul Islam, Farhad M. Hossain, Peter Brooks, William N. Rowlands, John Tulloch, Zoran D. Ristovski, and Richard J. Brown. 2015. "Fuel Characterisation, Engine Performance, Combustion and Exhaust Emissions with a New Renewable Licella Biofuel." *Energy Conversion and Management.* doi:10.1016/j.enconman.2015.02.085.

Nas, B., and A. Berktay. 2007. "Energy Potential of Biodiesel Generated from Waste Cooking Oil: An Environmental Approach." *Energy Sources, Part B: Economics, Planning and Policy.* doi:10.1080/15567240500400903.

Odoom, William, and Vida Opoku Edusei. 2015. "Evaluation of Saponification Value, Iodine Value and Insoluble Impurities in Coconut Oils from Jomoro District in the Western Région of Ghana." *Asian Journal of Agriculture and Food Sciences* 5 (3): 494–499.

Ogunkunle, Oyetola, and Noor A. Ahmed. 2019. "A Review of Global Current Scenario of Biodiesel Adoption and Combustion in Vehicular Diesel Engines." *Energy Reports.* doi:10.1016/j.egyr.2019.10.028.

Oliveira, Adriana Paiva de, Ricardo D. Villa, Keila Cristina Pinheiro Antunes, Aparecida de
   Magalhães, and Edinaldo Castro e. Silva. 2009. "Determination of Sodium in Biodiesel
   by Flame Atomic Emission Spectrometry Using Dry Decomposition for the Sample
   Preparation." *Fuel.* doi:10.1016/j.fuel.2008.10.006.
Parameswaran, Meenakshi H. N., Anisha Anand, and Shyamala R. Krishnamurthy. 2013. "
   A Comparison of Corrosion Behavior of Copper and Its Alloy in Pongamia Pinnata Oil
   at Different Conditions ." *Journal of Energy.* doi:10.1155/2013/932976.
Phan, Anh N., and Tan M. Phan. 2008. "Biodiesel Production from Waste Cooking Oils."
   *Fuel.* doi:10.1016/j.fuel.2008.07.008.
Pramanik, K. 2003. "Properties and Use of Jatropha Curcas Oil and Diesel Fuel Blends in
   Compression Ignition Engine." *Renewable Energy.* doi:10.1016/S0960-1481(02)00027-7.
Pranab, K. Barua. 2011. "Biodiesel from Seeds of Jatropha Found in Assam, India."
   *International Journal of Energy, Information and Communications* 2 (1): 53–65.
Ramos, María Jesús, Carmen María Fernández, Abraham Casas, Lourdes Rodríguez, and
   Ángel Pérez. 2009. "Influence of Fatty Acid Composition of Raw Materials on Biodiesel
   Properties." *Bioresource Technology.* doi:10.1016/j.biortech.2008.06.039.
Sahabdheen, Abdul Basheer, and Anitha Arivarasu. 2019. "Synthesis and Characterization
   of Reusable Heteropoly Acid Nanoparticles for One Step Biodiesel Production from
   High Acid Value Waste Cooking Oil – Performance and Emission Studies." *Materials
   Today: Proceedings.* doi:10.1016/j.matpr.2019.07.249.
Salaheldeen, Mohammed, M. K. Aroua, A. A. Mariod, Sit Foon Cheng, Malik A. Abdelrahman,
   and A. E. Atabani. 2015. "Physicochemical Characterization and Thermal Behavior of
   Biodiesel and Biodiesel-Diesel Blends Derived from Crude Moringa Peregrina Seed
   Oil." *Energy Conversion and Management.* doi:10.1016/j.enconman.2014.12.087.
Sani, S., M. U. Kaisan, D. M. Kulla, A. I. Obi, A. Jibrin, and B. Ashok. 2018. "Determination
   of Physico Chemical Properties of Biodiesel from Citrullus Lanatus Seeds Oil and
   Diesel Blends." *Industrial Crops and Products.* doi:10.1016/j.indcrop.2018.06.002.
Santhoshkumar, A., and Anand Ramanathan. 2020. "Recycling of Waste Engine Oil through
   Pyrolysis Process for the Production of Diesel like Fuel and Its Uses in Diesel Engine."
   *Energy* 197 (April). Elsevier Ltd. doi:10.1016/j.energy.2020.117240.
Santos, Eder J. Dos, Amanda B. Herrmann, Eduardo S. Chaves, Wellington W. D.
   Vechiatto, Adrielle C. Schoemberger, Vera L. A. Frescura, and Adilson J. Curtius.
   2007. "Simultaneous Determination of Ca, P, Mg, K and Na in Biodiesel by Axial
   View Inductively Coupled Plasma Optical Emission Spectrometry with Internal
   Standardization after Multivariate Optimization." *Journal of Analytical Atomic
   Spectrometry.* doi:10.1039/b702563g.
Saravanan, A., M. Murugan, M. Sreenivasa Reddy, and Satyajeet Parida. 2020. "Performance
   and Emission Characteristics of Variable Compression Ratio CI Engine Fueled with
   Dual Biodiesel Blends of Rapeseed and Mahua." *Fuel.* doi:10.1016/j.fuel.2019.116751.
Silitonga, A. S., H. H. Masjuki, T. M. I. Mahlia, H. C. Ong, W. T. Chong, and M. H. Boosroh.
   2013. "Overview Properties of Biodiesel Diesel Blends from Edible and Non-Edible
   Feedstock." *Renewable and Sustainable Energy Reviews.* doi:10.1016/j.rser.2013.01.055.
Silitonga, A. S., H. H. Masjuki, Hwai Chyuan Ong, F. Kusumo, T. M. I. Mahlia, and A. H.
   Bahar. 2016. "Pilot-Scale Production and the Physicochemical Properties of Palm and
   Calophyllum Inophyllum Biodiesels and Their Blends." *Journal of Cleaner Production.*
   doi:10.1016/j.jclepro.2016.03.057.
Singh, Digambar, Dilip Sharma, S. L. Soni, Sumit Sharma, Pushpendra Kumar Sharma, and
   Amit Jhalani. 2020. "A Review on Feedstocks, Production Processes, and Yield for
   Different Generations of Biodiesel." *Fuel.* doi:10.1016/j.fuel.2019.116553.
Souza, Roseli Martins De, Luiz Gustavo Leocádio, and Carmem Lucia P. Da Silveira. 2008.
   "ICP OES Simultaneous Determination of Ca, Cu, Fe, Mg, Mn, Na, and P in Biodiesel
   by Axial and Radial Inductively Coupled Plasma-Optical Emission Spectrometry."
   *Analytical Letters.* doi:10.1080/00032710802122248.

Tan, Shiou Xuan, Steven Lim, Hwai Chyuan Ong, Yean Ling Pang, Kusumo Fitranto, Brandon Han Hoe Goh, and Cheng Tung Chong. 2019. "Two-Step Catalytic Reactive Extraction and Transesterification Process via Ultrasonic Irradiation for Biodiesel Production from Solid Jatropha Oil Seeds." Chemical Engineering and Processing – Process Intensification. doi:10.1016/j.cep.2019.107687.

Tang, Haiying, Anfeng Wang, Steven O. Salley, and K. Y. Simon Ng. 2008. "The Effect of Natural and Synthetic Antioxidants on the Oxidative Stability of Biodiesel." Journal of the American Oil Chemists' Society. doi:10.1007/s11746-008-1208-z.

Thangarasu, Vinoth, and R. Anand. 2019. "Physicochemical Fuel Properties and Tribological Behavior of Aegle Marmelos Correa Biodiesel." Advances in Eco-Fuels for a Sustainable Environment. doi:10.1016/b978-0-08-102728-8.00011-5.

Thangarasu, Vinoth, B. Balaji, and Anand Ramanathan. 2019. "Experimental Investigation of Tribo-Corrosion and Engine Characteristics of Aegle Marmelos Correa Biodiesel and Its Diesel Blends on Direct Injection Diesel Engine." Energy. doi:10.1016/j. energy.2019.01.079.

Thangarasu, Vinoth, R. Siddharth, and Anand Ramanathan. 2020. "Modeling of Process Intensification of Biodiesel Production from Aegle Marmelos Correa Seed Oil Using Microreactor Assisted with Ultrasonic Mixing." Ultrasonics Sonochemistry. doi:10.1016/j.ultsonch.2019.104764.

Toscano, G., G. Riva, E. Foppa Pedretti, and D. Duca. 2012. "Vegetable Oil and Fat Viscosity Forecast Models Based on Iodine Number and Saponification Number." Biomass and Bioenergy. doi:10.1016/j.biombioe.2012.07.009.

Veny, Harumi, Saeid Baroutian, Mohamed Kheireddine Aroua, Masitah Hasan, Abdul Aziz Raman, and Nik Meriam Nik Sulaiman. 2009. "Density of Jatropha Curcas Seed Oil and Its Methyl Esters: Measurement and Estimations." International Journal of Thermophysics. doi:10.1007/s10765-009-0569-3.

Wakil, M. A., M. A. Kalam, H. H. Masjuki, A. E. Atabani, and I. M. Rizwanul Fattah. 2015. "Influence of Biodiesel Blending on Physicochemical Properties and Importance of Mathematical Model for Predicting the Properties of Biodiesel Blend." Energy Conversion and Management. doi:10.1016/j.enconman.2015.01.043.

Yaşar, Fevzi. 2020. "Comparision of Fuel Properties of Biodiesel Fuels Produced from Different Oils to Determine the Most Suitable Feedstock Type." Fuel. doi:10.1016/j. fuel.2019.116817.

Yesilyurt, Murat Kadir, Cüneyt Cesur, Volkan Aslan, and Zeki Yilbasi. 2020. "The Production of Biodiesel from Safflower (Carthamus Tinctorius L.) Oil as a Potential Feedstock and Its Usage in Compression Ignition Engine: A Comprehensive Review." Renewable and Sustainable Energy Reviews. doi:10.1016/j.rser.2019.109574.

Zhou, Dan, Baoquan Qiao, Gen Li, Song Xue, and Jianzhong Yin. 2017. "Continuous Production of Biodiesel from Microalgae by Extraction Coupling with Transesterification under Supercritical Conditions." Bioresource Technology. doi:10.1016/j.biortech.2017.04.097.

# 4 Effect of Biodiesel and Additives on Diesel Engine Efficiency and Emission

## 4.1 INTRODUCTION

Energy requirements are increasing along with an increase in industrialization and technological development (Sajjiadi et al. 2016). The world is presently facing the problems of fossil fuel depletion and environmental degradation due to the total dependence on fossil fuels (Sukjit et al. 2013). The more fossil fuels are used in automotive industries, agricultural activities, and power machinery, the greater the presence of exhaust emissions like carbon monoxide, carbon dioxide, unburned hydrocarbon (UBHC), and oxides of nitrogen, which cause global warming, ozone layer depletion, and acid rain (Tan et al. 2011). The depletion of fossil fuel resources with increase in energy consumption has triggered research interest to search for new alternative sources to fulfill the world's present and future energy demand. Its purpose is to replace fossil fuel and to reduce emissions (Rehan et al. 2018;Liu et al. 2015). According to International Outlet 2016, world energy consumption will increase by 48% from 2012 to 2040, due to population growth and economic development (Gebremaniam and Marchetti 2018). By 2040, world energy demand will reach 1094 million barrels per day (World Oil Outlook 2016). Petroleum diesel demand faces two challenges, one being a scarcity of sources and the being other environmental pollution (Olkiewicz et al. 2016). These challenges are the driving force in the search fora new alternative or a long-lasting substitute for diesel.

Biodiesel is one of the best alternative energy sources compared to the other renewable sources such as wind energy, solar energy, tidal energy, etc., because of low installation and maintenance costs and environmental friendliness (Anand 2017). Biodiesel, a monoalkyl ester of long-chain fatty acid, is derived from edible and non-edible sources like sunflower oil, palm, soybean, jatropha, and pungamia oil (Roy et al. 2013). Biodegradability, environmental friendliness, and a non-toxic nature make biodiesel a predominant alternative source for diesel engines (Esteves et al. 2017). Recently, biodiesel has emerged as a promising alternative due to it being free from aromatic and sulfur content, with properties similar to diesel (Aghbashlo et al. 2016).

Throughout the world, 350 crops have been identified as potential feedstock for biodiesel production. Edible oils are also the first and major source for biodiesel production. Approximately 90% of biodiesel is synthesized from edible oil sources.

71

Various edible oils contribute to biodiesel production, but the major contribution of 84% was from rapeseed oil, 3% from sunflower, and 1% from palm oil followed by soybean and other sources at 2% (Sajjadi et al. 2016). Several researchers have extracted biodiesel from edible oils such as camelina oil (Moser et al. 2010), cotton seed oil (corn oil (Demïrbas 2003), coconut oil (Jiang and Tan2012), groundnut oil (Bello and Daniel 2015), palm oil (Nongbe et al. 2017), rapeseed oil (Malins et al. 2016), and soybean oil (Nasreen et al. 2015).

Shahid and Jamal (2011)studied conventional and non-conventional sources for biodiesel production. Conventional sources are soybean oil, ground nut, sunflower, safflower, rice bran oil, palm oil, coconut oil, rapeseed oil, jatropha oil, cotton seed oil, and karanja oil, while non-conventional sources are lard oil, tallow, fish oil, algae, fungi, and microalgae. Earlier literature studies reported that the use of vegetable oil in biodiesel production caused environmental problems like deforestation, destruction of soil, and consumption of more arable land. Also, the price of vegetable oil will increase significantly, which will affect the economic viability of biodiesel. Further, it creates a food-versus-fuel conflict in society (Baskar and Aiswarya 2016). Vegetable oil–based biodiesel costs about 1.5 times more than conventional diesel due to feedstock cost. Feedstock should be cheap and easily available (Ma and Hanna 1999). Hence, selection of edible oil as a feedstock for biodiesel production is not a good long-term choice.

To overcome these problems, biodiesel is produced from non-edible oil crops, such as *Pongamia pinnata* oil (Goembira and Saka 2015), karanja oil (Agarval et al. 2015), jatropha oil (Agarval et al. 2015), rubber seed oil (Aravind, Joy, and Prabakaran 2015), and mahua oil (Puhan et al. 2005). Non-edible oil plant can be planted in wastelands throughout the world to reduce deforestation, and they are also efficient and environmentally friendly. However, it requires more area for planting, which leads to reduced land area for food crops. Hence, non-edible oil feedstock is not a good choice for biodiesel synthesis. Recently, many researchers have extracted biodiesel from waste frying oil (WFO) (Al-Hamamre and Yamin 2014;Chakraborty and Das 2012, Hamze et al. 2015;Kannan and Anand 2012; Kannan and Anand2011;Moazeni et al. 2019), due to it being inexpensive and not causing any environmental issues. Canacki and Van Gerpen (2001) reported on biodiesel production from WFO as a measure to prevent environmental pollution. It is a cheap feedstock for biodiesel production compared to edible and non-edible oils because of its worldwide availability and low cost (Kannan and Anand 2012;Moazeni et al. 2019;Sneha et al. 2015).

Direct injection of raw vegetable oil or WFO in engines may reduce engine durability and decrease engine performance due to its inferior fuel properties (Jaichandar et al. 2012). Many researchers have recommended transesterification, esterification, pyrolysis, micro-emulsion, and dilution methods to diminish the viscosity of oil. Among all methods, transesterification is the most favored method to reduce the viscosity of fuel. Several researchers have achieved maximum biodiesel yields using homogeneous catalysts on biodiesel production (Kulkarni and Dalai 2006;Abdullah et al. 2017;Efavi et al. 2018;Al-Hamamre and Yamin 2014). However, the major drawbacks of using homogeneous catalysts in biodiesel production are biodiesel separation from glycerol, soap formation, and the need for more water for washing. Heterogeneous catalysts overcome the problems of homogeneous catalysts and have the advantage of being simple in nature, free from separation and purification problems (Obadiah et al. 2012).

Although biodiesel's properties are similar to those of conventional diesel, it has some demerits like higher viscosity and density and lower calorific value, which lead to poor atomization and vaporization, in turn resulting in partial combustion that causes lower brake thermal efficiency (BTE) and higher engine exhaust emission levels (Kannan and Anand 2012). Huang and Wu (2008) stated that biodiesel had poor low-temperature properties and inferior engine characteristics compared to conventional diesel. These are major technical barriers for biodiesel use in diesel engines. Simultaneous improvements in fuel properties and engine characteristics are possible by fuel modification and engine modification (Agarwal et al. 2015; Kannan et al. 2013; Qi et al. 2011). Several approaches have been tried to enhance the low-temperature properties of biodiesel as well as the engine characteristics. To overcome the above problems, biodiesel is blended with oxygenated additives like lower alcohol, higher alcohol, and metal additives (Hwanam and Byungchul 2008; Moreau et al. 2012).

## 4.2 METAL ADDITIVES AND THEIR DRAWBACKS

Fuel additives are mainly used to enhance combustion efficiency and decrease the specific fuel consumption. Fuel additives should be selected based on the physicochemical properties, additive solubility, and feasibility from the economic and synthesis points of view. Different types of additives have been tried by researchers to improve combustion phenomena in the combustion ignition engine. Metal-based additives (Keskin, Gürü, and Altiparmak 2007), antioxidant additives (Ryu 2010), oxygenated additives (Cheung, Zhu, and Huang 2009), and cetane improvers have been tried in order to increase combustion efficiency and decrease engine emissions. Ganesh and Gowrishankar (2011) investigate the combustion characteristics of a 4.4-kW diesel engine powered with jatropha oil combined with magnalium (AL–Mg) and cobalt oxide ($Co_3O_4$) additives. They report that brake-specific fuel consumption (BSFC) was decreased by 3% and 2% with additions of AL–Mg and $Co_3O_4$, respectively. Higher BTE was found for AL–Mg additives, which is 1% higher than diesel. As for emissions, $NO_x$, HC, and CO were decreased for both metal additives. Babu, Rao, and Rao (2012) tried the addition of diethyl ether (DEE) in raw mahua oil at 3%, 5%, 10%, and 15% for viscosity reduction. They reported a decrease in BTE and an increase in BSFC with an increasing DEE ratio. Also, $NO_x$ emission significantly decreased with increment in the additive percentage due to a reduction in the combustion temperature. Varatharajan, Cheralathan, and Velraj (2011) investigated using L-ascorbic acid blended with raw jatropha oil at the ratio of 0.025%. They reported that the increase in BSFC and the HC and CO emission increment with the blending of L-ascorbic acid was due to improper combustion.

One of the most attractive methods is the addition of alcohol with biofuels for the reduction of viscosity and density. Also, alcohol improves the volatility of the biofuel. However, the blending of lower alcohols with raw oil is difficult because of phase separation and low solubility. Some surfactants can be used in the blends to overcome solubility and miscibility problems. Meanwhile, higher alcohols like butanol, n-pentanol have received increased research attention due to their higher miscibility with raw oil (Laza and Bereczky 2011). Singh et al. have investigated CI engine emission and performance characteristics using crude and virgin coconut oil blends in a diesel engine. To improve the miscibility, they used ethanol and 1-butanol (B) as the additive and the surfactant, respectively. They reported that engine emissions are significantly reduced with the addition of butanol and ethanol into coconut oil blends. The specific fuel consumption was higher for coconut oil blends for the entire range of operation. The BTE values of both 54CCO23E23B and 53VCO23E24B blends were identical with that of diesel (Singh, Khurma, and Singh 2010).

Overall, metal-based additives reduce the viscosity and pour point and increase the flashpoint of the blend. Due to the catalytic effect of metal additives, BTE was increased and BSFC reduced. Also, reduction in exhaust gas emissions was achieved by adding metallic additives in small quantities. However, metallic additives decrease the lifetime of the fuel injection system due to the higher wear rate. The addition of antioxidant additives like L-ascorbic acid improves the oxygen stability and cetane number of the raw bio-oil. Also, antioxidant additives effectively controlled $NO_x$ formation. However, BTE decreased and BSFC increased with the addition of antioxidant additives (Jiaqiang et al. 2018). Raw oil–alcohol blends exhibit higher BSFC compared to baseline reading. The BSFC significantly decreased for 50% alcohol blend s and it increased with increasing the alcohol blend above 50%. A higher alcohol ratio in the fuel decreases the viscosity and improves the volatility, which leads to better atomization. BTE decreased with an increase in the alcohol percentage due to the lower calorific value. Smoke emission was significantly decreased with the addition of short-chain alcohol in the ratio of 30% as compared to raw oil. However, smoke emission was slightly higher than the baseline value. Compared to lower alcohol, higher alcohol decreases the smoke emission effectively because of the higher cetane number (Datta and Mandal, 2017).

## 4.3 LOWER ALCOHOL ADDITIVES

Biodiesel is derived from vegetable oil, animal fat, and WFO. The properties of these fuels are similar to those of petroleum diesel fuel. Biodiesel is a promising substitute for petroleum diesel fuel in diesel engines. However, the obstacles to biodiesel use in diesel engines are the low-temperature properties (Kannan and Anand 2011;;Huang and Wu 2008) and inferior engine's performance, combustion, and emission characteristics (Huang and Wu 2008). Biodiesel has the drawbacks of poorer low-temperature properties, greater NO emissions, and inferior engine characteristics compared to conventional diesel. These must be rectified before the commercialization of biodiesel. Researchers have been making efforts from this perspective to improve fuel properties to attain better performance and lower emissions in diesel engines through fuel modification like biodiesel being blended with higher alcohol and conventional diesel or ternary blends.

Various additives such as alcohols, cetane improvers, water emulsions, oxygenates, and nano-additives were blended with biodiesel to reduce exhaust emission in diesel engines (Qi et al. 2011). Boretti (2012) reported that alcohol blended with biodiesel reduced exhaust emissions like smoke, unburned hydrocarbon, and carbon monoxide. Agarwal (2007) mentioned that alcohols are renewable energy sources with merits like lower viscosity and density and higher oxygen content. These are highly miscible with biodiesel and diesel fuel and enhance the properties of biodiesel blends. Alcohol blended with biodiesel improves the cold flow properties and reduces poisonous gas emission. The addition of more alcohol in biodiesel blends increased the latent heat of evaporation (LHE) of the fuel, causing reduction the in-cylinder temperature, leading to lower NO emission formation. Biodiesel and biodiesel blends produced lower NO emissions due to their lower iodine values related to neat diesel fuel. Carbon monoxide, sulfur oxide (SO), NO, and PM emissions significantly decreased in biodiesel–alcohol blends. Alcohol blends improve the cloud and pour point properties and reduce the viscosity and density of the blends (Ren et al. 2008).

Nadir and Sanchez (2012) investigated the effect of biodiesel–diesel–ethanol and –methanol blends on diesel engine emission characteristics under various load conditions. Biodiesel–diesel–methanol blends have higher CO and UBHC emissions compared to biodiesel–diesel–ethanol blends due to the higher latent heat of vaporization of methanol compared to ethanol, which suppressed in-cylinder temperature and pressure due to a cooling effect caused by adding more methanol into biodiesel blends. Mwangi et al. (2015) evaluated the influence of green fuel in diesel engines. Blends of biodiesel or acetone–butanol–ethanol solution reduced uncontrolled emissions like polycyclic aromatic hydrocarbons, persistent organic components, carbonyl and CO, UBHC, and particulate matter, whereas NO emission increased.

Hwanam and Byungehul (2008) confirmed that an additive like ethanol improved chemical properties; however, it negatively affected other properties like the flash and fire points, calorific value, and cetane number. So it is difficult to prepare a perfectly suitable blend to replace conventional diesel for use in diesel engines. Yao et al. (2010) observed that the drawbacks like poor blending solubility and poor miscibility of adding lower alcohols such as methanol and ethanol with diesel at low temperatures. Similarly, Moreau et al. (2012) noticed that the alcohols containing more carbons could be used as a resource because of their physical and thermodynamic properties. Sivalakshmi and Balusamy (2011) concluded that higher alcohols have higher cetane numbers and better miscibility with diesel. However, Giakoumis et al. (2013) found that using butanol used as a fuel additive in diesel engines resulted in favorable exhaust emission changes.

Arbab et al. (2015) investigated the effect of diesel–n-butanol binary blends on engine performance and emissions. The experimental study showed that diesel–n-butanol blends will be a next-generation fuel. Atmanli et al. (2015) optimized the diesel–n-butanol–cotton seed oil ternary blends to better emission and performance. The optimum concentration was 65.5% vol. diesel, 23.1% vol. n-butanol, and 11.4% vol. cotton seed oil. The results show that the BTE, braking power, and braking torque decreased, while BSFC increased for the above-mentioned blends compared to conventional diesel. Nitric oxide, carbon monoxide, and unburned hydrocarbon emissions decreased for the corresponding blends as 11.33%, 45.17%, and 81.45%, respectively, compared to conventional diesel. Sharon et al. (2013) examined diesel–palm oil–n-butanol blends for diesel engine as an alternative fuel. Increase of butanol in ternary blends resulted in a reduction of $NO_x$, CO, $CO_2$, and smoke emissions while UBHC emission and BTE increased.

Ileri (2016) investigated n-butanol–diesel–vegetable oil blends in diesel engine. It was noticed that blends having 60% of n-butanol content reduced CO and UBHC emissions. Kumar et al. (2013) examined alcohol properties compared to those of conventional diesel and gasoline fuel, and also evaluated the influence of bio-diesel–diesel–alcohol blends on diesel engine performance characteristics. Engine studies clearly showed that butanol was superior to methanol and ethanol due to its superior fuel properties. Experimental results showed that the alcohol blends along with a cetane improver reduced engine emissions. Jeevahan et al. (2018) studied the influence of 1-butanol-biodiesel blends on diesel engine characteristics. The results revealed that biodiesel-1-butanol blends reduced engine emissions like CO, UBHC, and NO and increased engine performance. They reported that biodiesel-1-butanol blends could replace conventional diesel.

## 4.4   HIGHER ALCOHOL ADDITIVES

Nour et al. (2019) carried out research on a common rail direct injection (CRDI)-assisted diesel engine fueled with biodiesel–higher alcohol blends. The study reported that biodiesel–higher alcohol blends had high stability with no phase separation. NO and smoke emissions were reduced for higher alcohol biodiesel blends, whereas CO and UBHC emissions increased for biodiesel–higher alcohol blends. Ashok et al. (2019) investigated the effect of n-biodiesel–n-pentanol blends on diesel engine performance, combustion, and emission characteristics. The addition of n-pentanol to biodiesel improved thermal efficiency and reduced UBHC and CO emission by about 15–43%, or 33–50% when compared to pure biodiesel. The addition of n-pentanol above 40% had a negative effect on engine performance. The study concluded that higher alcohols were better alternatives than lower alcohols.

Yilmaz et al. (2018) examined the effect of ternary blends on diesel engine characteristics. This study exposed that biodiesel, diesel, and alcohol are the most important fuel sources for diesel engines. The direct use of biodiesel in diesel engines is limited due to its inferior characteristics like higher viscosity and density and lower heating value, which leads to poor atomization and air-fuel mixture, in turn resulting in partial combustion, lower BTE, and higher exhaust emission. Zhi-Hui and Rajasekahar (2016) noticed that the higher percentage of biodiesel in diesel may cause poor atomization, carbon deposits, clogging of fuel lines, and thickening of lubricating oil due to its poor volatility and higher viscosity. In order to reduce the disadvantages of biodiesel, it is blended with alcohol and diesel. Popa et al. (2001) reported that the alcohols are extracted from renewable sources using fermentation process of biomass feedstock like corn and sugar. Atmanli et al. (2015) reported that alcohols are clean fuel because of the hydroxyl (OH) presence in the molecular structure. They can be easily mixed with organic components like vegetable oil and biodiesel, which reduces the viscosity and improves the phase stability at low temperatures. While alcohol can't be used directly in diesel engines, some of their properties make them suitable additives for biodiesel and diesel as noticed by Kumar and Saravanan (2016).

When Zaharin et al. (2017) reported that biodiesel and alcohol are promising alternative fuels for diesel engines, it gained much attention from researchers and scientific community. This study concluded that alcohol is a potential additive that can be blended with biodiesel and diesel fuel. The additives can improve the spray characteristics of biodiesel–diesel blends and also the air-fuel mixing process, which promotes complete combustion. The lower viscosity and higher volatility of alcohol when blended with biodiesel reduces the viscosity of biodiesel. Also, with the addition of alcohol to biodiesel, the oxygen content is further improved, which improves the combustion efficiency and lowers engine emissions. The low viscosity of alcohol improves the spray characteristics of blends and decreases fuel resistance during fuel injection as noticed by Popa et al. (2001). Lower alcohols allow biodiesel to be used in diesel engines, improving microemulsion, but by themselves they have limited use in diesel engines due to their inferior fuel properties like lower cetane numbers and calorific values. Especially, methanol and ethanol are not good alternative fuels for diesel engine.

Lower alcohol can be effortlessly mixed with diesel and biodiesel, which enhances the combustion efficiency and decreases emissions. Even though lower alcohols can increase the oxygenation of fuels, they have some limitations such as poor miscibility, lower heating values, lower cetane numbers, poor lubrication, and stability problems when mixed with biodiesel and conventional diesel as noticed by Yilmaz et al. (2014). Lower alcohols having low flashpoints are considered as class I fuel, while conventional diesel fuel is classified as class II fuel. Lower alcohol blended with diesel falls down to class I. Higher alcohol has been gaining much attention from researches due to its fuel properties like cetane number, less hygroscopic nature, higher energy density, and better blending stability when compared to lower alcohols like ethanol and methanol as noticed by Ren et al. (2008). Kumar and Saravanan (2016) observed that higher alcohol–biodiesel–diesel blends have the potential to replace fossil fuels wholly or partially. Higher alcohol can mix with biodiesel-diesel without any phase separation. Therefore, no co-solvent or emulsifying agent is required to maintain blend stability when higher alcohol is blended with biodiesel-diesel.

Higher alcohols exhibit less corrosion on materials and have high flashpoints; these properties make them safe to store, handle, and transport. Also, alcohol fuel reduced the delay period and improved the air-fuel mixing process, which contributed to clean combustion as observed by Xing-Cai et al. (2004). Balamurugan and Nalini et al. (2014) noticed that alcohol with increased carbon number can be mixed with organic components more easily and also reach a high cetane number and heat of combustion. Thus, higher alcohols have higher potential to mix with biodiesel than lower alcohols. Biodiesel–diesel–alcohol blends improve diesel engine characteristics and diminish hazardous exhaust emissions by physicochemical properties compared to biodiesel-diesel and biodiesel–alcohol blends. Biodiesel–diesel–alcohol blends can meet energy demands and regulate emissions as noticed by Zaharin et al. (2017).

Imdadul et al. (2016) reported that alcohol–diesel blends reduce viscosity, lubricity, cetane number, and ignitability. However, alcohol blended with biodiesel can provide higher oxygen content and increase the LHE, resulting in reduced NO and PM emissions and improved cold flow properties of the blends. The combination of biodiesel–diesel–alcohol blends can improve the cold flow properties, cetane number, and calorific value compared to raw biodiesel, and enhance the combustion efficiency, resulting in lower exhaust emissions and better thermal efficiency. The above literature studies clearly stated that a biodiesel–diesel–alcohol blend is a better option as fuel in diesel engines compared to biodiesel–diesel, biodiesel–alcohol, and diesel–alcohol blends. It can be a promising alternative fuel for diesel engines.

## 4.5  CONCLUSION

- Biodiesel production from WFO eliminated the conflict between food and fuel production. Further, it reduced the biodiesel production cost, the waste treatment cost, and environmental pollution compared to conventional diesel. It also has the advantage of low-cost feedstock and is readily available throughout the world.
- Major process parameters like reaction time, reaction temperature catalyst concentration, molar ratio of oil to alcohol, and stirrer speed significantly affected transesterification in the biodiesel process. Transesterification is a proven method to reduce the viscosity of vegetable oil compared to other methods like direct blending, micro-emulsification, and pyrolysis.
- Metal-based additives reduce the viscosity and pour point and increase the flashpoint of the blend. Due to the catalytic effect of metal additives, BTE was increased and BSFC reduced. Also, reduction in the exhaust gas emissions was identified by adding metallic additives in small quantities.
- Metallic additives decrease the lifetime of the fuel injection system due to the higher wear rate. Addition of antioxidant additives improves the oxygen stability and cetane number of the raw bio-oil. Also, antioxidant additives effectively controlled $NO_x$ formation.
- Poor low-temperature properties of biodiesel could be rectified by fuel modification without engine modification. Improvements in diesel engine performance, combustion, and emission characteristics without engine modification were achieved by biodiesel–diesel–alcohol blends.
- One of the most attractive methods is the addition of alcohol with biofuels for reductions of viscosity and density. Also, alcohol improves the volatility of the biofuel. However, the blending of lower alcohols with raw oil is difficult because of phase separation and low solubility. Some surfactants can be used in the blends to overcome solubility and miscibility problems. Meanwhile, higher alcohols like butanol, n-pentanol have received increased research attention due to their higher miscibility with raw oil.
- Higher alcohols have the potential to overcome drawbacks of lower alcohols like methanol and ethanol due to the higher cetane number and better miscibility with conventional diesel. Several researchers have stated that higher alcohols like n-pentanol and n-hexanol have the potential to become future-generation fuels, as their fuel properties are similar to those of conventional diesel.

## REFERENCES

Abdullah, R., N.Rahmawati Sianipar, D.Ariyani and I.Fatyasari Nata2017. "Conversion of palm oil sludge to biodiesel using alum and KOH as catalysts." *Sustainable Environment Research*, 291–95.

Agarwal, Avinash Kumar. 2007. "Biofuels (Alcohols and Biodiesel) Applications as Fuels for Internal Combustion Engines." *Progress in Energy and Combustion Science* 33 (3): 233–71. doi: 10.1016/j.pecs.2006.08.003.

Agarwal, A.K., A.Dhar, J.G.Gupta, W.I.Kim, C.S.Lee and S.Park2014. "Effect of fuel injection pressure and injection timing on spray characteristics and particulate size-number distribution in a biodiesel fuelled common rail direct injection diesel engine."*Applied Energy*, 130, 212–21.

Agarwal, A.K., A.Dhar, J.G.Gupta, W.I.Kim, K.Choi, C.S.Lee and S.Park2015. "Effect of fuel injection pressure and injection timing of Karanja biodiesel blends on fuel spray, engine performance, emissions and combustion characteristics."*Energy Conversion and Management*, 91, 302–14.

Aghbashlo, M., M.Tabatabaei, P.Mohammadi, M.Mirzajanzadeh, M.Ardjmand and A.Rashidi2016. "Effect of an emission-reducing soluble hybrid nano catalyst in diesel/biodiesel blends on energetic performance of a DI diesel engine."*Renewable Energy*, 93, 353–68.

Al-Hamamre, Z. and J.Yamin2014. "Parametric study of the alkali catalyzed transesterification of waste frying oil for Biodiesel production."*Energy Conversion and Management*, 79, 246–54.

Anand, R.2017. "Simultaneous control of oxides of nitrogen and soot in CRDI diesel engine using split injection and Cool EGR fuelled with waste frying oil biodiesel and its blends." In *Air Pollution and Control*, Springer, 11–44.

Aravind, A., M.L.Joy and K.Prabhakaran Nair2015. "Lubricant properties of biodegradable rubber tree seed (Hevea brasiliensis Muell. Arg) oil."*Industrial Crops and Products*, 74, 14–19.

Arbab, M.I., M.Varman, H.H.Masjuki, M.A.Kalam, S.Imtenan, H.Sajjad and I.M.Rizwanul Fattah2015. "Evaluation of combustion, performance and emissions of optimum palm-coconut blends in turbocharged and non-turbocharged conditions of a diesel engine."*Energy Conversion and Management*, 90, 111–20.

Ashoka, B., A.K.Jeevanantham, K.Nanthagopal, B.Saravanan, M.Senthil Kumar, AjithJohny, AravindMohan, Muhammad UsmanKaisan and ShituAbubaka2019. "An experimental analysis on the effect of n-pentanol – Calophyllum Inophyllum Biodiesel binary blends in CI engine characteristics."*Fuel*, 173, 290–305.

Atmanli, A., B.Yuksel, E.Ileri and A. DenizKaraoglan2015. "Response surface methodology based optimization of diesel – n-butanol – cotton oil ternary blends ratios to improve engine performance and exhaust emission characteristics."*Energy Conversion and Management*, 90, 383–94.

Babu, P Ramesh, K Prasad Rao, and B V Appa Rao. 2012. "The Role of Oxygenated Fuel Additive (DEE) along with Mahuva Methyl Ester to Estimate Performance and Emission Analysis of DI-Diesel Engine." *International Journal of Thermal Technologies*, 2(1), 119–123.

Balamurugan, T., and R. Nalini. 2014. "Experimental Investigation on Performance, Combustion and Emission Characteristics of Four Stroke Diesel Engine Using Diesel Blended with Alcohol as Fuel." *Energy* 78: 356–63. doi: 10.1016/j.energy.2014.10.020.

Baskar, G. and R.Aiswarya2016. "Trends in catalytic production of biodiesel from various feedstocks."*Renewable and Sustainable Energy Reviews*, 57, 496–504.

Bello, E.I. and F.Daniel2015. "Optimization of groundnut oil biodiesel production and characterization." Applied Science Report, 9, 172–80.

Boretti, Alberto. 2012. "Advantages of Converting Diesel Engines to Run as Dual Fuel Ethanol-Diesel." *Applied Thermal Engineering* 47: 1–9. doi: 10.1016/j.applthermaleng.2012.04.037.

Canakci, M. and J.Van Gerpen2001. "Biodiesel production via acid catalyst."*Transactions of the American Society of Agricultural Engineers*, 42, 1203–10.

Chakraborty, R. and S.K.Das2012. "Optimization of biodiesel synthesis from waste frying soyabean oil using fish scale-supported Ni catalyst."*Industrial and Engineering Chemistry Research*, 51, 404–14.

Cheung, C.S., Lei Zhu, and Zhen Huang. 2009. "Regulated and Unregulated Emissions from a Diesel Engine Fueled with Biodiesel and Biodiesel Blended with Methanol." *Atmospheric Environment* 43 (32): 4865–72. doi: 10.1016/j.atmosenv.2009.07.021.

Datta, A, and B K Mandal. 2017. "Effect of Alcohol Addition to Diesel on Engine Performance, Combustion and Emission Characteristics of a CI Engine." In *2017 International Conference on Advances in Mechanical, Industrial, Automation and Management Systems (AMIAMS)*, 110–14. doi: 10.1109/AMIAMS.2017.8069198.

Demirbaş, Ayhan. 2003. "Chemical and Fuel Properties of Seventeen Vegetable Oils." *Energy Sources* 25 (7): 721–28. doi: 10.1080/00908310390212426.

Demirbas, A.2009. "Production of biodiesel fuels from linseed oil using methanol and ethanol in non-catalytic SCF conditions."*Biomass and Bioenergy*, 33, 113–18.

Efavi, J.K., D.Kanbogtah, V.Apalangya, E.E.K.Nyankson Tiburu, D.Dodoo-Arhin, B.Onwona-Agyeman and A.Yaya2018. "The effect of NaOH catalyst concentration and extraction time on the yield and properties of Citrullus vulgaris seed oil as a potential biodiesel feed stock."*South African Journal of Chemical Engineering*, 25, 98–102.

Esteves, V.P.P., E.M.M.Esteves, D.J.Bungenstab, G.L.D.Feijo, O.D.Q.F.Araújo and C.D.R.V.Morgado2017. "Assessment of greenhouse gases (GHG) emissions from the tallow biodiesel production chain including land use change (LUC)."*Journal of Cleaner Production*, 151, 578–91.

Ganesh, D., and G. Gowrishankar. 2011. "Effect of Nano-Fuel Additive on Emission Reduction in a Biodiesel Fuelled CI Engine." In *2011 International Conference on Electrical and Control Engineering, ICECE 2011 - Proceedings*, 3453–59. doi: 10.1109/ICECENG.2011.6058240.

Gebremariam, S.N. and J.M.Marchetti2018. "Economics of biodiesel production: review."*Energy Conversion and Management*, 168, 74–84.

Giakoumis, Evangelos G. 2013. "A Statistical Investigation of Biodiesel Physical and Chemical Properties, and Their Correlation with the Degree of Unsaturation." *Renewable Energy* 50: 858–78. doi: 10.1016/j.renene.2012.07.040.

Goembira, F. and S.Saka2015. "Advanced supercritical methyl acetate method for biodiesel production from Pongamia pinnata oil."*Renewable Energy*, 83, 1245–49.

Green, Terrence R., and Radu Popa. 2011. "Endpoint Fragmentation Index: A Method for Monitoring the Evolution of Microbial Degradation of Polysaccharide Feedstocks." *Applied Biochemistry and Biotechnology* 163 (4): 519–27. doi: 10.1007/s12010-010-9058-1.

Hamze, H., M.Akiaa and F.Yazdani2015. "Optimization of biodiesel production from the waste cooking oil using response surface methodology."*Process Safety and Environmental Protection*, 94, 1–10.

Huang, Y.H. and J.H.Wu2008. "Analysis of biodiesel promotion in Taiwan."*Renewable and Sustainable Energy Reviews*, 12, 1176–86.

Ileri, E.2016. "Experimental study of 2-ethylhexyl nitrate effects on engine performance and exhaust emissions of a diesel engine fuelled with n-butanol or 1-pentanol diesel–sunflower oil blends."*Energy Conversion and Management*, 118, 320–30.

Imdadul, H K, H H Masjuki, M A Kalam, N W M Zulkifli, Abdullah Alabdulkarem, M Kamruzzaman, and M M Rashed. 2016. "A Comparative Study of C4 and C5 Alcohol Treated Diesel–Biodiesel Blends in Terms of Diesel Engine Performance and Exhaust Emission." *Fuel* 179: 281–88. doi: 10.1016/j.fuel.2016.04.003.

Jaichandar, S., P.Senthil kumar and K.Annamalai2012. "Combined effect of injection timing and combustion geometry on the performance of a biodiesel fueled diesel engine."*Energy*, 47, 388–94.

Jan, Ban, Jorge León Arellano, Amal Alawami, Roberto F. Aguilera, and Martin Tallett Contributors. 2006. *World Oil Outlook. Organization of the Petroleum Exporting Countries*. Organization of the Petroleum Exporting Countries. doi: 10.1002/9783527628148.hoc054.

Jeevahan, J., R.B.Durairaj, G.Sriramanjaneyulu and G.Mageshwaran2018. "Experimental investigation of the suitability of 1-butanol blended with biodiesel as an alternative biofuel in diesel engines."*Biocatalyst and Agricultural Biotechnology*, 15, 72–77.

Jiang, J.J. and C.S.Tan2012. "Biodiesel production from coconut oil in supercritical methanol in the presence of co-solvent."*Journal of the Taiwan Institute of Chemical Engineers*, 43, 102–07.

Jiaqiang, E., Zhiqing Zhang, Jingwei Chen, Minh Hieu Pham, Xiaohuan Zhao, Qingguo Peng, Bin Zhang, and Zibin Yin. 2018. "Performance and Emission Evaluation of a Marine Diesel Engine Fueled by Water Biodiesel-Diesel Emulsion Blends with a Fuel Additive of a Cerium Oxide Nanoparticle." *Energy Conversion and Management* 169: 194–205. doi: 10.1016/j.enconman.2018.05.073.

Kannan, G.R. and R.Anand2011. "Experimental evaluation of DI diesel engine operating with diestrol at varying injection pressure and injection timing."*Fuel Processing Technology*, 92, 2252–63.

Kannan, G.R. and R.Anand2012. "Effect of fuel injection pressure and fuel injection timing on DI diesel engine fuelled with biodiesel from waste cooking oil."*Biomass and Bioenergy*, 46, 343–52.

Keskin, Ali, Metin Gürü, and Duran Altiparmak. 2007. "Biodiesel Production from Tall Oil with Synthesized Mn and Ni Based Additives: Effects of the Additives on Fuel Consumption and Emissions." *Fuel* 86 (7–8): 1139–43. doi: 10.1016/j.fuel.2006.10.021.

Kim, Hwanam N., and Byungchul C. Choi. 2008. "Effect of Ethanol-Diesel Blend Fuels on Emission and Particle Size Distribution in a Common-Rail Direct Injection Diesel Engine with Warm-Up Catalytic Converter." *Renewable Energy* 33 (10): 2222–28. doi: 10.1016/j.renene.2008.01.002.

Kulkarni, M.G. and A.K.Dalai2006. "Waste cooking oils an economical source for biodiesel."*Industrial and Engineering Chemistry Research*, 45, 2901–13.

Kumar, S., J.H.Cho, J.Park and I.Moon2013. "Advances in diesel alcohol blends and their effects on the performance and emissions of diesel engines."*Renewable and Sustainable Energy Reviews*, 22, 46–72.

Laza, T., and Á Bereczky. 2011. "Basic Fuel Properties of Rapeseed Oil-Higher Alcohols Blends." *Fuel* 90 (2): 803–10. doi: 10.1016/j.fuel.2010.09.015.

Liu, X., M.Yao, Y.Wang, Z.Wang, H.Jin and I.Wei2015. "Experimental and kinetic modelling study of a rich and a stoichiometric low-pressure premixed laminar 2, 5-dimethyl furan/oxygen/argon flames."*Combustion and Flames*, 162, 4586–97.

Ma, F. and M.A.Hanna1999. "Biodiesel production: a review."*Bioresource Technology*, 70, 1–15.

Malins, K., J.Brinks, V.Kampars and I.Malina2016. "Esterification of rapeseed oil fatty acids using a carbon-based heterogeneous acid catalyst derived from cellulose."*Applied Catalyst A: General*, 519, 99–10.

Moazeni, F., Y.C.Chen and G.Zhang2019. "Enzymatic transesterification for biodiesel production from used cooking oil, a review."*Journal of Cleaner Production*, 216, 117–28.

Moreau, Alejandro, M. Carmen Martín, César R. Chamorro, and José J. Segovia. 2012. "Thermodynamic Characterization of Second Generation Biofuels: Vapour-Liquid Equilibria and Excess Enthalpies of the Binary Mixtures 1-Pentanol and Cyclohexane or Toluene." *Fluid Phase Equilibria* 317: 127–31. doi: 10.1016/j.fluid.2012.01.007.

Moser, Bryan R. 2010. "Camelina (*Camelina sativa* L.) Oil as a Biofuels Feedstock: Golden Opportunity or False Hope?" *Lipid Technology* 22 (12): 270–73. doi: 10.1002/lite.201000068.

Mwangi, John Kennedy, Wen Jhy Lee, Yu Cheng Chang, Chia Yang Chen, and Lin Chi Wang. 2015. "An Overview: Energy Saving and Pollution Reduction by Using Green Fuel Blends in Diesel Engines." *Applied Energy*, 159, 214–236. doi: 10.1016/j.apenergy.2015.08.084.

Nasreen, S., H.Liu, D.Skala, A.Waseem and L.Wan2015. "Preparation of biodiesel from soybean oil using La/Mn oxide catalyst."*Fuel Processing Technology*, 131, 290–96.

Nongbe, M.C., T.Ekou, L.Ekou, K.B.Yao, E.Le Grognec and F.X.Felpin2017. "Biodiesel production from palm oil using sulfonated graphene catalyst."*Renewable Energy*, 106, 135–41.

Nour, M., A.M.A.Attiaa and S.A.Nadaa2019. "Combustion, performance and emission analysis of diesel engine fuelled by higher alcohols (butanol, octanol and heptanol)/diesel blends."*Energy Conversion and Management*, 185, 313–29.

Obadiah, A., G.A.Swaroopa, S.V.Kumar, K.R.Jeganathan and A.Ramasubbu2012. "Biodiesel production from palm oil using calcined waste animal bone as catalyst."*Bioresource Technology*, 116, 512–16.

Olkiewicz, M., C.M.Torres, L.Jiménez, J.Font and C.Bengoa2016. "Scale-up and economic analysis of biodiesel production from municipal primary sewage sludge."*Bioresource Technology*, 214, 122–31.

Puhan, S., N.Vedaraman, V.B.Boppana, G.Sankarnarayanan and K.Jeychandran2005. "Mahua oil (Madhuca Indica seed oil) methyl ester as biodiesel-preparation and emission characteristics."*Biomass and Bioenergy*, 28, 87–93.

Qi, D.H., H.Chen, L.M.Geng and Y.Z.Bian2011. "Effect of diethyl ether and ethanol additives on the combustion and emission characteristics of biodiesel diesel blended fuel engine."*Renewable Energy*, 36, 1252–58.

Rajesh Kumar, B., and S. Saravanan. 2016. "Use of Higher Alcohol Biofuels in Diesel Engines: A Review." *Renewable and Sustainable Energy Reviews*, 60, 84–115. doi: 10.1016/j.rser.2016.01.085.

Rehan, M., J.Gardy, A.Demirbas, U.Rashid, W.Budzianowski, D.Pant and A.Nizami2018. "Waste to biodiesel: a preliminary assessment for Saudi Arabia."*Bioresour Technology*, 250, 17–25.

Ren, Y., Z.H. Huang, D.M. Jiang, W. Li, B. Liu, and X.B. Wang. 2008. "Effects of the Addition of Ethanol and Cetane Number Improver on the Combustion and Emission Characteristics of a Compression Ignition Engine." *Proceedings of the Institution of Mechanical Engineers, Part D: Journal of Automobile Engineering* 222 (6): 1077–87. doi: 10.1243/09544070JAUTO516.

Roy, M., W.Wang and J.J.Bujold2013. "Biodiesel production and comparison of emissions of a DI diesel engine fuelled by biodiesel-diesel and canola oil-diesel blends at high idling operations."*Applied Energy*, 106, 198–208.

Ryu, Kyunghyun. 2010. "Effect of Antioxidants on the Oxidative Stability and Combustion Characteristics of Biodiesel Fuels in an Indirect-Injection (IDI) Diesel Engine." *Journal of Mechanical Science and Technology* 23 (11): 3105–13. doi: 10.1007/s12206-009-0902-6.

Sajjadi, B., A.A. AbdulRaman and H.Arandiyan2016. "A comprehensive review on properties of edible and non-edible vegetable oil-based biodiesel: composition, specifications and prediction models."*Renewable and Sustainable Energy Reviews*, 63, 62–92.

Shahid, E.M. and Y.Jamal2011. "Production of biodiesel: a technical review."*Renewable and Sustainable Energy Reviews*, 15, 4732–45.

Sharon, H., P. JaiShiva Ram and K.Jenis Fernando2013. "Fuelling a stationary direct injection diesel engine with diesel-used palm oil-butanol blends – an experimental study."*Energy Conversion and Management*, 73, 95–105.

Singh, Pranil J., Jagjit Khurma, and Anirudh Singh. 2010. "Preparation, Characterisation, Engine Performance and Emission Characteristics of Coconut Oil Based Hybrid Fuels." *Renewable Energy* 35 (9): 2065–70. doi: 10.1016/j.renene.2010.02.007.

Sivalakshmi, S., and T. Balusamy. 2012. "Influence of Ethanol Addition on a Diesel Engine Fuelled with Neem Oil Methyl Ester." *International Journal of Green Energy* 9 (3): 218–28. doi: 10.1080/15435075.2011.621477.

Sneha, E., R.Anand, K.M.Meera, S.Begam and N.Anantharaman2015. "Biodiesel production from waste cooking oil using KBr impregnated CaO as catalyst."*Energy Conversion and Management*, 91, 442–50.

Sukjit, E., J.M.Herreros, J.Piaszyk, K.D.Dearn and A.Tsolakis2013. "Finding synergies in fuels properties for the design of renewable fuels-hydroxylated biodiesel effects on butanol-diesel blends."*Environmental Science Technology*, 47, 3535–42.

Tan, K.T., K.T.Lee and A.R.Mohamed2011. "Potential of waste palm cooking oil for catalyst free biodiesel production."*Energy*, 36, 2085–88.

Varatharajan, K., M. Cheralathan, and R. Velraj. 2011. "Mitigation of NOx Emissions from a Jatropha Biodiesel Fuelled Di Diesel Engine Using Antioxidant Additives." *Fuel* 90 (8): 2721–25. doi: 10.1016/j.fuel.2011.03.047.

Xing-Cai, Lü, Yang Jian-Guang, Zhang Wu-Gao, and Huang Zhen. 2004. "Effect of Cetane Number Improver on Heat Release Rate and Emissions of High Speed Diesel Engine Fueled with Ethanol-Diesel Blend Fuel." *Fuel* 83: 2013–20. doi: 10.1016/j. fuel.2004.05.003.

Yao, Mingfa, Hu Wang, Zunqing Zheng, and Yan Yue. 2010. "Experimental Study of N-Butanol Additive and Multi-Injection on HD Diesel Engine Performance and Emissions." *Fuel* 89 (9): 2191–2201. doi: 10.1016/j.fuel.2010.04.008.

Yilmaz, Nadir, Alpaslan Atmanli, and Francisco M. Vigil. 2018. "Quaternary Blends of Diesel, Biodiesel, Higher Alcohols and Vegetable Oil in a Compression Ignition Engine." *Fuel* 212: 462–69. doi: 10.1016/j.fuel.2017.10.050.

Yilmaz, Nadir, and Tomas M. Sanchez. 2012. "Analysis of Operating a Diesel Engine on Biodiesel-Ethanol and Biodiesel-Methanol Blends." *Energy* 46 (1): 126–29. doi: 10.1016/j.energy.2011.11.062.

Yilmaz, Nadir, and Francisco M. Vigil. 2014. "Potential Use of a Blend of Diesel, Biodiesel, Alcohols and Vegetable Oil in Compression Ignition Engines." *Fuel* 124: 168–72. doi: 10.1016/j.fuel.2014.01.075.

Zhang, Zhi-Hui, and Rajasekhar Balasubramanian. 2016. "Investigation of Particulate Emission Characteristics of a Diesel Engine Fueled with Higher Alcohols/Biodiesel Blends." *Applied Energy* 163: 71–80. doi: 10.1016/j.apenergy.2015.10.173.

# 5 Recent Advanced Injection Strategy on Biodiesel Combustion

## 5.1 INTRODUCTION

Many researchers have tried biodiesel–diesel blends to improve performance and reduce emissions. However, this method often results in reduced engine performance and can damage various automotive components (Mohsin et al. 2014). Hence, there is a need for investigation of injection parameters, namely fuel injection pressure (FIP) and fuel injection timing (FIT) effects on biodiesel performance, combustion, and emission characteristics. Many researchers have found that injection pressure and timing are vital operating parameters that are found to affect engine characteristics.

Suryawanshi and Deshpande (2005) investigated the effect of advanced and retarded fuel injection timing on diesel engine characteristics. They reported that delayed injection timing effectively decreased the $NO_x$, CO, and HC emission with a negligible rise in the total fuel consumption. At a delayed injection time, BTE and exhaust temperature followed a similar trend as baseline fuel. Nevertheless, ignition delay and peak cylinder pressure were decreased. Nwafor and Ogbonna (2000) described that advanced injection timing increases the ignition delay period of rapeseed oil in a CI engine with a penalty of higher fuel consumption. Also, they found that $NO_x$ emissionwas enhanced for the fuel injection advancement by 3 °CA, and it decreased significantly for the retardation of fuel injection by 3 °CA from the standard injection timing. Gumus Sayin, and Canakci (2012) found that standard injection timing gained a better performance result in the brake thermal efficiency (BTE),brake-specific fuel consumption (BSFC), and brake-specific energy consumption (BSEC). Both retarded and advanced injection timing slightly deteriorate the performance characteristics. The overall investigation reported that engine performance and emissions vary in terms of the fuel properties; higher emissions were also reported for both advanced and retarded fuel injection timing. Hence, further investigation is needed to obtain suitable fuel injection timing for different kinds of fuels.

Sayin and Gumus (2011) studied the impact of nozzle opening pressure (NOP) on the CI engine combustion characteristics for the biodiesel–diesel blends. The increase in NOP provides higher BTE and lower BSFC and BSEC compared to lower NOP due to the breakup of fuel into smaller droplets. The smaller diameter of the fuel droplets provides higher homogeneity in the air-fuel ratio, and the higher surface area of the fuel droplets also leads to better combustion. Smoke, HC, and CO emissions are significantly reduced with a penalty of higher NOx emission at higher NOP for all fuel blends. Sukumar et al. studied the influence of higher NOP ranges from

220 bar to 240 bar for the linseed biodiesel used in a Kirloskar diesel engine. They found that smoke and CO emission reduced, and $NO_x$ and HC emissions increased with the boost in NOP. Also, the study reports that too-high NOP improves the fuel spray movement at a higher velocity and increases the vaporization rate, which significantly reduces the NOx emission (Puhan et al. 2009). Purushothaman and Nagarajan (2009) studied the combustion behavior of orange skin powder biodiesel blend (B30) with varying NOPs in the range of 215 bar to 255 bar. The results exhibited that CO emission decreased up to 235 bar and it further increased when increase in the NOP above 255 bar. At higher NOPs up to 235 bar, the BTE and heat release rate (HRR) were increased due to the higher homogeneity of the air-fuel charge, and also it resulted in lower CO emission due to complete combustion. At a too-large NOP, the combustion duration was increased, which led to a lower heat release rate and a longer combustion duration, resulting in higher CO emission.

The simultaneous lowering of diesel engine soot and NOx emissions is needed to meet rigorous exhaust emission, but it seems very difficult to reduce the NOx emissions without increasing soot emissions through varying fuel injection timing. The explanation is that NOx and soot emissions often contradict each other when the timing of the fuel injection is retarded or advanced. An optimized split injection strategy proved to be an effective method for simultaneously reducing soot and NOx emissions for diesel engines. It is known as splitting the main single injection profile between the injections into two or more injection pulses with a definite delay dwell.

In diesel engines, multiple injection strategies typically include one main injection and two pilot injections of a small quantity of fuel before the main injection and after top dead center. More than three injections are usually used in car engines. Mendez and Thirouard (2009) studied the effect of the double injection on diesel engine noise and fuel economy and found that it is possible to simultaneously decrease the noise and fuel consumption using a double injection strategy. Moreover, the pilot injection strategy raises the in-cylinder temperature at the premixed combustion and enhances the atomization and evaporation (d'Ambrosio and Ferrari 2015).

The pilot injection tends to minimize the noise of combustion, but it can lead to an increase in particulate matter (PM), as it decreases the ignition delay of the diesel that occurs during the main injection (Yadav and Ramesh 2018). Splitinjection eliminates noise, pollution, and the BSFC with its versatile injection approach (Mohan, Yang, and Chou 2013). It was found that pilot injection reduces the smoke emission because of increased air-fuel mixing at premixed combustion in the cylinder (Hotta et al. 2005). The air-fuel mixing is improved by post-pilot injection, which leads to rises the in-cylinder temperature, thus enhancing soot oxidation (Yao et al. 2009). Besides, the post-injection reduces the PM emission with a large effect on fuel consumption and hydrocarbon emission (Chen 2000). Further research also suggested the efficacy of post-injection in soot reduction, with no effect on NOx emissions (Benajes, Molina, and García 2001). Optimization of post-injection timing was important to minimize smoke emission (Yao et al. 2009). Higher smoke and $NO_x$ emission were observed when post-injection was below 8°CA offset to the main injection. This is due to elevated temperature during the combustion (Desantes et al. 2007). However, it was found in another study that above 13°CA after top dead center post-injection did not help to lower the smoke level either. Furthermore, the

NOx emission was significantly reduced with three pulse injections compared with a single injection (Desantes et al. 2007).

## 5.2 SINGLE INJECTION ON CRDI-ASSISTED DIESEL ENGINE

Conventional diesel produces more exhaust emission, which does not meet emission norms. To reduce emission and improve engine performance, a modern electronic injection system was used in the diesel engine instead of a conventional fuel injection system (Agarwal et al. 2014). Agarwal et al. (2015) stated that combustion duration significantly diminished with an increase in NOP due to better atomization. Higher NOP was preferred for higher-viscosity fuel to ensure better atomization. PM was reduced at higher NOP and advanced FIT. More delay and a better air-fuel mixture led to complete combustion. Gumus et al. (2012) found that maximum NO emission occurred at higher NOP due to better atomization and evaporation. At higher NOP, droplet size was very small, resulting in faster combustion, leading to higher NO emission. Mohamed Shameer and Ramesh (2018) examined the influence of FIT and NOP on diesel engine performance and combustion characteristics. Experimental results infer that advanced FIT and higher NOP amplified biodiesel combustion phenomena, resulting in higher performance and lower emissions.

Sayin et al. (2012) examined the effects of NOP on CI diesel engine combustion characteristics. Results showed that higher NOP increased in-cylinder temperature and pressure, causing higher NO emission. This was due to higher NOP enhancing atomization and reducing fuel droplet size. It resulted in complete combustion, which leads to an increase in cylinder pressure and temperature. Ryu (2013) studied the effect of nozzle opening pressure on diesel engine combustion and emission characteristics. It was noticed that at higher load, unburned hydrocarbon (UBHC) and CO emissions decreased. The reason was that increase in NOP reduced droplet size and improved combustion efficiency. Rajesh Kumar et al. (2016) investigated the effect of NOP on diesel engine performance and combustion characteristics fueled by biodiesel and its blends. It was noticed that BTE increased and BSFC decreased when NOP increased from 180 to 220 bar due to better atomization, leading to faster combustion. Yesilyurt (2019) investigated the influence of fuel injection pressure on diesel engine characteristics fueled by biodiesel blends. Results indicated that biodiesel had lower emissions and performance compared to conventional diesel. Fuel injection pressure improved BTE and lowered emissions, but NO and $CO_2$ emissions increased.

Jindal et al. (2010) examined the influence of NOP on diesel engine performance and emission characteristics fueled by Jatropha biodiesel. It was observed that BTE increased with an increase in NOP, while BSFC decreased with increased NOP from 200 to 250 bar. Kannan and Anand (2011) investigated the effect of NOP and FIT on diesel engine characteristics fueled by biodiesel and its blends. The use of pure biodiesel in engines causes issues like injector choking, piston ring sticking, and poor spray characteristics. Results showed that higher NOP and advanced FIT enhanced engine characteristics. Kshirsagar and Anand (2017) suggested higher NOP and advanced FIT for higher-viscosity fuel to achieve better performance characteristics and lower emissions. Higher NOP and advanced FIT improved fuel spray characteristics, resulting in complete combustion. Harish et al. (2019) investigated the effect of fuel injection timing on diesel engine characteristics fueled by bio-diesel–diesel–ethanol blends. Maximum BTE and minimum CO and UBHC emission were revealed at advanced FIT.

## 5.3 SPLIT INJECTION STRATEGY

Minimum engine emission and better performance were achieved by employing a split injection strategy in diesel engines. Lee and Reitz (2003) investigated the effect of the split injection strategy on a high-speed direct injection (HSDI) diesel engine equipped with a common rail direct injection (CRDI) system. Various parameters like exhaust gas recirculation (EGR), start of injection, inlet boost pressure, nozzle opening pressure, and dwell period were involved in the optimization of HSDI engine combustion. Results showed that the oxides of nitrogen ($NO_x$) and PM emissions were reduced in the split injection strategy, whereas BSFC increased. Yehliu et al. (2010) examined the single and split injection strategies on a CRDI diesel engine fueled by vegetable oil–derived biodiesel, ultra-low sulfur diesel, and Fischer–Tropsch fuel. It was observed that fuel ignition characteristics affected the start of combustion. Fischer–Tropsch fuel obtained minimum $NO_x$ and PM emissions compared to other tested fuels under single and split injection modes.

How et al. (2018) used split injection on a diesel engine to investigate engine characteristics fueled by biodiesel blends. The results showed that the split injection strategy drastically reduced NO emission and improved engine performance. Edara et al. (2019) studied the influence of single retarded injection and split injection strategy with cooled EGR on a CRDI-assisted light-duty diesel engine. Test results revealed that retarded injection with 5% cooled EGR showed higher BTE compared to split injection with 5% EGR at a higher injection pressure of 350 bar. Both single and split injection strategies are recommended for higher-viscosity fuel to obtain better engine characteristics. Sindhu et al. (2018) numerically investigated the effect of the split injection strategy on diesel engine characteristics. Results showed that split injection with a small quantity of pre-injection reduced NO emission for the tested fuels. Park et al. (2018) investigated the effect of single and split injection strategies on diesel engine combustion and emissions characteristics. Results revealed that higher injection pressure with the split injection strategy decreased CO, UBHC, and NO emissions. Higher injection pressure with advanced injection timing lowered the BSFC and CO emissions but increased NO emissions.

Li et al. (2016) carried out research on the split injection strategy with swirl ratio on a single-cylinder diesel engine. The results revealed that split injection with swirl ratio enhanced the air-fuel mixture and accelerated combustion efficiency, causing lower engine emissions and minimum BSFC. Teoh et al. (2018) examined the impact of the start of injection and dwell angle on a turbocharged diesel engine's performance, combustion, and emissions characteristics. Advanced start of injection and long dwell period influenced combustion efficiency and increased NO emission formation. Proper start of injection and dwell angle under two-stage split injection reduced emissions and improved performance characteristics. Plamondon and Seers (2019) studied the impact of single and double injection strategies on a diesel engine fueled by waste frying oil biodiesel. The results revealed that pilot injection enhanced main injection combustion by providing a better air-fuel mixture. Also, emissions were reduced during double injection when compared to the single injection strategy.

Imperato et al. (2016) noticed that splitting of fuel into small quantities for pilot injection reduced premixed combustion and lowered cylinder pressure. Also, pilot injection reduced ignition delay. An increase in the dwell period reduced NO emission, and increased fuel consumption. Jain et al. (2017) examined the influence of split injection and EGR on diesel engine characteristics. Results showed that premixed charge compression ignition combustion relatively lowered in-cylinder temperature when compared to conventional diesel engines. Advancement of main and post-injection timing improved the performance of the engine and reduced emissions. However, retarded main and post-injection timing reduced performance due to lower in-cylinder temperature, leading to partial combustion. Similarly, a higher EGR percentage reduced the performance of the engine due to lower in-cylinder temperature and pressure. Fayad et al. (2017) studied the effect of a post-injection strategy on a modern CRDI system–assisted diesel engine equipped with diesel oxidation catalyst and fueled by butanol–diesel blends. The results showed that the post-injection strategy reduced particulate matter and NO emission, but CO and UBHC emissions increased. This modern injection strategy reduced emissions and improved engine fuel economy.

## 5.4 MULTIPLE INJECTION STRATEGY

Currently, advanced injection strategies, including split and multiple injection strategies, are used in diesel engines to further reduce engine exhaust emissions and improve engine performance characteristics (Mohan et al. 2013). Li et al. (2015) examined the effect of the multiple injection strategy on diesel engine characteristics under different modes. The results revealed that minimum premixed and diffusion combustions were observed in the pilot-main injection strategy. In addition, main-post injection favors soot oxidation by decreasing particle diameter. Xining et al. (2015) employed a multiple injection strategy on diesel engine under different modes. Various injection modes are single main injection, pilot-main injection, and main-post injection. Minimum premixed and diffusion combustion was observed in the pilot-main injection strategy compared to the single main injection strategy. In addition, main-post injection favors soot oxidation, leading to decreased particle diameter.

Suh (2014) investigated the impact of the twin-pilot injection strategy on a diesel engine. The results showed that the exhaust emissions of CO, UBHC, and soot were reduced in the twin-pilot injection strategy compared to the single injection and the one-pilot injection strategy. FIT of 20–10°CA proved to be the best timing for lower CO emission formation due to the shorter ignition delay and the higher in-cylinder pressure, which enhanced the CO oxidation rate. Jing et al. (2016) reported the effect of the two injection strategy associated with pilot injection on diesel engine performance and emissions. It was observed that faster flame development was identified for the double injection strategy compared to the single injection strategy. Soot and NO emissions were reduced in the double injection strategy compared to the single injection strategy due to low combustion temperature. Jaykumar and Ramesh (2018) examined diesel engine performance and combustion characteristics fueled by butanol–diesel blends. Test results showed improvement in BTE and lower smoke emission in the main injection plus post injection (MPI) modes. However, there was a drop in BTE and an increase in smoke emission in the pilot injection plus main injection (PMI) mode.

Park et al. (2011) observed the influence of the multiple injection strategy on a diesel engine. The results showed that CO and UBHC emissions decreased but NO emission increased due to a shorter dwell period. Vanegas et al. (2008) observed the influence of the multiple injection strategy on a CRDI diesel engine's performance and emission. In-cylinder pressure and temperature were reduced due to minimum fuel being burnt at premixed combustion, resulting in lower CO, UBHC, and NO emissions and lower BSFC. Hiwasea et al. (2013) studied the effect of split multiple stage fuel injection on diesel engine performance and emissions characteristics. Split multiple stage fuel injection showed strong combustion and provided controlled temperature and pressure inside the combustion chamber. This strategy significantly reduced NO emission and reduced SFC compared to a conventional continuous fuel injection system. Choi and Reitz (1999) investigated the influence of the multiple injection strategy on diesel engine characteristics fueled by oxygenated fuels. Soot emission was lower in the split injection strategy compared to the single injection strategy. Split injection reduced particulate emission at low loads, particularly with advanced fuel injection timing.

## 5.5 COMBINATION OF SPLIT INJECTION AND EGR ON BIODIESEL COMBUSTION

The $NO_x$ and PM trade-off is the major drawback in CI engine combustion. The dilemma between $NO_x$ and PM in CI engine combustion was first reported by Akihama et al. (2001). They conducted experiments using a higher volume of EGR for the trade-off between the $NO_x$ and PM. The study concluded that the addition of EGR with a higher percentage decreases the $NO_x$ and PM significantly. However, a higher volume percentage of EGR deteriorates engine performance and combustion characteristics. The combustion duration increased by 39.8% for increasing the EGR volume percentage from 0 to 30% because of the higher equivalence ratio and temperature (Yu et al. 2013). Hence, a few works have reported that EGR along with advanced injection strategy enhances the combustion and performance characteristics with regard to the trade-off between the $NO_x$ and PM. In the previous study, the single-stage fuel injection with advanced fuel injection was adopted for the low-temperature combustion technology to reduce the $NO_x$ and PM. Advanced fuel injection creates the wall impingement problem, which results in higher HC emission and lower brake thermal efficiency (Hashizume et al. 1998; Neely, Sasaki, and Leet 2004). Hence, the researcher employed the split injection strategy for regulating the in-cylinder condition during the time of the main injection. Fang et al. reported that split injection along with EGR was used for a heavy-duty engine. The study concludes that higher pilot injection quantity with lower EGR effectively controlled $NO_x$ emissions (Fang et al. 2012). Engelmayer et al. (2015) conducted experiments using a 300-kW test engine by adopting the cooled EGR along with multiple fuel injections. The study reported that soot emissions were effectively controlled in the partial load and rated load conditions using high-pressure multiple injections with addition of EGR.

CRDI has been used for high-pressure injection systems since it better meets the demand for lower emissions and specific fuel consumption. Also, CRDI enhances the specific power output and torque, and it injects the fuel at multiple times; the time separation between the injections can be reduced to 0 microseconds. Dober et al. (2008) studied the effect of multiple injections using CRDI along with EGR addition for diesel engines. The study reported that HC and CO emissions are increased in the advanced injection with EGR condition due to low combustion temperature and leanburn conditions. The cooled EGR is preferable for the emission reduction technology due to the higher density of exhaust gas, which occupies a low volume in the combustion chamber. This creates a higher volume for the fresh air entering the combustion chamber, which digests a higher quantity of EGR as compared to hot EGR. Mehrotra et al. reported that around 16.4% and 23.3% reductions of $NO_x$ and PM, respectively, were achieved for the cooled EGR in the same quantity of burnt fuel (Mehrotra et al. 2014). Another study reported that cooled EGR with delayed injection timing improves thermal efficiency even at a low fuel injection pressure (Brijesh, Chowdhury, and Sreedhara 2013). Sarangi et al. (2013) investigates the combined effect of EGR, and 50% of pilot and main injection quantity requires the minimum quantity of EGR as compared to single injection.

Gautam Edara et al. (2019) investigated the effect of single and multiple injections with 5% and 10% EGR additions in light-duty diesel engines. They reported that multiple injections with and without EGR of 5% improve the brake thermal efficiency. The exhaust gas temperature decreased for the multiple injections as compared to single injection, and the addition of 10% EGR decreases the exhaust gas temperature by 10°C compared to a retarded single injection. Bhowmick et al. (2019) investigated the combined effect of cold EGR and split injection for a *Calophyllum inophyllum* biodiesel blend (B10)–powered diesel engine. They varied the split injection quantity by 10% pilot and 90% main injection quantities and adjusted the EGR rate by 10% and 20%. The study reported that brake thermal efficiency decreased with EGR addition in the split injection strategy. $NO_x$ emissions significantly decreased with the addition of 20% EGR along with split injection as compared to split injection without EGR. The addition of EGR decreases the heat release rate and combustion temperature in the split injection condition due to a dilution effect. Because of this, higher CO, HC, and smoke emissions were noted in the EGR addition.

## 5.6 CONCLUSION

The above literature studies discussed details about advanced injection strategies, including the single injection strategy, the split injection strategy, and multiple and split combined with EGR technique on diesel engine performance, emission, and combustion characteristics fueled with biodiesel and conventional diesel. The main conclusions drawn from the literature survey are summarized as given below.

- Delayed injection timing decreases the $NO_x$, CO, and HC emissions with a negligible rise in the total fuel consumption. Advanced injection timing increases the ignition delay period and enhances $NO_x$ emission formation. Retarded injection timing slightly deteriorates the performance characteristics.
- At too-high NOP, the combustion duration was increased, which leads to a lower heat release rate and longer combustion duration, resulting in higher CO emission. At higher NOP, droplet size was very small, resulting in faster combustion and hence more NO emission.
- Splitinjection eliminates noise, pollution, and the BSFC with its versatile injection approach. Pilot injection reduces smoke emission because of the increased air-fuel mixing at premixed combustion. Split injection with swirl ratio enhanced the air-fuel mixture and accelerated combustion efficiency, causing lower engine emissions and minimum BSFC.
- The air-fuel mixture is improved by post pilot injection, in turn leading to rises in the in-cylinder temperature that enhance the soot oxidation. Oxides of nitrogen ($NO_x$) and PM emissions were reduced in the split injection strategy, whereas BSFC increased.
- Retarded injection with 5% cooled EGR showed higher BTE compared to split injection with 5% EGR at a higher injection pressure. An increase in the dwell period reduced NO emissions and increased fuel consumption. The post injection strategy reduced particulate matter and NO emission. But CO and UBHC emissions increased.
- A higher volume percentage of EGR deteriorates engine performance and combustion characteristics. The combustion duration increased by increasing the EGR volume percentage from 0% to 30% because of the higher equivalence ratio and temperature.
- A higher pilot injection quantity with lower EGR effectively controlled $NO_x$ emission. HC and CO emissions are increased in the advanced injection with the EGR condition due to low combustion temperature and leanburn conditions. The cooled EGR is preferable for the emission reduction technology due to the higher density of exhaust gas.

## REFERENCES

Agarwal, A. K., A.Dhar, J. G.Gupta, W. I.Kim, C. S.Lee, and S.Park. 2014. "Effect of Fuel Injection Pressure and Injection Timing on Spray Characteristics and Particulate Size-Number Distribution in a Biodiesel Fuelled Common Rail Direct Injection Diesel Engine."*Applied Energy*130: 212–21.

Akihama, Kazuhiro, YoshikiTakatori, KazuhisaInagaki, ShizuoSasaki, and Anthony M.Dean. 2001. "Mechanism of the Smokeless Rich Diesel Combustion by Reducing Temperature." In *SAE Technical Papers*. doi:10.4271/2001-01-0655.

Benajes, Jesús, SantiagoMolina, and José M.García. 2001. "Influence of Pre- and Post-Injection on the Performance and Pollutant Emissions in a HD Diesel Engine." In *SAE Technical Papers*. doi:10.4271/2001-01-0526.

Bhowmick, Pathikrit, A. K.Jeevanantham, B.Ashok, K.Nanthagopal, D.Arumuga Perumal, V.Karthickeyan, K. C.Vora, and AatmeshJain. 2019. "Effect of Fuel Injection Strategies and EGR on Biodiesel Blend in a CRDI Engine."*Energy*181: 1094–113. doi:10.1016/j. energy.2019.06.014.

Brijesh, P., A.Chowdhury, and S.Sreedhara. 2013. "Effect of Ultra-Cooled EGR and Retarded Injection Timing on Low Temperature Combustion in CI Engines." In *SAE Technical Papers*. doi:10.4271/2013-01-0321.

Chen, S. Kevin. 2000. "Simultaneous Reduction of NOx and Particulate Emissions by Using Multiple Injections in a Small Diesel Engine." In *SAE Technical Papers*. doi:10.4271/2000-01-3084.

Choi, C. Y., and R. D.Reitz. 1999. "An Experimental Study on the Effects of Oxygenated Fuel Blends and Multiple Injection Strategies on DI Diesel Engine Emissions."*Fuel*78: 1303–17.

d'Ambrosio, S., and A.Ferrari. 2015. "Potential of Double Pilot Injection Strategies Optimized with the Design of Experiments Procedure to Improve Diesel Engine Emissions and Performance."*Applied Energy*. doi:10.1016/j.apenergy.2015.06.050.

Desantes, José M., JeanArrègle, J. JavierLópez, and AntonioGarcía. 2007. "A Comprehensive Study of Diesel Combustion and Emissions with Post-Injection." In *SAE Technical Papers*. doi:10.4271/2007-01-0915.

Dober, Gavin, SimonTullis, GodfreyGreeves, NebojsaMilovanovic, MartinHardy, and StefanZuelch. 2008. "The Impact of Injection Strategies on Emissions Reduction and Power Output of Future Diesel Engines." In *SAE Technical Papers*. doi:10.4271/2008-01-0941.

Edara, Gautam, Y. V. V.Satyanarayana Murthy, JayashriNayar, MerigalaRamesh, and PaletiSrinivas. 2019. "Combustion Analysis of Modified Light Duty Diesel Engine under High Pressure Split Injections with Cooled EGR."*Engineering Science and Technology, an International Journal*22 (3): 966–78. doi:10.1016/j.jestch.2019.01.013.

Engelmayer, Michael, AndreasWimmer, GertTaucher, GernotHirschl, and ThomasKammerdiener. 2015. "Impact of Very High Injection Pressure on Soot Emissions of Medium Speed Large Diesel Engines."*Journal of Engineering for Gas Turbines and Power*. doi:10.1115/1.4030096.

Fang, Qiang, JunhuaFang, JianZhuang, and ZhenHuang. 2012. "Influences of Pilot Injection and Exhaust Gas Recirculation (EGR) on Combustion and Emissions in a HCCI-DI Combustion Engine."*Applied Thermal Engineering*. doi:10.1016/j. applthermaleng.2012.03.021.

Fayad, M. A., A.Tsolakis, D.Fernández-Rodríguez, J. M.Herreros, F. J.Martos, and M.Lapuerta. 2017. "Manipulating Modern Diesel Engine Particulate Emission Characteristics through Butanol Fuel Blending and Fuel Injection Strategies for Efficient Diesel Oxidation Catalysts."*Applied Energy*190: 490–500.

Gumus, Metin, CenkSayin, and MustafaCanakci. 2012. "The Impact of Fuel Injection Pressure on the Exhaust Emissions of a Direct Injection Diesel Engine Fueled with Biodiesel – Diesel Fuel Blends."*Fuel*95 (x): 486–94. doi:10.1016/j.fuel.2011.11.020.

Harish, V., D.Raju, and L.Subramani. 2019. "Combined Effect of Influence of Nano Additives, Combustion Chamber Geometry and Injection Timing in a DI Diesel Engine Fuelled with Ternary (Diesel-Biodiesel-Ethanol) Blends."*Energy*174: 386–406.

Hashizume, Takeshi, TakeshiMiyamoto, AkagawaHisashi, and KinjiTsujimura. 1998. "Combustion and Emission Characteristics of Multiple Stage Diesel Combustion." In *SAE Technical Papers*. doi:10.4271/980505.

Hiwasea, S. D., S.Moorthy, H.Prasad, M.Dumpad, and R. M.Metkare. 2013. "Multidimensional Modeling of Direct Injection Diesel Engine with Split Multiple Stage Fuel Injections."*Procedia Engineering*51: 670–75.

Hotta, Yoshihiro, MinajiInayoshi, KiyomiNakakita, KiyoshiFujiwara, and IchiroSakata. 2005. "Achieving Lower Exhaust Emissions and Better Performance in an HSDI Diesel Engine with Multiple Injection." In *SAE Technical Papers*. doi:10.4271/2005-01-0928.

How, H. G., H. H.Masjukia, M. A.Kalam, and Y. H.Teoh. 2018. "Influence of Injection Timing and Split Injection Strategies on Performance, Emissions, and Combustion Characteristics of Diesel Engine Fuelled with Biodiesel Blended Fuels."*Fuel*213: 106–14.

Imperato, M., O.Kaario, T.Sarjovaara, and M.Larmi. 2016. "Split Fuel Injection and Miller Cycle in a Large-Bore Engine."*Applied Energy*162: 289–97.

Jain, A., A. P.Singh, and A. K.Agarwal. 2017. "Effect of Split Fuel Injection and EGR on NOx and PM Emission Reduction in a Low Temperature Combustion (LTC) Mode Diesel Engine."*Energy*122: 249–64.

Jaykumar, Y., and A.Ramesh. 2018. "Injection Strategies for Reducing Smoke and Improving the Performance of Butanol-Diesel Common Rail Dual Fuel Engine."*Applied Energy*212: 1–12.

Jindal, S., B. P.Nandwana, N. S.Rathore, and V.Vashistha. 2010. "Experimental Investigation of the Effect of Compression Ratio and Injection Pressure in a Direct Injection Diesel Engine Running on Jatropha Methyl Ester."*Applied Thermal Engineering*30: 442–48.

Jing, W., Z.Wu, W. L.Roberts, and T.Fang. 2016. "Spray Combustion of Biomass-Based Renewable Diesel Fuel using Multiple Injection Strategy in a Constant Volume Combustion Chamber."*Fuel*181: 718–28.

Kannan, G. R., and R.Anand. 2011. "Experimental Evaluation of DI Diesel Engine Operating with Diestrol at Varying Injection Pressure and Injection Timing."*Fuel Processing Technology*92: 2252–63.

Kshirsagar, C. M., and R.Anand. 2017. "Artificial Neural Network Applied Forecast on a Parametric Study of Calophyllum inophyllum Methyl Ester-Diesel Engine Out Responses."*Applied Energy*189: 555–67.

Lee, T., and R.Reitz. 2003. "The Effects of Split Injection and Swirl on a HSDI Diesel Engine Equipped with a Common Rail Injection System."*SAE Tech Paper 2003-01-0349*, 2003.

Li, X., C.Guan, Y.Luo, and Z.Huang. 2015. "Effects of Multiple-Injection Strategies on Diesel Engine Exhaust Particle Size and Nanostructure."*Journal of Aerosol Science*89: 69–76.

Li, X., H.Zhou, L. M.Zhao, L.Su, H.Xu, and F.Liu. 2016. "Effect of Split Injections Coupled with Swirl on Combustion Performance in DI Diesel Engines."*Energy Conversion and Management*129: 180–88.

Mehrotra, A., S.Juttu, S.Ravishankar, G.Rambhaji, et al.2014. "Simultaneous Reduction of NOx and PM Emissions through Low Temperature EGR Cooling in Diesel Engines."*SAE Technical Paper*01: 2803.

Mendez, Sylvain, and BenoistThirouard. 2009. "Using Multiple Injection Strategies in Diesel Combustion: Potential to Improve Emissions, Noise and Fuel Economy Trade-off in Low CR Engines."*SAE International Journal of Fuels and Lubricants*. doi:10.4271/2008-01-1329.

Mohamed Shameer, P., and K.Ramesh. 2018. "Assessment on the Consequences of Injection Timing and Injection Pressure on Combustion Characteristics of Sustainable Biodiesel Fuelled Engine."*Renewable and Sustainable Energy Reviews* 81: 45–61.

Mohan, B., W.Yang, and S. K.Chou. 2013. "Fuel Injection Strategies for Performance Improvement and Emissions Reduction in Compression Ignition Engines – A Review."*Renewable and Sustainable Energy Reviews* 28: 664–76. doi:10.1016/j. rser.2013.08.051.

Mohsin, R., Majid, Z. A., Shihnan, A. H., Nasri, N. S., and Sharer, Z. 2014. "Effect of biodiesel blends on engine performance and exhaust emission for diesel dual fuel engine". *Energy Conversion and Management*. 88. 821–828. https://doi.org/10.1016/j.enconman. 2014.09.027

Neely, Gary D., ShizuoSasaki, and Jeffrey A.Leet. 2004. "Experimental Investigation of PCCI-DI Combustion on Emissions in a Light-Duty Diesel Engine." In *SAE Technical Papers*. doi:10.4271/2004-01-0121.

Nwafor, O. M. I., G.Rice, and A. I.Ogbonna. 2000. "Effect of Advanced Injection Timing on the Performance of Rapeseed Oil in Diesel Engines."*Renewable Energy*. doi:10.1016/S0960-1481(00)00037-9.

Park, S., H. J.Kim, D. H.Shin, and J. T.Lee. 2018. "Effects of Various Split Injection Strategies on Combustion and Emissions Characteristics in a Single-Cylinder Diesel Engine."*Applied Thermal Engineering*140: 422–31.

Park, S. H., S. H.Yoon, and C. S.Lee. 2011. "Effects of Multiple-Injection Strategies on Overall Spray Behaviour, Combustion and Emissions Reduction Characteristics of Biodiesel Fuel."*Applied Energy*88: 2976–87.

Plamondon, E., and P.Seers. 2019. "Parametric Study of Pilot–Main Injection Strategies on the Performance of a Light-Duty Diesel Engine Fueled with Diesel or a WCO Biodiesel–Diesel Blend."*Fuel*236: 1273–81.

Puhan, Sukumar, R.Jegan, K.Balasubbramanian, and G.Nagarajan. 2009. "Effect of Injection Pressure on Performance, Emission and Combustion Characteristics of High Linolenic Linseed Oil Methyl Ester in a DI Diesel Engine."*Renewable Energy*34 (5): 1227–33. doi:10.1016/j.renene.2008.10.001.

Purushothaman, K., and G.Nagarajan. 2009. "Effect of Injection Pressure on Heat Release Rate and Emissions in CI Engine Using Orange Skin Powder Diesel Solution."*Energy Conversion and Management*. doi:10.1016/j.enconman.2008.12.030.

Rajesh Kumar, B., Saravanan, S., Rana, D., and Nagendran, A. 2016. "A comparative analysis on combustion and emissions of some next generation higher-alcohol/diesel blends in a direct-injection diesel engine". *Energy Conversion and Management*. 119. 246–256. https://doi.org/https://doi.org/10.1016/j.enconman.2016.04.053

Ryu, K.2013. "Effects of Pilot Injection Pressure on the Combustion and Emissions Characteristics in a Diesel Engine Using Biodiesel-CNG Dual Fuel."*Energy Conversion and Management*76: 506–16.

Sarangi, Asish K., Colin P.Garner, Gordon P.McTaggart-Cowan, Martin H.Davy, EmadWahab, and MarkPeckham. 2013. "The Effects of Split Injections on High Exhaust Gas Recirculation Lowtemperature Diesel Engine Combustion."*International Journal of Engine Research*. doi:10.1177/1468087412450987.

Sayin, C., M.Gumus, and M.Canakci. 2012. "Effect of Fuel Injection Pressure on the Injection, Combustion and Performance Characteristics of a DI Diesel Engine Fuelled with Canola Oil Methyl Esters-Diesel Fuel Blends."*Biomass and Bioenergy*46: 435–46.

Sayin, Cenk, and MetinGumus. 2011. "Impact of Compression Ratio and Injection Parameters on the Performance and Emissions of a Di Diesel Engine Fueled with Biodiesel-Blended Diesel Fuel."*Applied Thermal Engineering*. doi:10.1016/j.applthermaleng.2011.05.044.

Sindhu, R., G.Amba Prasad Rao, and K. MadhuMurthy. 2018. "Effective Reduction of NOx Emissions from Diesel Engine Using Split Injections."*Alexandria Engineering Journal*57: 1379–92.

Suh, H. K.2014. "Study on the Twin-Pilot-Injection Strategies for the Reduction in the Exhaust Emissions in a Low-Compression-Ratio Engine."*Proceedings of the Institution of Mechanical Engineers, Part D: Journal of Automobile Engineering*228: 335–43.

Suryawanshi, J. G., and N. V.Deshpande. 2005. "Effect of Injection Timing Retard on Emissions and Performance of a Pongamia Oil Methyl Ester Fuelled CI Engine." In *SAE Technical Papers*. doi:10.4271/2005-01-3677.

Teoh, Y. H., H. H.Masjuki, H. G.How, M. A.Kalam, K. H.Yu, and A.Alabdul Karem. 2018. "Effect of Two-Stage Injection Dwell Angle on Engine Combustion and Performance Characteristics of a Common-Rail Diesel Engine Fueled with Coconut Oil Methyl Esters-Diesel Fuel Blends."*Fuel*234: 227–237.

Vanegas, A., H.Won, C.Felsch, M.Gauding, and N.Peters. 2008. "Experimental Investigation of the Effect of Multiple Injections on Pollutant Formation in a Common-Rail DI Diesel Engine."*SAE Tech Paper 2008-01-1191*, 2008.

Xinling, Li, G.Chun, Y.Luo, and Z.Huang. 2015. "Effects of Multiple-Injection Strategies on Diesel Engine Exhaust Particle Size and Nanostructure."*Journal of Aerosol Science*89: 69–76.

Yadav, Jaykumar, and A.Ramesh. 2018. "Injection Strategies for Reducing Smoke and Improving the Performance of a Butanol-Diesel Common Rail Dual Fuel Engine."*Applied Energy*. doi:10.1016/j.apenergy.2017.12.027.

Yao, Mingfa, HuWang, ZunqingZheng, and YanYue. 2009. "Experimental Study of Multiple Injections and Coupling Effects of Multi-Injection and EGR in a HD Diesel Engine." In *SAE Technical Papers*. doi:10.4271/2009-01-2807.

Yehliu, K., A. L.Boehman, and O.Armas. 2010. "Emissions from Different Alternative Diesel Fuels Operating with Single and Split Fuel Injection."*Fuel*89: 423–37.

Yesilyurt, M. K.2019. "The Effects of the Fuel Injection Pressure on the Performance and Emission Characteristics of a Diesel Engine Fuelled with Waste Cooking Oil Biodiesel-Diesel Blends."*Renewable Energy*132: 649–66.

Yu, Byeonghun, Sung MinKum, Chang EonLee, and SeungroLee. 2013. "Effects of Exhaust Gas Recirculation on the Thermal Efficiency and Combustion Characteristics for Premixed Combustion System."*Energy*. doi:10.1016/j.energy.2012.10.057.

# 6 Low-Temperature Combustion Technology on Biodiesel Combustion

## 6.1 INTRODUCTION

The modern world carries out its daily energy activities in different forms and the demand for energy continues to increase for obvious reasons such as demographic growth and the introduction of new technologies, which once again depend on the external energy source. Most of the energy used in day-to-day activities is extracted from fossil fuels (Nanthagopal et al. 2019). The energy extracted from fossil resources is non-renewable and anticipated to last for only a few years depending on the current use; also fossil fuels emit pollutants such as carbon dioxide. $CO_2$ and other toxic gases are the major causes of pollution and global warming. According to forecast statistics, there will be an increase in $CO_2$ emissions of 1.7% per year until 2030 (Laha and Chakraborty 2017).

In a developing country like India, most of the energy extracted from fossil fuel sources is through diesel engine combustion. Major fossil fuel resources are used in off-road and on-road vehicles, power generators, the agricultural sector, the industrial sector, and marine transportation (Vigneswaran et al. 2018). Diesel engines are the most prominent users of fossil fuels due to their better brake thermal efficiency, more affordable fuel consumption, lower brake-specific energy consumption, more rigid structure, and higher power delivery compared to gasoline engines (Dhinesh et al. 2016). Continuous increases in the diesel engine population and in the quantum of $CO_2$ emissions from diesel engines are primary reasons for global warming, acid rain, ozone layer depletion, and nitrogen oxides emission (Thiyagarajan et al. 2019). As per the Paris agreement, India as one of the parties has agreed to reduce its carbon footprint in order to maintain the global surface temperature rise below 2°C from the pre-industrial level (Schleussner et al. 2016). Hence, carbon neutral biofuels can be used in CI engine to reduce engine emissions and decrease fossil fuel usage.

A biofuel is a different alternative to crude oil–based fuels, and it can be extracted from many sources such as vegetable oils, diluted oils, and bio fats. In fact, its biodegradable, sulfur-free, and non-toxic properties may serve beneficial roles in engine combustion. Particular attention has been paid to adjusting and replacing the conventional fuel resources in different countries (Elumalai, Annamalai, and Dhinesh 2019). Biofuels derived from biomass release exhaust gas that contains lower carbon emissions compared to fossil fuels, and it also produces lower hydrocarbons emission and particulate emissions due to its non-toxic and sulfur-free constituents. Despite the known benefits of using biofuels instead of fossil fuels, it is important to evaluate possible negative impacts on the environment during biofuel production (Naik et al. 2010). During the research and development (RD) stage of processes, evaluation of the biofuel synthesis pathways is essential to eliminate the undesired processes according to predetermined criteria and objectives (Jayswal et al. 2011). To date, several analyses on the production of biofuels have been carried out around the world based on different aspects, such as economic performance, environmental impact, and social development (Jaeger and Egelkraut 2011). That information provides an in-depth understanding of the present research gap in biofuel production.

In addition, extensive research has started to continually improve technologies and assessment tools for cleaner biofuel production (Klein-Marcuschamer et al. 2010). During the evaluation of biofuel synthesis processes, some fuel upgrade processes have been identified as increasing production costs and global warming potential due to higher energy and catalyst consumption at the production and processing stage (Sims et al. 2010). Due to this fact, some biofuels are used in CI engines without further refinement and processing, which results in poor physicochemical properties and combustion behavior. Poor kinematic viscosity and higher density were identified in biodiesel due to greater molecular strength between the bonds and higher molecular weight. These factors cause poor atomization as well as deposition and formation of coke in the throat of the piston, piston segments, and injectors (Hoang et al. 2019).

Different research has reported that the use of highly viscous biofuels like straight vegetable oils and biodiesel in a diesel engine emits higher smoke, CO, and HC emissions due to improper atomization (Che Mat et al. 2018). In general, fuel modification with diesel or cetane improver blending is the most common method for reducing viscosity. Previous works have reported that biofuel blended with diesel up to 20% decreased engine emissions, but beyond that limit, this type of blending affects fuel injection systems and increases engine emissions (Ramalingam, Rajendran, and Ganesan 2017; Abed et al. 2018). Advanced high-pressure injection systems like the CRDI system have been adopted for highly viscous fuel to improve the combustion efficiency and decrease the smoke and hydrocarbon emissions. The CRDI system improves the atomization rate and decreases the surface tension of the biofuel but also produces higher $NO_x$ emissions (Babu and Anand 2019). Diesel engine combustion is generally a lean-burn process. However, the diffused combustion phase is mainly stoichiometric combustion microscopically due to flame initiate and tends to propagate to towards the downstream of spray. Diffused phase combustion improves the combustion temperature and heterogeneous nature of the biofuel combustion in diesel engines but also increases NO emissions (Zheng, Reader, and Hawley 2004).

Different research has reported use of different additives, water injection (WI), emulsification technology (ET), exhaust gas recirculation (EGR), fuel injection pattern modification, geometrical modification in the combustion chamber, and low-temperature combustion strategy (LTC) for simultaneous reduction of smoke and $NO_x$ emissions for biofuel combustion. Among these various technologies, LTC mode is most efficient to decrease $NO_x$ and smoke emissions by 95% and 98%, respectively (Jiaqiang et al. 2017). Hence, this review elaborates on the highly viscous biofuels used in CI engines with different emission reduction techniques previously used for emission reduction. Also, biofuels used in CI engines under low-temperature combustion are evaluated in terms of the $NO_x$ and smoke trade-off.

## 6.2 HISTORY AND DIFFERENT METHODS USED FOR EMISSION REDUCTION

Researchers have attempted biofuel synthesis from lignocellulose biomass (second-generation biofuels) using different techniques like catalyst cracking, hydrotreating, and pyrolysis for biofuel production. The second-generation biofuels have decreased emission percentages compared to first-generation biofuels (Sims et al. 2010). However, these second-generation biofuels do not meet the essential fuel properties like viscosity and density and also contain water during the pyrolysis (Lee and Lavoie 2013). Hydrocracking and hydrotreating of vegetable oil, which produce comparable viscosity to diesel, are better compared to the other methods. However, consumption of catalyst and hydrogen increases the production cost (Kohli et al. 2019). The higher viscosity of the raw vegetable oil restricted the diesel engine operation to within certain limits. Incomplete combustion and higher smoke emissions were identified in the combustion of biofuel, due its higher kinematic viscosity and surface tension (Hellier, Ladommatos, and Yusaf 2015).

In current practice, raw vegetable oil has undergone a transesterification process to synthesize biodiesel, and the viscosity of the fuel is closer to that of diesel. In some respects, this transesterification process is time-consuming and expensive. Moreover, the major drawback for biodiesel is fuel stability. Fuel stability is influenced by different factors—temperature, light, and oxidation reactions—and biodiesel is more hygroscopic and highly corrosive compared to diesel. The maximum storage stability of the biodiesel was reported to be 6 months; after that it lost stability (Jiaqiang et al. 2017). Also, biodiesel production forms undesirable by-products like glycerol. Hence, preheating the raw oil before introducing it into the diesel engine will decrease the production cost and increase energy recovery. With preheating, biodiesel was found to be comparable to diesel fuel (Ndayishimiye and Tazerout 2011). Similarly, different additives and combustion chamber modifications are also done improving biofuel combustion efficiency and reducing emissions (Cheung, Zhu, and Huang 2009). In recent days, the CRDI system has been adopted for improving the combustion efficiency of biofuels (Jaliliantabar et al. 2020). Detailed emission reduction technologies commonly used in CI engine combustion are given in Sections 6.2.1 to 6.2.7.

### 6.2.1 BLENDING WITH DIESEL

The main disadvantage of using straight vegetable oil on an existing diesel engine it that it decreases engine performance because of higher kinematic viscosity, which leads to improper atomization. The raw vegetable oil is blended with diesel to decrease the kinematic viscosity. Twenty percent raw oil mixed with 80% diesel may be the best option for running CI engines without any major modifications (Chauhan et al. 2016). However, increased raw oil percentages in the blends have also been studied by several researchers. Ramadhas, Jayaraj, and Muraleedharan (2005) conducted an engine study using 20%, 40%, 60%, and 80% rubber seed oil blends with diesel. They reported a blending of 20% and 40% rubber seed oil performed well and comparable with diesel. The brake thermal efficiency was highest for 80% rubber seed oil. The smoke density of the 40% and 60% blends were similar to that of the diesel and the 20% blend was highest.

Another engine study reported that a CI engine was operated with preheated and unpreheated raw rapeseed oil blended with diesel at 20% and 50%. They reported raw fuel and preheated fuel show lower brake power compared to diesel. The unpreheated blend shows higher smoke density compared to the baseline value due to higher viscosity. In contrast, preheated oil emitted lower smoke compared with diesel because the viscosity decreased with increasing preheated temperature (Hazar and Aydin 2010). The earlier works prove that the brake-specific fuel consumption (BSFC) of raw oil–diesel blends indicates the incremented BSFC compared to diesel. This is because the lower heating value of raw oil–diesel blends increases the fuel consumption to maintain the same power output. Also, brake thermal efficiency (BTE) decreased due to the increase in BSFC. The blend of 50% raw oil shows higher BTE at partial load and on par with diesel at full load (Hazar and Aydin 2010). Some research exhibits higher exhaust gas temperature due to higher viscosity and poor atomization, which undergoes delayed combustion. The emission characteristics of raw oil–biodiesel blends show increased CO and HC emissions due to a rich air-fuel mixture because of undesirable spray characteristics. The blend with less than 50% exhibits lower smoke value; however, smoke increases with increase in blend ratio due to poor atomization (Agarwal, Kumar, and Agarwal 2008). The increase in blend ratio decreases the $NO_x$ emission significantly due to the decrease in peak cylinder temperature and the lower calorific value of the raw oil–diesel blend.

## 6.2.2 Exhaust Gas Recirculation System

In recent days, the exhaust gas recirculation (EGR) technique is commonly used for in-cylinder $NO_x$ reduction for biofuels. Much research work has described that EGR is the most preferable potential $NO_x$ mitigation technique (Gill et al. 2012). This technology works by recirculating 10% to 40% of exhaust gas and bringing it back to the engine cylinder to reduce oxygen availability. About 15% of EGR reduced the $NO_x$ by about 80% (Agarwal, Singh, and Agarwal 2011). The $NO_x$ controlling mechanism works based on two actions, which are dilution and chemical effects. During the dilution effect, the portion of EGR in the suction air acts as an inert gas in the combustion and absorbs the in-cylinder heat to reduce the peak cylinder temperature. Also, the dilution effect increases the non-combustible mass portion (Song et al. 2012). During the chemical effect, dissociation occurs due to an increase in molecular complexity (Ladommatos, Abdelhalim, and Zhao 2000).

EGR is an effective method for $NO_x$ reduction even though there are some persistent disadvantages such as higher hydrocarbon, smoke, and CO emissions. Also, increased EGR decreased the BTE unless the quantity was optimized. Saleh (2009) reported that 12% of EGR was the optimized level for biodiesel combustion, which decreases the $NO_x$ by 36%. However, it increases the HC and CO emissions slightly and also increases the BSFC by 9%. Kass et al. (2009) identified that 27% of EGR effectively reduced the $NO_x$ emission by 87% at a 68 Nm load for raw soybean biodiesel. EGR technology increases the BSFC due to altering the air-fuel ratio, reducing the oxygen concentration, creating a dilution effect, and lowering the burn fall,

all of which make achieving stable combustion difficult (Qi et al. 2010). Further, EGR increases the inlet air temperature, which affects the volumetric efficiency due to a damnatory effect at a higher load (Yoon, Suh, and Lee 2009). At a rated load, the disassociation of $CO_2$ to CO can also cause higher CO formation without varying the HC emission (Tsolakis et al. 2007). Many research works suggest that EGR is the most suitable technology for combustion due to its low volume requirement and design simplicity. The oxygen content in the biofuel compensates for the oxygen deficiency during the biofuel combustion. Hence, EGR with less than 25% can effectively reduce the $NO_x$ without increasing the other emissions.

## 6.2.3 WATER INJECTION

An important technique for controlling $NO_x$ emissions from the diesel engine is to inject the water into the cylinder. The water injection was carried out in two different ways: one is direct injection into the cylinder during combustion, and another one is injection (water fumigated) during the suction stroke through intake manifold (Bedford et al. 2000). A major benefit for injecting the water into the cylinder is the greater potential to decrease the $NO_x$ for the entire load range with a minimal detrimental effect on smoke emissions (Tauzia, Maiboom, and Shah 2010). The evaporation of injected water droplets absorbs the surrounding heat due to high latent heat, which significantly reduces the local adiabatic flame temperature, which results in low $NO_x$ emission (Park, Huh, and Park 2000). Fumigation is inducting or injecting the water to the upstream of the intake manifold. The water injection technology effectively decreases the $NO_x$ emission, although it has some disadvantages like incremental increase in HC and CO emission. Also, the BSFC value increased along with an increase in water injection rate. Water injection improves the low combustion temperature for the entire load range (Jiaqiang et al. 2017).

Tauzia, Maiboom, and Shah (2010) reported the effect of the water injection rate on the automotive direct injection diesel engine. They reported that an increase in the water flow rate extended the ignition delay period and lowered the $NO_x$ emissions, but with a penalty of higher CO and HC emissions. Tesfa et al. (2012) examined the influence of water injection on the 4-cylinder turbocharged diesel engine. They found water injection at 3 kg/h resulted in 50% lower $NO_x$ emissions without increasing specific fuel consumption. However, CO emissions increased by 40% compared to baseline. Overall, water injection effectively decreased the $NO_x$ emissions but with the penalty of higher CO and HC emissions. Hence, water injection combined with other HC and CO emission technology is a better option for biofuel combustion in the CI engine.

## 6.2.4 EMULSION TECHNOLOGY

Fuel emulsion technology has been used to introduce water during the combustion of biofuels. Emulsion technology has some advantages compared to water injection, which improves combustion efficiency and simultaneously reduces the $NO_x$, smoke, and other emissions (Lif and Holmberg 2006). A mixture of two or three immiscible fuels and liquids is called an emulsion. For example, water droplets are immersed in the oil phase emulsion, which helps distribute the water droplets uniformly in the combustion chamber. The fuel droplets are distributed uniformly everywhere using an adequate surfactant by mechanical or ultrasonic forces. Two main types of emulsion technology, namely two-phase emulsion and three-phase emulsion, have been used for biofuel combustion. Two-phase emulsion has been further divided into two types, namely water–oil emulsions and oil–water emulsions. Similarly, three-phase emulsion has been divided into water–oil–water and oil–water–oil. The important criteria for the selection of an emulsion are that it should possess characteristics similar to those of the fuel and that the presence of water does not overly influence the physicochemical characteristics of the biofuel (Matsumoto and Kang 1989).

The major disadvantage in emulsion technology is that water and oil separate over time and become stable in the dispersed phase. To maintain equal composition in the dispersed phase in the emulsion, the proper surfactants are added in the emulsion. The surfactants encase the water droplets in the oil phase emulsion and distribute them equally in the dispersed phase. Also, the surfactants prevent the water droplets from coalescing. During combustion, the dispersed phase water droplets are dissociated and absorb the heat of evaporation. Also, the specific capacity of fuel gas increases during the water dissociation from the emulsion. The emulsification technology increases the momentum of injected fuel. This is due to higher water content, which enhances the atomization and air entrainment (Hountalas, Mavropoulos, and Zannis 2007). Additionally, premixed combustion increases due to the higher atomization rate, which significantly reduces smoke emission. Further, the formation of OH radicals due to dissociation of water decreases the smoke and $NO_x$ emission (Nazha, Rajakaruna, and Wagstaff 2001).

Davis (2012) investigated the influence of 10% water emulsified in diesel and biodiesel combustion on a single-cylinder variable-speed diesel engine. They found that $NO_x$ emission decreased by 10% and 25%, respectively, for the emulsified B20 and B100 fuels. Compared to raw biodiesel, emulsified biodiesel (Eu-B100) performed well without reducing the engine power. This is due to the micro explosion of water droplets, higher vapor pressure, increase in velocity, and atomization of the fuel jet. The exhaust gas temperature of the B100 and Eu-B100 remains the same for the entire operation. Hence, the $NO_x$ reduction was mainly based on OH radical formation during combustion. However, a 15% increase in specific fuel consumption and a 14% drop in BTE were found for the emulsified B20 fuel (Davis 2012). Basha and Anand (2011) investigated the effect of 15% water and 2% surfactant added in jatropha emulsified biodiesel combustion. They reported 22% and 15% drops in $NO_x$ and PM, respectively; however, HC increased by 45% compared to raw biodiesel. Also, BTE and BSFC were improved by 2.4% and 2.7%, respectively. The improvements in the performance parameters have differed, and the reason for this is not clear: several researchers have found an increase in engine power and a few have observed a reduction. Hence, further research work is needed for utilizing the emulsion technology for the low-temperature combustion application.

### 6.2.5 Combustion Geometry Modification

An alternative method for overcoming the problems of a diesel engine powered by biodiesel is to make the correct modification in the combustion chamber and fuel injection. The injection time modification and nozzle opening pressure (NOP) play an essential role in the combustion. The pressure and timing should be optimized depending on the combustion chamber configuration. The modification in the combustion chamber should not affect the combustion efficiency and emission characteristics. Improving the air movement and swirling effect in the combustion chamber is essential to increase the homogeneity in the air-fuel mixture (Jiaqiang et al. 2017). The combustion chamber geometry creates a swirling effect of air, which creates the turbulence with biodiesel. Increasing turbulence and swirling effect increases the BTE and reduces the BSFC. New rotating grooves have been provided at the top of the piston to improve the turbulence of the air-biodiesel mixture by improving rotational movement.

Jaichandar and Annamalai (2012) examined the impact of three types of combustion chamber such as shallow depth combustion chamber (SCC), toroidal combustion chamber (TCC), and hemispherical combustion chamber (HCC) for the Pongamia biodiesel blend (B20) combustion. They found BTE increased for TCC compared to the other two types. Emission characteristics like PM, HC, and CO significantly decreased for TCC compared to other combustion chambers. However, $NO_x$ emission was slightly increased due to higher swirl motion, which increases the heat release rate and combustion temperature. Isaac et al. investigated the effect of turbulence inducer piston and injection pressure on Adelfa biodiesel blend (A20) combustion. They reported that smoke, CO, and HC emission were decreased due to improvement in the swirling movement, which enhances the premixed combustion. However, premixed combustion significantly enhances the $NO_x$ emission due to the increase in combustion rate (Isaac JoshuaRamesh Lalvani et al. 2016). Hence, combustion chamber modification will effectively work when injection pattern modification is adopted for biodiesel combustion.

## 6.2.6 Different Nozzle Opening Pressure and Timing

Governmental policies around the world have imposed the call to reduce pollution. This challenges engine manufacturers to establish an optimal position between engine performance and emissions. However, with emerging technologies like advanced fuel injection systems, the task is increasingly feasible. In recent years, the improvement of combustion in CI engines has been a hot topic for attaining lowest emissions. Choosing between different injection pressures and timings is an effective technique for reducing engine emissions because injection characteristics have a significant impact on the combustion process (Mohan, Yang, and Chou 2013). Different research work has reported that variation of fuel injection pressure and injection time effectively decreases emissions while increasing engine performance. In existing work, optimization of the fuel injection timing (FIT) and nozzle opening pressure (NOP) or fuel injection pressure (FIP) are effectively described for diesel fuel. Still, optimization of FIT and NOP is required for biofuel combustion to attain the lowest emissions with higher combustion efficiency (Mohamed Shameer et al. 2017). The following section explains how the FIT and NOP influence the diesel engine combustion characteristics using different kinds of fuel.

### 6.2.6.1  Effect of Fuel Injection Timing

Fuel injection timing is one of the essential parameters which significantly changes engine performance and emissions (Tat and Gerpen 2003). In general, $NO_x$ emission for the biofuels was higher due to higher density, viscosity, and bulk modulus and lower compressibility. Retarding and advancing the fuel injection timing from the standard value has been proven to decrease and increase $NO_x$ emissions. The combustion process advanced when the fuel injection timing was advanced and retarded when the injection timing was retarded. Retarded injection timing was mainly used for $NO_x$ reduction (Saroj 2013). However, fuel injection retardation has some disadvantages like increased HC and smoke emission and decreased BTE due to reduced peak pressure. Advanced injection timing improves the homogeneity of the air-fuel mixture, which promotes the premixed phase combustion. Increases in BTE and decreases in smoke and HC emissions were noted in the early injection strategy due to increment in the peak pressure and cylinder temperature. The higher $NO_x$ emission was noted in the early injection due to higher cylinder temperature. Surplus advancement in the injection timing causes higher CO, HC, smoke, and $NO_x$ emissions due to wall wetting. Also, the peak pressure attained during the combustion was earlier in excess advancement in the fuel injection timing, which resulted in power loss and lowered BTE (Caresana 2011). Hence, fuel injection timing should be optimized for obtaining higher brake thermal efficiency with lower emissions.

### 6.2.6.2  Effect of Fuel Injection Pressure (FIP) or Nozzle Opening Pressure (NOP)

Nozzle opening pressure (NOP) is one of the most important parameters which significantly affects the spray characteristics and fuel atomization rate and also contributes to the CI engine performance and emission characteristics. Generally, an increase in NOP leads to earlier combustion because of better atomization and proper air-fuel mixing. As a result, combustion temperature increased rapidly due to higher premixed stage combustion, which facilitates higher $NO_x$ and lowers HC formation. Also, the ignition delay period decreased at higher NOP, which resulted in better engine performance and BSFC; in contrast, lower NOP leads to poor BTE and BSEC due to longer ignition delay period, deteriorated air-fuel mixing, and poor atomization (Gumus, Sayin, and Canakci 2012). However, too long of an ignition delay period retarded the combustion process and decreased the peak pressure developed in the cylinder, which resulted in higher heat transfer loss and low power output. Similarly, too short of an ignition delay period advanced the combustion, which resulted in power loss due to peak pressure attained at compression stroke. However, too-high NOP causes inhomogeneity in the air-fuel mixture, which resulted in lower combustion efficiency. Increasing NOP increased the premixed combustion heat release rate (HRR) and early stage of controlled combustion HRR, which increases the steady-state heat transfer, resulting in lower BTE (Çelikten 2003).

The fuel droplets move at a higher velocity in the higher NOP, which hits the cylinder wall (wall wetting) without proper mixing, leading to higher HC emission (Puhan et al. 2009; Purushothaman and Nagarajan 2009). Also, $NO_x$ emission was decreased at higher NOP due to a decrease in HRR at the premixed phase stage and extension in the combustion duration, which resulted in a lower in-cylinder temperature (Purushothaman and Nagarajan 2009). The study also reported that smoke emission decreases up to 235 bar and further increased up to 255 bar. A higher NOP of up to 235 bar decreased the enriched fuel area, which resulted in lower smoke emission. However, further increment in NOP leads to a delayed and long diffused combustion phase, which favors higher smoke emission (Purushothaman and Nagarajan 2009).

Similarly, Hwang et al. (2014) reported that smoke emissions are significantly decreased while increasing NOP for waste cooking oil biodiesel. The soot and carbon particles get evaporated at a higher temperature in the premixed phase combustion, resulting in lower smoke emission in the higher NOP. The primary literature review shows that too-high and too-low NOP values increase the CO, HC, $NO_x$, and smoke emission with poor engine performance characteristics. Hence, the optimization of NOP for the different fuels is essential to obtain the lowest emissions with higher engine performance.

### 6.2.7 Influence of Fuel Injection Timing and Fuel Injection Pressure (FIP) on the Spray Characteristics

The spray properties of highly viscous fuels like biodiesel are enhanced with increments in fuel injection pressure. In general, spray tip penetration was extended, owing to the boost in the fuel injection pressure at low ambient density. A spray developing at a higher ambient density will tend to penetrate less and become a wider spray angle. The ambient density for the injected fuel depends on the fuel injection timing. Retarded injection timing increases the ambient density of the gas, which increases the spray angle and dispersion of fuel droplets equally (Arai 2012). Delacourt, Desmet, and Besson (2005) reported that a spray angle increased with an increase in pressure drop across the orifice. Also, the study reported that spray tip penetration increased with a decrease in ambient density with a higher pressure drop across the orifice. The spray angle mainly depends on the ambient density, and it increases with increases in ambient pressure

Parrish et al. (2010) reported that the largest spray angle was formed at the lowest ambient pressure, and it was decreased with an increase in ambient pressure up to 120 kPa due to reduced radical penetration. The spray angle increased again when the ambient pressure was between 120 kPa and 500 kPa. This is due to a toroidal vortex formed peripherally to the spray during the interaction of spray and high-density gas. At ambient pressures between 500 kPa and 1000 kPa, the spray angle decreased due to excessive loss of penetration. An approximately 20 bar increase in combustion chamber pressure decreased the spray angle from 74° to 50°.

The impingent and the higher concentration of the fuel at the piston bowl are preferable to the concentration of fuel at the squish region and the linear surface of the piston. Narrow spray angles (25° to 35°) are formed when the piston and spray direction are perpendicular to each other. Under the perpendicular spray condition, fuel cannot mix with air properly as a result of a low heat release rate, and higher levels of HC and CO are formed. The maximum and minimum gas temperatures have been observed at 35° and 74° spray angles, respectively. The ringing intensity was lower than the critical value of 5 MW/m$^2$ for the spray angle of 25° to 74°. The lowest ringing intensity was noticed when the spray angle became wider. The lowest and highest brake thermal efficiencies were noticed for spray angles of 55° and 35° due to better and poor atomization, respectively (Imtenan et al. 2014). Hence, optimization of nozzle opening time and pressure is most important to obtain better engine performance, lower emissions, and lower ringing intensity of the engine.

The fuel penetration length and velocity of fuel injection decide the turbulence, air-fuel mixing rate, and air utilization for the combustion. Increase in fuel penetration and higher velocity of injection improve the air-fuel mixing and swirl effect in some combustion chamber designs; however, the probability of spray hitting on the cold region may lead to increases in unburnt hydrocarbon emissions. Lower penetration length leads to poor air utilization and improper air-fuel mixing, which causes higher HC, CO, and smoke emissions (Mohan, Yang, and Chou 2013). Thus, the spray penetration of the biofuel is an essential parameter for biofuel combustion, and it should be carefully optimized by adjusting the fuel injection pressure for different loads and speeds.

The prediction of penetration length is essential for emission prediction. Dent (1971) predicted the spray penetration length ($L_{tip}$) using a gas jet mixing model, which is given in Equation 6.1:

$$L_{tip} = 3.07 \left( \frac{\nabla p}{\rho_g} \right)^{1/4} (tD_n)^{1/2} \left( \frac{294}{T_g} \right)^{1/4} \qquad (6.1)$$

Where
$\nabla p$ = pressure drop across the nozzle
$\rho_g$ = ambient gas density
$t$ = time after injection start
$D_n$ = diameter of the injector hole
$T_g$ = ambient gas temperature

This equation is only applicable (and predicts the penetration length at best) for injector nozzle length-to-diameter ratios of between 2 to 4. Also, this equation effectively works at higher ambient pressures ($p \geq 100$ atm) and times ($t \geq 0.5$ ms). Hiroyasu, Kadota, and Arai (1980) tried different equations to predict spray tip penetration length. The spray tip penetration is a function of injection pressure, ambient conditions, and time. It depends on the time ($t$) that the penetration length of the fuel jet is changed until the jet breaks up. According to the Hiroyasu model, the penetration length varied depends on the $\sqrt{t}$ after the jet breaks up. The Hiroyasu model is represented in Equations 6.2 and 6.3.

$$t < t_{break} : L_{tip} = 0.39 \left( \frac{2\nabla p}{\rho_g} \right)^{1/2} t \qquad (6.2)$$

$$t > t_{break} : L_{tip} = 2.95\left(\frac{\nabla p}{\rho_g}\right)^{\frac{1}{4}}\left(tD_n\right)^{\frac{1}{2}} \qquad (6.3)$$

Where

$$t_{break} = \frac{29\rho_l D_n}{\left(\rho_g \nabla p\right)^{\frac{1}{2}}}$$

$\rho_l$ and $\rho_g$ = liquid fuel density and ambient gas density

In general, the earlier start of combustion occurs when boosting the fuel injection pressure, because of proper atomization and increased penetration length, which results in higher homogeneity in the air-fuel mixture. As a result, the premixed combustion phase increased, which facilitates the increase in temperature and pressure in the combustion chamber, which causes higher $NO_x$ emission. Increase in NOP leads to lowering the ignition delay period; as a result, the engine power output and BSFC significantly improve. At lower NOP, poor BTE and BSFC result due to deteriorated atomization and air-fuel mixing.

## 6.3 LOW-TEMPERATURE COMBUSTION (LTC)

Around the world, CI engines fueled with conventional petroleum emit higher amounts of $NO_x$ and PM. Improvements in environmental impacts and CI engine efficiency can be achieved by encompassing low-temperature combustion technology within engine research. The methods for attaining low-temperature combustion and the individual merits and demerits are explained in the subsequent section. LTC combustion has been attained by combining the different emissions reduction techniques like blending with diesel and cetane improver, fuel additives, fuel emulsion, EGR, compression geometry modification, higher fuel injection pressure, and earlier and delayed injection timing. The individual effects of these techniques have been described in the previous sections.

### 6.3.1 DIFFERENT METHODS FOR ATTAINING LTC

Several techniques, such as homogenous charged compression ignition (HCCI), premixed charged compression ignition (PCCI), and reactive controlled compression ignition (RCCI), have been used for advanced low-temperature combustion strategies. Throughout the LTC combustion modes, the whole air and fuel mixture is premixed before combustion. The combustion is regulated by the predefined equivalence ratio and cylinder temperature in the LTC mode for reducing the formation of soot, PM, and $NO_x$ emissions at the same time. The LTC mode maintains the combustion temperature close to or below 1800 K–2200 K; at this temperature $NO_x$ pollutants are not produced in the rich mixing region and soot does not form below 1800 K in the lean mixing region. Like traditional diesel combustion, LTC techniques generally improve the pre-combustion and also reduce locally fuel dense areas and lower peak combustion temperatures, resulting in lower $NO_x$ and soot pollution (Hoekman and Robbins 2012). The LTC strategy is to combine different techniques such as different EGR volumes of 10% to 50%, increasing fuel injection pressure, split injection, and multiple injections with retarded injection timing closer to TDC.

Recently, reactivity controlled compression ignition (RCCI) has been implemented by different researchers to overcome the drawbacks of HCCI and PCCI (Nieman, Dempsey, and Reitz 2012; Splitter, Hanson, Kokjohn, and Reitz 2011). In the LTC strategy, the ignition delay of the fuel increases, which increases the premixed combustion phase without increasing the overall cylinder temperature and $NO_x$. Also, diffused combustion duration is decreased in LTC combustion (Jiaqiang et al. 2017). The presence of a homogeneous lean fuel mixture in the combustion chamber also decreases PM. A higher fuel injection pressure enables the fuel atomization, and biodiesel oxygen content guarantees full soot oxidation. However, HC and CO emissions are affected and vary significantly compared to conventional diesel combustion due to too-early and too-retarded injection timing and higher injection pressure, operating load, injection style (Venkataramana 2013), air-fuel temperature (Bunting et al. 2009), and combustion phasing and temperature (Petersen, Ekoto, and Miles 2010). In the premix combustion, early and late fuel injection raises the HC and CO as compared with traditional diesel and biodiesel combustion. HC is enhanced by the long ignition interval in the LTC mode, which creates excessive lean fuel mixture zones and increases the amount of fuel species existing beyond the flammability limits. Increasing the premixed combustion phase without increasing the combustion temperature reduces the HC and CO emissions as related to conventional diesel and biodiesel combustion, which are the major drawbacks for applying HCCI, PCCI or RCCI (Jiaqiang et al. 2017).

### 6.3.2 Importance of Advanced Injection Strategy in the LTC

First and foremost, mechanical fuel injection was adopted for operating CI engines with a high NOP of 200–300 bar. The higher NOP pressure is attained by adjusting the spring tension in the injector. Fuel injection by the mechanical system allowed only one injection per cycle. Also, the mechanical injection system decreases the atomization rate due to single injection and higher nozzle diameter, which resulted in weak air-fuel mixture and formation of fuel clouds in multiple zones with different temperatures in the cylinder. The low-temperature cloud region flame produces more soot and smoke, and the high-temperature leaner region produces higher $NO_x$ (Mohan, Yang, and Chou 2013). To overcome the drawbacks in the mechanical injection system, the electronic fuel injection system is being adopted in CI engines. The electronic fuel injection system known as the common rail direct injection (CRDI) system can be operated at high NOP. The nozzle diameter in the CRDI fuel injector was smaller than that in a mechanical injection system. Also, the CRDI fuel injector contains more holes per injector. The smaller diameter and multiple injection holes promote air-fuel mixing, and fuel cloud formations are smaller than in the mechanical injection system. Also, the temperature difference between the fuel clouds and air is minimal, which leads to better utilization of air in the cylinder, resulting in lower emissions (Khandal et al. 2017). Hence, the CRDI system has replaced the mechanical injection system recently for high-speed DI diesel engines to achieve lower emissions. The most important benefits of the CRDI system are listed below.

- Ultra-high NOP up to 2500 bar has been used to atomize the fuel droplets into fine droplets, which significantly reduces the evaporation time.
- The fuel droplets utilize the air entirely due to the higher velocity of the fuel droplets, which penetrate in the combustion chamber within a short period.
- The injection quantity can be precisely controlled by adjusting the injection duration, which results in controlled power output with lower smoke.
- The cycle-to-cycle disparity in the injection fuel quantity is drastically decreased.
- The initial rate of injection can be controlled, which helps to decrease the noise and emissions.
- The sharp end of the injector can control the smoke and hydrocarbon emissions by eliminating the nozzle dribble.
- Fuel injection shaping is also possible in the CRDI, which helps to suppress the heat release rate during the premixed and diffused combustion phase for simultaneous decrements in smoke and $NO_x$ emissions.
- Injection splitting also possible to suppress the rapid premixed combustion rate for a decrement in $NO_x$ emission.

Therefore, the replacement of a conventional mechanical injection system with common rail direct injection may help to attain higher performance and combustion efficiency with lower emissions (Mohan, Yang, and Chou 2013; Khandal et al. 2017). Also, a comparison of mechanical injection system with the CRDI system for different biofuel combustions is needed to find improvements in the engine emission and performance characteristics. Also, the comparison may help to analyze the feasibility of the CRDI system for highly viscous fuels.

## 6.3.3 Effect of Higher Injection Pressure in the Electronic Injection Strategy

Combustion, emission, and performance characteristics of the diesel engine are specifically influenced by many variables such as the start of fuel injection (SOI) timing, nozzle opening pressure (NOP), fuel injection quantity, the number of fuel injection pulses, the configuration of the combustion chamber, the number of holes in the injector, and the spray pattern. However, many of the parameters often indirectly influence engine efficiency and heat transfer. High injection pressure is a realistic solution to increase efficiency and decrease pollution after the advent of electronic fuel injection devices. In the early years, the maximum possible NOP of 1000 bar was attained using an inline or rotary pump system. In recent days, the NOP has been increased up to 2000 bar using CRDI (Mohan, Yang, and Chou 2013).

Pierpont and Reitz reported that the effect of NOP ranges from 720 bar to 1600 bar on engine performance and emission. NOP increased from 720 to 1220 bar decreased the PM emission and reduced the BSFC. A further increase in injection pressure from 1220 bar to 1600 bar decreased the PM emission and BSFC due to the higher driving torque required for increasing the fuel pressure. Also, this study reported that an increase in $NO_x$ emission was substantially higher at higher NOP. In contrast, reductions in PM and BSFC were observed at higher NOP (Pierpont and Reitz 1995).

Ye and Boehman (2012) examined the effect of NOP in a CRDI diesel engine fueled with ultra-low sulfur diesel (ULSD) and B40 of soybean methyl ester (SME). The study reported that increase in NOP enhances the $NO_x$ emissions for both diesel and B40 blend. The premixed combustion phase HRR was increased when the start of injection retraced to the after top dead center (TDC) side. Also, higher HRR and earlier start of combustion were identified at higher NOP due to better air-fuel mixing. Yamane, Ueta, and Shimamoto (2001) found a significant difference in the fuel injection characteristics of diesel and biodiesel at higher loads due to variation in the bulk density. Agarwal et al. (2013) found that the effect of NOP ranges from 300 to 750 bar with advanced and retarded injection timing on the particulate matter concentration of the CRDI single-cylinder research engine. The study reported that increasing NOP and advanced fuel injection timing decreased the mass concentration of PM emissions. This is because mixing of air and fuel droplets takes more time at advanced fuel injection time and higher NOP.

Ashok et al. (2018) evaluate the combustion characteristics of CI engine using lemon and orange-peel steam-extracted biofuel in a CRDI diesel engine. The study was carried out by adjusting the NOP between 400 bar and 600 bar, adjusting the pilot injection quantity from 10% to 30%. The study reported that increasing NOP increases the BTE and lowers the BSFC with lower smoke, HC, and CO emissions. The increment in NOP leads to higher $NO_x$ emission for both lemon and orange-peel biofuels. Orange-peel oil performed well at a higher pilot injection quantity; in contrary, the lemon peel oil was not suitable for injecting a higher quantity of pilot injection due to higher viscosity. The higher quantity of pilot injection quenches the possible ignition sources, resulting in poor combustion characteristics. Hence, they reported that 10% and 30% pilot injections improve the combustion characteristics for the lemon and orange-peel biofuels, respectively

Giakoumis (2012) carried out a statistical investigation using best-fit quadratic regression curves for the combustion of biodiesel blends. The collected data were highly scattered, but the general trend showed decreasing HC, CO, and PM emissions. The best-fit approximation gives accurate values for biodiesel blends up to 50%. Due to the higher viscosity and bulk density of the biodiesel, the data was highly scattered, and prediction accuracy decreased at higher blending ratio. Most of the injection methods have a different effect on the formation of fuel spray in the combustion chamber. Hence, the formation of fuel spray at different conditions plays a significant role during the combustion in the CI engines. Several researchers follow different injection techniques for reducing emissions and improving engine efficiency. Hence, the optimization of fuel injection pressure, fuel injection timing, and pilot and main injection fuel quantity are important parameters to enhance performance characteristics and decrease the tailpipe emissions simultaneously for the different kind of fuels.

## 6.3.4 Effect of Split Injection Strategy

In multiple injection strategies, the main injection is divided into two to enhance the reaction zone between the flame periphery and air in the cylinder, which results in a decreased fuel-rich region. Therefore, multiple injection strategies decrease smoke formation during combustion and decrease the peak pressure and temperature. Hence, $NO_x$ emission also decreases due to lower peak pressure and temperature compared to single injection. Split injection techniques in diesel engines typically consist of a small amount of initial pilot injection accompanied by the main injection and finally a small amount of post-injection. Three injection blasts are commonly used for automotive vehicles. Pilot injection is mainly used for combustion noise suppression; however, it also contributes to the rise of particulate matter emissions due to decrement in the ignition delay of the fuel which is injected during the main injection (Park, Kook, and Bae 2004). Post-injection enhances the air-fuel mixer due to the higher kinetic energy of the molecules at higher temperatures, which improves the diffusion phase combustion and results in increased in-cylinder temperature.

The higher temperature at diffused phase combustion decreases soot formation due to improvements in the oxidation of carbon particles (Hotta et al. 2005; Vanegas et al. 2008). Chen reported that post-injection decreases the PM emission without much effect on the HC emission and fuel consumption (Chen 2000). Also, another study reported that post-injection decreases the PM emission without an increase in $NO_x$ emission (Herfatmanesh et al. 2013; Montgomery and Reitz 1996). This is due to a drop in the premixed combustion rate and flame temperature at the beginning of combustion. Also, air-fuel homogeneity has been found to increase due to the higher kinetic energy of the molecules, which results in lower soot and $NO_x$ emission. The optimization of post-injection timing is essential to reduce smoke emissions (Yao et al. 2009). The dwell time between the main injection and post-injection is less than 8°CA, which results in higher smoke emission and elevated temperature during the combustion, which in turn contributes to $NO_x$ formation (Desantes et al. 2007; Martin et al. 2016). Also, another study reported that post-injection timing did not influence smoke emission when it retreaded above 13°CA (Vanegas et al. 2008).

Babu et al. (2018) explore the effect of split injection of waste cooking biodiesel on CI engine emission and performance characteristics. The study was conducted using main-post injection (MPI) strategy with varying induction quantities of 90%–10%, 80%–20%, and 75–25% and retarded main injection timing between 25°CA and 15°CA and post-injection timing between –5°CA and 5°CA. The study reported that retreading both injection timings decreases NO emissions and slightly increases smoke emissions. Also, premixed combustion and diffused combustion heat release rate occur at a single peak when the dwell time between the main injection and post-injection decreased. Another study by Babu et al. (2020) reported that multiple injections with retarded injection timing decrease the NO emission effectively. The slight increment in the smoke emission was observed in the retarded injection timing of main and post-injection. The split injection quantity of 15%–70%–15% attained the lowest NO, HC, CO, and smoke emission (Babu et al. 2020). Thus, the total literature revealed that the MPI strategy with retarded injection timing effectively reduces the smoke and $NO_x$ emission. The optimum injection quantity, dwell time, and retarded

timing have varied depending on the properties of the fuel and the engine specifications. Hence, further optimization of split injection quantity and timing is required
for attaining low-temperature combustion with minimum $NO_x$ and smoke emissions.

## 6.4   HOMOGENEOUS CHARGE COMPRESSION IGNITION (HCCI)

The combination of both SI engine and CI engine combustion characteristics in the
IC engine is called the homogeneous charge compression ignition (HCCI) engine.
In the HCCI strategy, fuel is well mixed, as for SI engines, and then auto-ignited to
initiate the combustion, as for CI engines. The fuel is vaporized and homogeneously
premixed with air before the combustion initiation. The HCCI has the potential to
suppress the $NO_x$ emission with higher brake thermal efficiency due to lean-burn
combustion. The lean-burn combustion suppresses the combustion temperature,
which causes lower $NO_x$ emission. Compared to traditional diesel combustion, HCCI
combustion increases the brake thermal efficiency up to 50% with lower smoke emission due to the increase in displacement volume. The higher compression ratio and
premixed combustion of the fuel increase the brake thermal efficiency with lower
smoke emissions in the HCCI engine (Komninos and Rakopoulos 2016; Desantes et
al. 2016). The homogeneous mixture and uniform equivalence ratio in the cylinder
promote the multi-zone auto-ignition and spontaneous combustion of the entire mixture. Flame propagation does not influence this, and some cases there is no evidence
for flame propagation during the combustion in HCCI mode (Yap, Megaritis, and
Wyszynski 2004).

The multi-zone combustion and unexpected ignition points increase the unpredictable pressure rise and cycle-to-cycle variation. The unpredictable pressure rise may
generate high oscillation frequency and high-amplitude oscillations, which leads to
knocking (Ganesh, Nagarajan, and Ganesan 2011; Contino et al. 2013). Early direct
injection, early multiple injection, water injection, port fuel injection, external cold
EGR, variable valve timing, variable compression ratio, air preheating, and alcohol
injection techniques are widely used to manage the combustion in the HCCI strategy
at moderate loading conditions. The appropriate working condition and compression
ratio have been chosen for the biodiesel blends. A higher compression ratio is suitable for the biofuel auto-ignition temperature, due to higher viscosity than diesel.
The combustion phasing can be controlled by adjusting the compression ratio for the
different loads (Yao, Zheng, and Liu 2009). Few research works have reported that
internal EGR with negative exhaust valve overlapping is the most preferred strategy
for combustion phase control and decreasing $NO_x$ emissions.

Combustion phasing and emission reduction were attained in the spark-assisted
HCCI engine by adjusting the spark time and spark plug place arrangement
(Wiemann et al. 2018). The combustion phenomenon in the HCCI engine is identified
by pressure rise rate, combustion noise, and ringing intensity. The ringing intensity is
mainly used to identify the combustion noise for the required cylinder pressure in the
real-time combustion application (Maurya and Saxena 2018). The higher compression ratio can be used to attain the fuel's auto-ignition temperature, which indirectly
increases the brake thermal efficiency due to the increase in stroke volume. However,
the higher compression ratio leads to higher peak pressure and $NO_x$ emission. To

overcome this problem, a number of either high- or low-octane fuels can be used as a port fuel. A minor variation of inlet charge temperature (5–10°C) can increase the combustion temperature and also lead to higher $NO_x$ emission. So, maintaining the inlet charge temperature in the HCCI engine is essential. Similarly, the compression ratio for the HCCI engine could be properly maintained from 10:1 to 28:1. For higher-cetane fuel like n-heptane, the compression ratio was preferred to be 10:1 and for high-octane fuel like iso-octane, the compression ratio was preferred to be 28:1. An intermediate compression ratio was preferred for biodiesel; this was chosen based on the cetane number of the fuel (Fukushima et al. 2015; Gan, Ng, and Pang 2011). The use of alcohol fuel like ethanol, n-butonal, and methanol is the alternative method for attaining lower emissions in the HCCI engine. Alcohol fuel increases the premixed burning and complete oxidization of fuel due to oxygen enrichment. Also, it decreases the combustion temperature because of the higher latent heat of vaporization, which enhances the quenching effect (Yao, Zheng, and Liu 2009). The equivalence ratio is the most important factor in the HCCI to decide the engine power output. The equivalence ratio should be maintained below 1 to obtain the maximum power output and attain low-temperature combustion. Therefore, its use poses significant challenges in HCCI engines at higher and lower loads due to higher knocking and misfire (Babu, Murthy, and Rao 2017).

### 6.4.1 Challenges of HCCI Combustion

Many kinds of research on the combustion of HCCI engine clearly state that obstacles need to be resolved before attaining the advantages of HCCI combustion, as observed from previous studies. Research clearly shows the difficulties which occur depending on the mode of combustion.

### 6.4.2 Factors Affecting Combustion Phasing Control

The primary task and obstacle for HCCI combustion are managing the combustion phasing. The auto-ignition chemistry and homogeneous air-fuel combination govern the starting of combustion. Auto-ignition is mainly subjected to the properties of the air-fuel mixture, and also it depends on the time and temperature. The combustion phasing of HCCI combustion is mainly influenced by factors like the fuel's auto-ignition properties along with the concentration of the fuel in the different zones of the combustion chamber, the reactivity rate, the homogeneity of the mixture, the intake charge temperature, the latent heat of fuel evaporation, the compression ratio, heat transfer, and other parameters, depending on the engine (Stanglmaier and Roberts 1999).

### 6.4.3 Higher Level of HC and CO along with Combustion Noise

The increase in the noise along with higher amounts of unburnt hydrocarbon and CO emissions are the second main problems faced by the HCCI engine. Most of the combustion volume is homogeneous, and there is a certain area in the cylinder where the fuel is shielded in crevices during the compression stroke and its escape at the end of combustion. However, the unburned fuel cannot be burned during the re-entering of the fuel into a cylinder in the expansion stroke due to lower temperature. This leads to a specified increment in HC and CO emissions compared to conventional combustion. CO cannot be converted into $CO_2$ at a minimum load condition, as the peak combusted gas temperature is quite low (sometimes less than 1400K or 1500K). Partial oxidation deteriorates the combustion efficiency (Dec 2002). The combustion efficiency deterioration and difficulties in the start of ignition may affect the HCCI combustion effectiveness at a minimum load condition. At higher engine loads, the noise of the combustion increases with increase in the pressure rate and if not rectified may cause damage to the engine components (Eng 2002).

### 6.4.4 Operation Range

The other fundamental obstacle in the development of HCCI, in addition to the above problems, is to expand the operating load limit and maintain maximum HCCI gain. Covering the operating range for the entire load by controlling the auto-ignition phase is essential. The usage of HCCI engine at low loads is often minimized in addition to expanding the HCCI cycle to a higher load since the thermal energy is inadequate to trigger the auto-ignition during the late compression stroke. Furthermore, high CO and HC emissions with low exhaust temperature make this HCCI mode less desirable in terms of exhaust emissions and combustion efficiency perspectives (Yao, Zheng, and Liu 2009).

### 6.4.5 Cold Start

Since the temperature of the combustion chamber is too low during the cold start, the heat transfer from the compressed charge to the walls is high during the initial starting condition, and the HCCI engine will experience great difficulty in the cold start due to the leaner mixture. The problem of cold start can be solved by starting the engine in traditional diesel mode by converting it to HCCI mode for a short heating-up period. Hence, the cold start problem persists in homogenous combustion, and this can be a real-time challenge. The cold start–related HCCI problems are to be overcome by more effort and development in the charge preparation methods (Yao, Zheng, and Liu 2009).

### 6.4.6 Homogeneous Mixture Preparation

In order to achieve lower HC and PM emission with higher fuel efficiency, oil dilution prevention and preparation of a homogeneous air-fuel mixture by restricting the wall impingement are necessary. Many disadvantages have been shown for HC emission when fuels like gasoline have impinged on the combustion chamber's surface (Stanglmaier, Li, and Matthews 1999). The combustion phasing of HCCI engines is also controlled by the effect of the auto-ignition reaction because of mixture homogeneity. Much evidence has identified lower amounts of $NO_x$ emission for the inhomogeneity mixture for the accepted extend. Creating the homogeneous mixture with low ambient temperature is the biggest challenge for low-volatility fuels like diesel.

The diesel requires high intake temperature to prepare the homogeneous charged mixture, which produces low-smoke emissions.

HCCI combustion has gained attention in recent years because of its potential to reduce $NO_x$ along with particulate emission (Harada et al. 1998). Also, it achieves high thermal efficiency at various loads. Three main fields, namely electronic control technology, optical technology, and numerical simulations, have made progress in establishing fundamental theories of HCCI, including its combustion in both gasoline-filled and diesel-filled engines. Research works have been made not only on the improvement of HCCI and fuel characteristics, but also regarding the development of the HCCI concept in real time.

## 6.5   PREMIXED CHARGE COMPRESSION IGNITION (PCCI)

Premixed charge compression ignition can be achieved by atomizing the fuel more finely using a higher-pressure fuel injection system with advanced injection timing. The time duration between the start of injection and start of combustion has been increased due to the early injection of fuel, which significantly improves the homogeneity of the air-fuel mixture prior to the combustion (Zhang et al. 2016). The impact of earlier injection and higher injection pressure has been discussed in previous sections. Minimum and too-lean air-fuel ratios used for the PCCI combustion are 34:1 and 80:1, respectively. A slightly higher intake charge temperature was maintained at 170°C for the lean combustion (Egüz et al. 2014). The PCCI combustion strategy follows the lean-burn technology, and it works on higher-compression-ratio engines. The ignition of PCCI started after the completion of fuel injection. Also, the combustion phenomena are mainly identified by chemical kinetics and do not follow the diffusion mixture combustion and speed of the combustion as followed in the conventional combustion. Therefore, the injection pattern and the combustion of fuel do not overlap, which decreases the chances of direct combustion control (Girish, Neeraj, and Suryawanshi 2017). A single-stage fuel injection pattern with an earlier start of injection has been used to attain the premixed charge in the PCCI combustion. However, too-early start of injection leads to wall impingement and wall wetting during the injection, which leads to incomplete combustion and higher HC and CO emissions. To overcome these problems, the fuel injection pattern has been modified with split and multiple injection strategies. Even in these strategies, though, the duration of the multiple injections is completed prior to the start of combustion. Also, controlling the auto-ignition due to early injection is an important task in the PCCI combustion. The auto-ignition and ignition delay period have been managed by adopting a higher quantity of EGR. Also, a higher EGR volume helps to decrease the combustion temperature and $NO_x$ formation because of the dilution of the fresh charge mixture (Ying et al. 2010; Drews et al. 2011).

Compared to HCCI combustion, PCCI combustion was better due to stability in the combustion instigated by partially premixed charge and controlled auto-ignition rage and temperature. The combustion phasing in the PCCI mainly depends on the chemical kinetics, and it also can be altered by adjusting the inlet charge temperature, the EGR rate, and the fuel injection timing and pressure. Different types of fuel pattern have been used in the PCCI combustion such as early single-pulse injection, port fuel injection, advanced multiple injections, and advanced injection with a low-quantity late injection. The effect of advanced and late injection timings are discussed in the previous section. The small quantity of late injection is mainly provided for decreasing the smoke emission (Jia et al. 2011). In order to eliminate wall wetting during the advanced injection, a spray angle of 70° has been used to atomize the fuel within the combustion chamber (Horibe et al. 2009; Jia et al. 2011).

The compression ratio of the PCCI engine maintained the same level of the conventional diesel engine to prevent the formation of HC and CO emissions. A few research works have revealed that higher flammability and volatility of the fuel create some limitations in the PCCI combustion. Uniform air-fuel preparation, combustion phasing control for the entire load range, and control of the UBHC emissions due to wall impingement are the major challenges for PCCI combustion (Pandey, Sarma Akella, and Ravikrishna 2018).

Spark-assisted PCCI combustion has been implemented for the low-volatility fuels like kerosene, diesel, and biofuels. The use of low-quality cetane fuel in the spark-assisted PCCI strategy engine has improved the engine performance compared to conventional CI combustion (Verma et al. 2016). In PCCI-DI dual-mode combustion, the partially premixed combustion mixture is formed by injecting a huge amount of fuel in the intake port or early pilot injection, and the same or another fuel is injected by conventional direct injection. The combustion phasing of the PCCI-DI dual-mode combustion can be achieved by adjusting the pilot fuel quantity because of variation in the ignition delay, and the combustion rate can be adjusted by varying the port fuel ratio (Parks et al. 2010). Single-fuel or dual-fuel port injection is used to form a premixed mixture with a suitable air-fuel ratio. Compared to single-fuel LTC, dual-fuel premixed LTC has higher brake thermal efficiency and attained better decrements in soot and $NO_x$ emissions. Undesirable cycle-to-cycle disparity was identified in the single-fuel premixed LTC due to a lean air-fuel ratio and lower temperature (Kocher et al. 2013; Li et al. 2012). The dual-fuel premixed LTC is known as reactivity controlled compression ignition (RCCI), and a detailed description of the RCCI is given in subsequent sections.

## 6.6 PARTIALLY PREMIXED CHARGE COMPRESSION IGNITION (PPCI)

Partially premixed charge compression ignition is also similar to the PCCI strategy, which is intermediate between conventional diesel combustion and HCCI combustion. However, PPCI combustion is mostly preferred for low-cetane-number fuels. The longer ignition delay period and enhanced air-fuel mixing can be achieved as for the PCCI combustion strategy. Blending of low-cetane fuel with diesel is one of the methods to obtain a longer ignition delay period in PPCI combustion. The low-cetane fuels increase the ignition delay period due to high resistance to auto-ignition and slow reactivity with air (Ilango and Natarajan 2014). A few research works have revealed that longer ignition delay in the PPCI combustion can be obtained by advanced and delayed injection techniques. Also, low and moderate compression ratio and moderate to high EGR dilution were used to attain a higher ignition delay (An et al. 2018). The effect of advanced and retarded injection and EGR dilution was discussed in previous sections. The main advantage of PPCI mode compared to HCCI mode is that it emits lower particulate matter and $NO_x$ emissions and exhibits better control in the combustion phasing. PPCI can be classified into two types, namely early injection PPCI and late injection PPCI. In early injection PPCI, the fuel is injected at the middle of the compression stroke, and in late injection PPCI, the fuel injection falls at the start of the expansion stroke or exact TDC (Ilango and Natarajan 2014). Due to the partial compression, the fuel-injected gases are denser and colder in the early injected PPCI model. Similarly, the fuel-injected gases are colder and denser in the late injected PPCI model since injection occurs on the expansion cycle, which decreases the temperature in the later stage (Singh and Agarwal 2018). The specific fuel consumption of the PPCI combustion was slightly higher than conventional diesel combustion due to incomplete oxidation and non-optimal combustion phasing (Liu et al. 2015). A higher EGR rate and retarded injection time decrease the power out of both low- and higher-power engines under low load conditions. To avoid deterioration of the power output, an advanced injection strategy has been adopted in EGR-assisted PPCI combustion. Another drawback to PPCI combustion is higher HC and CO emission due to unburnt fuel present in the high-pressure squish region and piston bowl (Fridriksson et al. 2017). Gasoline addition is the alternative method to EGR addition in PPCI combustion to minimize soot and $NO_x$ emissions (Mehregan and Moghiman 2018). The longer ignition delay period in the PCCI combustion increases the premixed phase heat release, resulting in higher cylinder pressure and noise levels. The diffused combustion phase occurs as the ignition delay periods become short, which leads to a similar state as seen in conventional diesel combustion (Han, Choongsik, and Choi 2012). Hence, the optimization of combustion phasing is the biggest task in PPCI combustion.

## 6.7 REACTIVE CONTROLLED COMPRESSION IGNITION (RCCI)

Recently, some advanced low-temperature combustion technologies such as HCCI, PCCI, and RCCI have been developed. Among these, RCCI has especially received increased research attention due to its easy adaptability. Earlier studies show that HCCI and PCCI improve engine efficiency and lower emissions by attaining low-temperature combustion. However, these two technologies have some limitations, and they are not suitable for low load and higher load conditions due to knocking, misfire, and higher pressure rise rate. To overcome the issues, fuel modification is required in the HCCI and PCCI combustion (Epping et al. 2002; Kanda et al. 2005). Biessonette et al. reported that gasoline and diesel had been used to improve the ignition quality of HCCI combustion in a heavy-duty engine. Also, they reported that combustion quality was improved over a wide range of the engine's operation (Bessonette et al. 2007) Li et al. (2015) studied the influence of a gasoline-diesel partially mixed charge under minimum to higher load conditions in the CI engine. They reported that raw diesel is preferred for the minimum load condition, and a higher percentage of gasoline blend is suitable for the higher load condition. This dual-fuel PCCI operation is named as RCCI combustion in a later stage. It is used to obtain the $NO_x$ to smoke trade-off and higher efficiency by adjusting the low to high reactive fuel ratio and adjusting the injection pattern of the high reactive fuels. Also, fuel properties like viscosity, volatility, and ignition ability can be involved in the reactivity stratification in the RCCI combustion.

### 6.7.1  A Fundamental Concept of RCCI

Dual-fuel combustion technology using different reactive fuels is called RCCI technology. The combustion phasing is controlled by blending of higher reactive fuel and lower reactive fuel in different proportions (Reitz and Duraisamy 2015). Higher cetane number fuels such as biofuels, diesel, and pyrolysis fuels are used as the pilot fuels. Higher octane number fuels like liquefied petroleum gas (LPG), compressed natural gas (CNG), natural gas, and alcohol fuels are used as the port fuels in the RCCI combustion. During the suction stroke, the low reactive port fuels are injected with air and premixed well due to the cylinder movement. During the compression stroke, the pilot fuels are injected directly into the cylinder after completion of low reactive air-fuel blending. The modification in the blending ratio, appropriate EGR rate, and multiple injections of the pilot fuel are adopted to manage the combustion phasing. Also, these parameters influence $NO_x$ and smoke emissions. The early injection of high reactive fuel increases the squish region and promotes $NO_x$ emission. Late injection suppresses the $NO_x$ emission due to lower peak cylinder temperature (Eichmeier, Reitz, and Rutland 2014).

RCCI combustion meets emission regulations by simultaneously decreasing the $NO_x$ and smoke emissions. RCCI combustion extended the operating region from the 4.6 bar engine load to 14.6 bar (Eichmeier, Reitz, and Rutland 2014; Kokjohn and Reitz 2013). Eichmeier, Reitz, and Rutland 2014) reported that RCCI combustion provides higher gross thermal efficiency compared to conventional diesel combustion. Curran, Hanson, and Wagner (2012) reported that compared to conventional diesel combustion, RCCI brake mean effective pressure improved by 7% and BTE increased up to 39% due to a reduction in the heat loss. Yang et al. (2014) reported that the formation of OH radical mitigates the peak pressure rise due to the staged heat generation.

Low-temperature combustion can be attained by decreasing the combustion rate in RCCI combustion. The HRR in the RCCI combustion consists of two stages. The first small peak, which occurs during compression, is known as low-temperature heat release (LTHR), and the bigger second peak, which occurs during combustion, is known as high-temperature heat release (HTHR). The LTHR peak is primarily due to high reactive fuel with a negative temperature coefficient (Li, Yang, and Zhou 2017). The HTHR peak relies on premixed low reactive fuel. Li, et al. 2015) simulated RCCI combustion using gasoline-biodiesel blended at different ratios. The study reported that the increase in gasoline ratio decreased the LTHR and delayed the start of HTHR and also increased the peak HRR values.

### 6.7.2  Importance of Fuel Reactivity

Fuel reactivity of the RCCI combustion is categorized into two types: the mixture's global reactivity and the reactivity stratification. The global reactivity mainly depends on the port fuel octane number and pilot fuel cetane number values. The reactivity stratification is unpredictable, mainly depending on the pilot fuel penetration length and air-fuel entrainment (Li et al. 2014; Li et al. 2015) numerically investigated how reactivity stratification can be attained by sustaining the same equivalence ratio throughout the combustion by fuel reactivity elimination. They reported that retarded injection timing of pilot fuel decreases the HRR and pressure rise. In RCCI combustion, the ignition starts in the pilot fuel region due to higher reactiveness and then the flame propagates towards the low reactive region (higher octane region) (Kokjohn, Musculus, and Reitz 2015). The direction of flame propagation mainly depends on the injection pattern, injection timing, spray angle, and engine dimensions. A few investigations have shown that the flame starts near the cylinder liner and propagates towards the center. It has also been shown that flame propagation starts from the cylinder bowl and propagates towards the cylinder liner (Kokjohn et al. 2012).

### 6.7.3 Low Reactive Fuel Management

Fuel management in the RCCI combustion is classified into two types: the single-fuel approach and the dual-fuel approach. The dual-fuel strategy requires two fuel tanks, and it works based on the reactive difference between them. In a single-fuel strategy, the additive can be mixed in the fuel, and these additives create a reactive difference during the combustion. Most RCCI-based vehicles work based on the two-fuel strategy due to the higher reactive difference compared to a single-fuel strategy (Li, Yang, and Chou 2017). Earlier studies reported the use of gaseous fuels like LPG, CNG, NG, and alcohol fuels used as the port fuel in the two-fuel strategy. LPG, a petroleum-based fuel, is readily available on the market. The major advantages of the LPG used as the port fuel (low reactive fuel) are greater reliability and the wide availability of this kind of RCCI engine in many areas. Some studies reported that LPG potentially improves the RCCI combustion. However, gasoline contains higher aromatic content, which leads to higher smoke emission (Li, Yang, and Zhou 2017).

Natural gas and compressed natural gas are also non-renewable conventional fossil fuels like gasoline. Natural gas primarily consists of methane and hydrocarbon mixtures like ethane and propane. Small fractions of nitrogen and carbon dioxide are also present in natural gas (Kakaee, Rahnama, and Paykani 2015). The primary advantage of natural gas is that it is free from aromatics which lead to lower smoke emission (Jia and Denbratt 2015). Also, the market price of natural gas is lower than gasoline in many countries. Compared to gasoline, the octane number of natural gas is higher, which is preferred for high-compression-ratio engines like CI engines. Moreover, the reactiveness with high reactive fuel is much slower than LPG when used under the same operating conditions (Walker et al. 2014). The heating value of natural gas is higher than that of gasoline by mass and lower than that of gasoline by volume, which leads to higher brake thermal efficiency and lower heat release rate. All the characteristics of natural gas make it a potential low reactive fuel for RCCI combustion (Li, Yang, and Zhou 2017).

Nieman, Dempsey, and Reitz 2012) investigated RCCI combustion fueled with natural gas–diesel under different intake pressures. Also, the results were compared with gasoline-diesel at the same intake conditions. The results infer that decreasing the intake pressure from 1.75 bar to 1.45 bar retarded the combustion event for the gasoline-diesel operation. However, the natural gas–diesel combustion event was unaffected by decreasing the intake pressure. Also, this study inferred that split injection with an EGR rate of 43% is required for the gasoline-diesel combustion at optimum operation. In the case of natural gas–diesel, EGR is not required for attaining low-temperature combustion. Compared to gasoline-diesel combustion, natural gas–diesel required 4% lower premixed fuel for attaining the optimum condition. At the optimum condition, the smoke, HC and CO emissions dropped effectively in the natural gas–diesel combustion compared to gasoline-diesel. However, 2% higher gross thermal efficiency was noticed in the gasoline-diesel operation with lower $NO_x$ emission

Traditional fuels such as gasoline are often supplemented with alcohol fuels like ethanol, methanol, and iso-butanol because one of the benefits of the latter is their renewability. These alcohol fuels are further characterized based on their octane numbers; thus, the alcohol fuel selected for the RCCI engine is based on cetane number in order to mitigate the knocking tendency, and it creates a larger reactivity gradient in the RCCI combustion mode. Alcohol fuels have higher heat of vaporization than do gasoline and natural gas. The higher heat of vaporization increases the cooling effect of the charger due to the absorption of higher heat by the mixture; accordingly, the temperature of the air-fuel charge is decreased drastically. The temperature drop would theoretically decrease $NO_x$ formation. In addition, alcohol fuels are supported with oxygen and are non-aromatic. This is thought to help eliminate soot. However, the heat content of alcohol fuels is almost half that of natural gas and gasoline. Due to the lower density of the alcohol fuel, the specific fuel consumption is increased for the generation of the same power in the engine (Han et al. 2015).

### 6.7.4 Biofuels Used in the RCCI Combustion

Around the world, different biodiesels have been examined in different kinds of engines under different working conditions. The $NO_x$ emission was elevated for the biodiesel-fueled engine due to the bonding of oxygen molecules in the fuel (Vallinayagam et al. 2013; Vedharaj et al. 2013). Li, Yang, An, and Zhao (2015) numerically analyzed the RCCI engine powered by gasoline-biodiesel. The result from the study revealed that lower $NO_x$ emissions were identified in the gasoline-biodiesel operation as compared to raw biodiesel. Therefore, utilization of biodiesel under the RCCI strategy can be a preferred option to decrease $NO_x$ pollution as compared to biodiesel-powered diesel engines. Hanson et al. (2013) investigated the RCCI combustion using a diesel and biodiesel blend (B20) as a direct-injected fuel along with gasoline, E85 (85% ethanol and 15% diesel blend), and E20 as a port fuel. The results of the E20-diesel blend in the RCCI combustion indicate that maximum pressure and HRR decreased, which allowed the peak load to increase by 2 bar. E85-B20 increased the peak load by an additional 1 bar (BMEP 11 bar). The use of E20 increases the combustion efficiency and decreases heat transfer and exhaust loss; in this case, the BTE increased by 1.33%. This study also found that an increased pilot fuel fraction decreased the reactivity of the E20, which causes increased $NO_x$ emissions. In the gasoline-B20 RCCI operation, decreased port fuel to pilot fuel fractions were observed; however, in contrast to E20, it reduced the $NO_x$ emissions. Higher combustion efficiency and lower UBHC emission were observed in the gasoline-B20 operation. The improvement in the combustion efficiency of fuels strongly suggests boosting the BTE by 1.68%. As for the gasoline-B20 operation, the reduced port fuel fraction elevated the pilot fuel fraction, which was similar to the findings of the maximum pressure rise rate in the E20. Compared with gasoline-diesel, the BTE of E85-B20 combustion increased from 40% to 43% (Hanson et al. 2013). Gharehghani et al. (2015) have observed that the use of biodiesel as pilot fuel provides greater stability for the cyclic operation of the RCCI engine fueled by natural gas–biodiesel. This is because biodiesel contains an oxygen component that increases the cetane number. However, the composition of natural gas–biodiesel produced 1.6% higher BTE compared to natural gas–diesel.

### 6.7.5 SINGLE-FUEL RCCI COMBUSTION

The intention of using a single-fuel strategy in the RCCI combustion is to reduce the need for multiple fuel tanks in automobiles. The fuel additives and cetane improvers were mixed with low reactive fuels. The cetane improver creates a reactivity gradient in the mixture and also acts in the high reactive fuel role in the RCCI combustion. Most common cetane improvers like 2-EHN (2-ethylhexyl nitrate) (Kaddatz et al. 2012) and DTBP (di-tert-butyl peroxide) (Dempsey, Walker, and Reitz 2013) have been tried by different researchers in the single-fuel strategy for RCCI combustion. Splitter, Reitz, and Hanson (2010) investigated the RCCI combustion under a single-fuel strategy using DTBP. They reported that a small quantity of DTBP significantly decreased the emission comparable with the gasoline-diesel dual-fuel combustion and increased the thermal efficiency Wang et al. (2014) conducted the experiment and simulation using iso-butanol and DTBP blend in the RCCI combustion. The study found that the reactivity between the DTBP and iso-butanol was low during the combustion. Hence, it increases the direct injection fuel quantity for initiation of the ignition. The iso-butanol–iso-butanol+DTBP engine performance was comparable with gasoline-diesel and iso-butanol–diesel operations. Dempsey, Walker, and Reitz (2013) investigated the effect of 2-EHN and DTBP cetane improver for increasing the fuel reactivity of gasoline, methanol, and ethanol. Compared to DTBP, 2-EHN effectively increased the fuel reactivity; nevertheless, 2-EHN increased the $NO_x$ emission as compared to DTBP due to the nitrogen content in the additives

## 6.8 LOW-TEMPERATURE COMBUSTION ADVANTAGES AND CHALLENGES

The LTC mode combustion temperature was always less than that of the conventional diesel combustion temperature. Two major ways to meet the low-temperature combustion are by operating the engine with higher EGR and by operating with an excess air ratio ($\lambda$) of higher than 1 (Huang et al. 2016). At stoichiometric operating conditions, fuel combusted and oxidized at a higher temperature reaction, which leads to the formation of higher $NO_x$. Also, higher soot emissions were identified at the stoichiometric condition as compared to conventional diesel combustion due to reduction in the oxygen availability in the fuel spray periphery (Goldsborough et al. 2017; Feng et al. 2014). Normally, higher fuel injection pressure is a viable option to overcome the issues mentioned above. The higher injection pressure creates better atomization, mixing, and vaporization, which indirectly increase the excess air region in the combustion chamber. However, the wall impingement of fuel occurs due to spray tip penetration at higher fuel injection pressure is the major task to rectified in the advanced injection technology in the low-temperature combustion (Wu et al. 2017). Also, advanced injection strategies such as high-pressure injection and CRDI techniques decrease the ignition delay and also promote the premixed phase combustion, which resulted in higher $NO_x$ emission. The ignition delay and combustion phasing will be lengthened by higher level of cold EGR addition, decreasing the compression ratio and advancing the exhaust valve opening by variable valve time control. The increasing ignition delay period promotes the air-fuel mixing, which causes higher homogeneity in the air-fuel mixture. Achieving a homogeneous diluted cylinder mixture is the biggest challenge before the start of combustion during the higher ignition delay period. The peak pressure and temperature of the cylinder were reduced by higher EGR rate and lower compression ratio, which significantly affected the engine performance due to higher fuel consumption. Also, it affects the combustion process and combustion efficiency due to higher CO and HC emissions.

Getting LTC mode on high engine loads in real-time situations faces challenges. Manufacturing the engines with a higher quantity of EGR is not possible. Also, the higher BTE of the engine should be compensated for with the addition of higher EGR. An external charge booster is required to achieve higher BTE in the LTC condition (Thangaraja and Kannan 2016; Harari 2017). At higher engine loads, moderate EGR with an intake charge booster increases the cylinder peak pressure. Depending on the engine load, the combustion process will change, and it is affected by the different equivalence ratios and different fuel mixing zones, which make it difficult for the engine to change the operating condition for each load (Imtenan et al. 2014). Dual-fuel technology with a multiple injection strategy and negative valve overlapping is currently being used in the modern diesel engine. However, these technologies are expensive, and it is also complicated to adopt all the needed equipment. These technologies decrease the fuel-rich region by increasing the premixed charge quantity with lower peak pressure and temperature (Girish, Neeraj, and Suryawanshi 2017; Singh and Agarwal 2018). The higher swirl ratio can be used for the homogeneous charge compression ignition, and a higher compression ratio and piston modification are required for the RCCI combustion. Adopting the proper technology decreased the $NO_x$ and particulate matter by 85% and 95% in the LTC combustion as compared to the conventional diesel combustion.

## 6.9   HIGH-EFFICIENCY CLEAN COMBUSTION

The high reactive fuel injection strategy is one of the factors which contributes to increasing RCCI engine performance. The technique of injection is associated with the high reactive fuel injection cycle. The injection technique involves many controllable parameters such as the use of single, double, and triple pulses to drive the fuel to the cylinder. Splitter et al. (2011) investigated split injection and single injection using sweep of start of injection timing in the RCCI combustion fueled with iso-octane and n-heptane. For split injection, a double pulse with a fixed dwell period of 25°CA and 55% fuel injecting in the first pulse was used. The study infers that advanced injection time in the split injection decreases the peak pressure as compared to the single injection. However, the lowest peak pressure was identified in the single injection between 50°CA bTDC to 30°CA bTDC. Retarded injection times in both single and double injection increased the peak pressure and also promoted $NO_x$ formation. Retarded injection time in the double injection increased the peak pressure as compared to the single injection, due to an increase in the equivalence ratio gradient, which resulted in higher reactivity difference in the cylinder. This is due to the first and second injections targeting the squish and bowl regions, respectively, which resulted in high-temperature combustion. Higher $NO_x$ and CO emissions were observed in the split injection, and PM and UBHC were same in both single and split injections. Nieman, Dempsey, and Reitz (2012) numerically investigated the natural gas–diesel RCCI combustion at an IMEP of 23 bar. They reported that the optimum start of injection was 81°CA bTDC. The double injection generates higher soot emission, and it maintained the same level of HC, $NO_x$, and HC emissions (Splitter et al. 2011).

Ma et al. (2013) investigated the RCCI combustion using two injection strategies: early second injection timing (35°CA bTDC) and late second injection timing (9°CA bTDC) with fixed first injection at 58°CA bTDC. This study reported that smoke and $NO_x$ emissions were decreased with improving first injection fraction in the retarded injection time due to lower peak pressure. The advanced second injection increased the UBHC and CO emissions, and the retarded second injection decreased them. This is because the advanced second injection promotes the reactive controlled combustion and the late second injection promotes the mixing controlled combustion. Also, the late second injection is mainly used to initiate the combustion process at the later stage. The reactive controlled mixture from the early injection enhances incomplete combustion, which increases the UBHC and CO emissions. However, the mixing controlled charge from the late injection decreases UBHC and CO emissions due to an improvement in the diffusion flame combustion (Ma et al. 2013). Kokjohn et al. (2012) used imaging techniques to identify the fuel distribution as well as to find the combustion process of the RCCI engine. The CRDI injection system with three starts of injection timing of 145°CA, 50°CA, and 15°CA bTDC was adopted in this study. The study reported that starting injection at 145°CA and 15°CA bTDC increased the peak pressure rise. Lower peak pressure was attained in 50°CA bTDC. The reactive gradient and equivalence ratio were uniform in the early injection time. A steep gradient was seen in the late injection and a moderate value was seen in the 50°CA bTDC, which decreased the peak pressure.

Liu et al. (2014) investigated the RCCI combustion under different low reactive fuel energy ratios with early injection (43°CA and 35°CA bTDC) and delayed injection (10°CA and 4°CA bTDC) strategies. The results obtained from the different energy ratios of the low reactive fuel trends are identical with what was seen for varying the high reactive fuel injection. Retarding the early start of injection increased the peak pressure, which resulted in higher soot and $NO_x$ emissions. Also, higher thermal efficiency and lower CO and HC emissions were observed. In contrast, retarded delayed injection decreased thermal efficiency due to incomplete combustion. The $NO_x$ and soot emissions were decreased, and HC and CO emissions were increased. Moreover, the single-stage HRR occurred in the earlier start of injection. However, in the delayed injection, HRR occurred in the two-stage process.

Hanson et al. (2010) investigated the RCCI combustion with a split injection of high reactive fuel. The study reported the influence of the first and second injection on the emission and combustion characteristics of the CI engine. The advanced injection of the first pilot pulse from 55°CA to 62°CA bTDC delayed the combustion event and reduced the peak pressure and $NO_x$ emission. The early start of injection only increased the CO emissions, and soot and HC emissions remained constant at the entire sweep of the first injection time. Similarly, advanced second injection time decreased the peak pressure, $NO_x$, and soot emissions, while HC and CO emissions remain unchanged at the entire sweep of the second injection. Also, the study reported that the increase in fuel fraction of the first injection from 0.36 to 0.62 decreased the peak pressure and HRR occurring at a single stage.

Curran et al. (2010) reported the influence of gasoline percentage and start of injection on the RCCI combustion. Advanced injection of 60°CA bTDC decreased the peak pressure and HRR, as in the earlier studies. The study also reported that retarded injection timing from 20°CA to 30°CA bTDC decreased the HRR and advanced the combustion phase. The injection time between 30°CA and 60°CA bTDC increased the HRR as the maximum, which promotes high-temperature combustion. The major advantage of the retarded injection is that it promotes the mixed controlled combustion, and the maximum HRR and peak pressures attained closer to TDDC, which leads to higher BTE. Also, this study reports on the influence of the increase in the gasoline percentage in constant valve timing and variable valve timing conditions. In variable valve timing, HRR increased by increasing the gasoline percentage. In contrast, the HRR decreased by increasing the gasoline percentage under a constant valve time condition.

The overall study infers that at the first start of injection retardation, there is less time for air-fuel mixing; therefore, the mixture will be prepared with a high gradient of reactivity. Subsequently, higher local equivalence ratio zone (i.e., a fuel-rich area) is formed inside the cylinder. The large reactivity gradient and difference in the equivalence ratio at the multiple zones advanced the ignition period and CA50; hence, peak pressure and temperature were increased. In split injection, for a delayed first injection time, the high reactive fuel can be burned immediately at the end of injection due to higher in-cylinder temperature, which leads to the higher first heat release state. As a result, delayed start of the second injection time will delay the ignition period and the CA50 due to the formation of combusted gas during the first stage combustion. Hence, maintaining the fuel injection time between the reactive region and mixing controlled region decreases the entire emissions with higher thermal efficiency.

## 6.10   CONCLUSION

This study provides a comprehensive overview of LTC experiments carried out to improve the reliability and fuel efficiency of the CI engine combustion cycle with low emissions and noise. High-pressure fuel injection, split and multiple injections, alcohol blending, and exhaust gas recirculation strategies are major factors which influence the engine performance and combustion of a diesel engine under low-temperature combustion. The findings presented in this review can be summarized as follows:

- In general, an increase in injection pressure leads to higher brake thermal efficiency and lower specific fuel consumption with higher $NO_x$ emission. However, it decreases the HC, CO, and smoke emissions.
- Advanced injection time decreases the smoke, HC, and CO emissions with the penalty of higher $NO_x$ emission. The performance characteristics of the engine were improved with an early injection time. However, early injection beyond the limit decreases the engine performance and increases the HC and CO emissions due to wall impingement.
- Retarded injection time decreases the $NO_x$ emission effectively. However, retarded injection slightly increases the HC and CO emission due to insufficient time for the air-fuel mixing. Also, too-retarded injection deteriorates the specific energy consumption and smoke emission.
- Multiple and split injections effectively decrease the $NO_x$ and smoke emission simultaneously. However, a short dwell period between the two injections increases the heat release rate and $NO_x$ emission. The small quantity of post-injection decreases the smoke emission without increasing the other emissions.
- Exhaust gas recirculation at 15% to 40% significantly decreases the $NO_x$ emission. Advanced high-pressure injection with higher EGR rate may help to decrease the $NO_x$ emission without affecting smoke, HC, and CO emissions. However, a higher EGR rate increases the HC emission at low load conditions.
- PCCI combustion efficiently decreases HC and CO emissions as compared to the HCCI engine, but $NO_x$ and PM emissions were increased significantly with increased premixed charge percentage. However, the smoke and $NO_x$ emissions were identified as being at a minimum level compared with conventional diesel engine combustion.
- Higher cycle-to-cycle variation, unpredictable pressure rise, combustion noise, and knocking occurred in the HCCI mode of combustion due to higher homogeneity and an unpredictable auto-ignition zone.
- RCCI combustion is preferable for higher load conditions due to combustion phase control and higher brake thermal efficiency than PCCI and HCCI modes. The use of natural gas as a reactive fuel extended the load limit and attained efficient, clean combustion which significantly decreased $NO_x$ and soot emissions as compared to other techniques.
- The double injection of high reactive fuel in the RCCI combustion decreases the peak pressure and ringing intensity, which efficiently decreases smoke and $NO_x$ emissions. The advanced second injection in the RCCI increases the reactive controlled combustion, and late second injection increases the mixed controlled combustion.

- The combustion efficiency was increased while using B20 as the high reactive fuel. The oxygen availability in the biodiesel promotes the oxidization process, which decreases the HC and CO emission. B20-gasoline decreases the amount of CO and HC as compared to diesel-gasoline RCCI combustion.

Many experiments have extensively demonstrated that there is a wide and unexploited scope for improving low-temperature combustion using different reactive fuel injection parameters. The overall study infers that the operating condition, engine configuration parameters, fuel injection mechanism, and fuel mixing method influenced the engine combustion, emission, and performance characteristics. Hence, further research work will be needed to examine the trade-off between PM and $NO_x$ emission and improvement in engine performance.

## REFERENCES

Abed, K. A., A. K. El Morsi, M. M. Sayed, A. A. El Shaib, and M. S. Gad. 2018. "Effect of Waste Cooking-Oil Biodiesel on Performance and Exhaust Emissions of a Diesel Engine." *Egyptian Journal of Petroleum.* doi:10.1016/j.ejpe.2018.02.008.

Agarwal, Deepak, Lokesh Kumar, and Avinash Kumar Agarwal. 2008. "Performance Evaluation of a Vegetable Oil Fuelled Compression Ignition Engine." *Renewable Energy.* doi:10.1016/j.renene.2007.06.017.

Agarwal, Deepak, Shrawan Kumar Singh, and Avinash Kumar Agarwal. 2011. "Effect of Exhaust Gas Recirculation (EGR) on Performance, Emissions, Deposits and Durability of a Constant Speed Compression Ignition Engine." *Applied Energy.* doi:10.1016/j.apenergy.2011.01.066.

Agarwal, A. K., Srivastava, D. K., Dhar, A., Maurya, R. K., Shukla, P. C., and Singh, A. P. 2013. "Effect of fuel injection timing and pressure on combustion, emissions and performance characteristics of a single cylinder diesel engine". *Fuel.* 111. 374–383. https://doi.org/10.1016/j.fuel.2013.03.016

An, Yanzhao, Mohammed Jaasim, Vallinayagam Raman, Francisco E. Hernández Pérez, Hong G. Im, and Bengt Johansson. 2018. "Homogeneous Charge Compression Ignition (HCCI) and Partially Premixed Combustion (PPC) in Compression Ignition Engine with Low Octane Gasoline." *Energy.* doi:10.1016/j.energy.2018.06.057.

Arai, Masataka. 2012. "Physics behind Diesel Sprays." *ICLASS 2012, 12th Triennial International Conference on Liquid Atomization and Spray Systems*, Heidelberg, Germany, September 2–6, 2012 Physics.

Ashok, B., K. Nanthagopal, Bhaskar Chaturvedi, Shivam Sharma, and R. Thundil Karuppa Raj. 2018. "A Comparative Assessment on Common Rail Direct Injection (CRDI) Engine Characteristics Using Low Viscous Biofuel Blends." *Applied Thermal Engineering* 145 (September): 494–506. doi:10.1016/j.applthermaleng.2018.09.069.

Babu, D., and R. Anand. 2019. "Influence of Fuel Injection Timing and Nozzle Opening Pressure on a CRDI-Assisted Diesel Engine Fueled with Biodiesel-Diesel-Alcohol Fuel." *Advances in Eco-Fuels for a Sustainable Environment.* doi:10.1016/b978-0-08-102728-8.00013-9.

Babu, D., R. Karvembu, and R. Anand. 2018. *Impact of Split Injection Strategy on Combustion, Performance and Emissions Characteristics of Biodiesel Fuelled Common Rail Direct Injection Assisted Diesel Engine. Energy.* Vol. 165. Elsevier B. V. doi:10.1016/j.energy.2018.09.193.

Babu, D., Vinoth Thangarasu, and Anand Ramanathan. 2020. "Artificial Neural Network Approach on Forecasting Diesel Engine Characteristics Fuelled with Waste Frying Oil Biodiesel." *Applied Energy* 263 (February). doi:10.1016/j.apenergy.2020.114612.

Bedford, F., C. Rutland, P. Dittrich, A. Raab, and F. Wirbeleit. 2000. "Effects of Direct Water Injection on Di Diesel Engine Combustion." In *SAE Technical Papers*. doi:10.4271/2000-01-2938.

Benajes, Jesús, Santiago Molina, and José M. García. 2001. "Influence of Pre- and Post-Injection on the Performance and Pollutant Emissions in a HD Diesel Engine." In *SAE Technical Papers*. doi:10.4271/2001-01-0526.

Bessonette, Paul W., Charles H. Schleyer, Kevin P. Duffy, William L. Hardy, and Michael P. Liechty. 2007. "Effects of Fuel Property Changes on Heavy-Duty HCCI Combustion." In *SAE Technical Papers*. doi:10.4271/2007-01-0191.

Bunting, Bruce G., Scott J. Eaton, Robert W. Crawford, Yi Xu, Les R. Wolf, Shankar Kumar, Don Stanton, and Howard Fang. 2009. "Performance of Biodiesel Blends of Different FAME Distributions in HCCI Combustion." In *SAE Technical Papers*. doi:10.4271/2009-01-1342.

Caresana, Flavio. 2011. "Impact of Biodiesel Bulk Modulus on Injection Pressure and Injection Timing. the Effect of Residual Pressure." *Fuel*. doi:10.1016/j.fuel.2010.10.005.

Çelikten, İsmet. 2003. "An Experimental Investigation of the Effect of the Injection Pressure on Engine Performance and Exhaust Emission in Indirect Injection Diesel Engines." *Applied Thermal Engineering* 23 (16): 2051–60. doi:10.1016/S1359-4311(03)00171-6.

Chauhan, Bhupendra Singh, Ram Kripal Singh, H. M. Cho, and H. C. Lim. 2016. "Practice of Diesel Fuel Blends Using Alternative Fuels: A Review." *Renewable and Sustainable Energy Reviews*. doi:10.1016/j.rser.2016.01.062.

Chen, S. Kevin. 2000. "Simultaneous Reduction of NOx and Particulate Emissions by Using Multiple Injections in a Small Diesel Engine." In *SAE Technical Papers*. doi:10.4271/2000-01-3084.

Contino, Francesco, Fabrice Foucher, Philippe Dagaut, Tommaso Lucchini, Gianluca D'Errico, and Christine Mounaïm-Rousselle. 2013. "Experimental and Numerical Analysis of Nitric Oxide Effect on the Ignition of iso-Octane in a Single Cylinder HCCI Engine." *Combustion and Flame*. doi:10.1016/j.combustflame.2013.02.028.

Curran, Scott, Vitaly Prikhodko, Kukwon Cho, Charles Sluder, James Parks, Robert Wagner, Sage Kokjohn, and Rolf Reitz. 2010. "In-Cylinder Fuel Blending of Gasoline/Diesel for Improved Efficiency and Lowest Possible Emissions on a Multi-Cylinder Light-Duty Diesel Engine." In *SAE Technical Papers*. doi:10.4271/2010-01-2206.

Curran, Scott J., Reed M. Hanson, and Robert M. Wagner. 2012. "Reactivity Controlled Compression Ignition Combustion on a Multi-Cylinder Light-Duty Diesel Engine." *International Journal of Engine Research*. doi:10.1177/1468087412442324.

Davis, J. A. 2012. "NOx Emissions and Performance of a Single-Cylinder Diesel Engine with Emulsified and Non-Emulsified Fuels." *Applied Engineering in Agriculture* 28 (X): 179–86.

Dec, John E. 2002. "A Computational Study of the Effects of Low Fuel Loading and EGR on Heat Release Rates and Combustion Limits in HCCI Engines." In *SAE Technical Papers*. doi:10.4271/2002-01-1309.

Delacourt, E., B. Desmet, and B. Besson. 2005. "Characterisation of Very High Pressure Diesel Sprays Using Digital Imaging Techniques." *Fuel* 84 (7–8): 859–67. doi:10.1016/j.fuel.2004.12.003.

Dempsey, Adam B., N. Ryan Walker, and Rolf Reitz. 2013. "Effect of Cetane Improvers on Gasoline, Ethanol, and Methanol Reactivity and the Implications for RCCI Combustion." *SAE International Journal of Fuels and Lubricants*. doi:10.4271/2013-01-1678.

Dent, J. C. 1971. "A Basis for the Comparison of Various Experimental Methods for Studying Spray Penetration." In *SAE Technical Papers*. doi:10.4271/710571.

Desantes, J. M., J. M. García-Oliver, W. Vera-Tudela, D. López-Pintor, B. Schneider, and K. Boulouchos. 2016. "Study of the Auto-Ignition Phenomenon of PRFs under HCCI Conditions in a RCEM by Means of Spectroscopy." *Applied Energy*. doi:10.1016/j. apenergy.2016.06.134.

Desantes, José M., Jean Arrègle, J. Javier López, and Antonio García. 2007. "A Comprehensive Study of Diesel Combustion and Emissions with Post-Injection." In *SAE Technical Papers*. doi:10.4271/2007-01-0915.

Dhinesh, B., J. Isaac JoshuaRamesh Lalvani, M. Parthasarathy, and K. Annamalai. 2016. "An Assessment on Performance, Emission and Combustion Characteristics of Single Cylinder Diesel Engine Powered by Cymbopogon Flexuosus Biofuel." *Energy Conversion and Management*. doi:10.1016/j.enconman.2016.03.049.

Drews, P., T. Albin, F. J. Heßeler, N. Peters, and D. Abel. 2011. "Fuel-Efficient Model-Based Optimal MIMO Control for PCCI Engines." In *IFAC Proceedings Volumes (IFAC-PapersOnline)*. doi:10.3182/20110828-6-IT-1002.01138.

Egüz, Ulaş, Niels Leermakers, Bart Somers, and Philip De Goey. 2014. "Modeling of PCCI Combustion with FGM Tabulated Chemistry." *Fuel*. doi:10.1016/j.fuel.2013.10.073.

Eichmeier, Johannes Ulrich, Rolf Reitz, and Christopher Rutland. 2014. "A Zero-Dimensional Phenomenological Model for RCCI Combustion Using Reaction Kinetics." *SAE International Journal of Engines*. doi:10.4271/2014-01-1074.

Elumalai, P. V., K. Annamalai, and B. Dhinesh. 2019. "Effects of Thermal Barrier Coating on the Performance, Combustion and Emission of DI Diesel Engine Powered by Biofuel Oil–Water Emulsion." *Journal of Thermal Analysis and Calorimetry*. doi:10.1007/s10973-018-7948-6.

Eng, J. A. 2002. "Characterization of Pressure Waves in HCCI Combustion." In *SAE Technical Papers*. doi:10.4271/2002-01-2859.

Epping, Kathi, Salvador Aceves, Richard Bechtold, and John Dec. 2002. "The Potential of HCCI Combustion for High Efficiency and Low Emissions." In *SAE Technical Papers*. doi:10.4271/2002-01-1923.

Feng, Hongqing, Chunhong Zhang, Meiying Wang, Daojian Liu, Xiaoxi Yang, and Chia Fon Lee. 2014. "Availability Analysis of N-Heptane/iso-Octane Blends during Low-Temperature Engine Combustion Using a Single-Zone Combustion Model." *Energy Conversion and Management*. doi:10.1016/j.enconman.2014.04.061.

Fridriksson, Helgi, Bengt Sundén, Martin Tunér, and Öivind Andersson. 2017. "Heat Transfer in Diesel and Partially Premixed Combustion Engines: A Computational Fluid Dynamics Study." *Heat Transfer Engineering*. doi:10.1080/01457632.2016.1255086.

Fukushima, Naoya, Makito Katayama, Yoshitsugu Naka, Tsutomu Oobayashi, Masayasu Shimura, Yuzuru Nada, Mamoru Tanahashi, and Toshio Miyauchi. 2015. "Combustion Regime Classification of HCCI/PCCI Combustion Using Lagrangian Fluid Particle Tracking." *Proceedings of the Combustion Institute*. doi:10.1016/j. proci.2014.07.059.

Gan, Suyin, Hoon Kiat Ng, and Kar Mun Pang. 2011. "Homogeneous Charge Compression Ignition (HCCI) Combustion: Implementation and Effects on Pollutants in Direct Injection Diesel Engines." *Applied Energy*. doi:10.1016/j.apenergy.2010.09.005.

Ganesh, D., G. Nagarajan, and S. Ganesan. 2011. "Performance and Emission Analysis on Mixed-Mode Homogeneous Charge Compression Ignition (HCCI) Combustion of Biodiesel Fuel with External Mixture Formation." In *SAE Technical Papers*. doi:10.4271/2011-01-2450.

Gharehghani, Ayatallah, Reza Hosseini, Mostafa Mirsalim, S. Ali Jazayeri, and Talal Yusaf. 2015. "An Experimental Study on Reactivity Controlled Compression Ignition Engine Fueled with Biodiesel/Natural Gas." *Energy*. doi:10.1016/j.energy.2015.06.014.

Giakoumis, Evangelos G. 2012. "A Statistical Investigation of Biodiesel Effects on Regulated Exhaust Emissions during Transient Cycles." *Applied Energy.* doi:10.1016/j.apenergy.2012.03.037.

Gill, S. S., D. Turner, A. Tsolakis, and A. P. E. York. 2012. "Controlling Soot Formation with Filtered EGR for Diesel and Biodiesel Fuelled Engines." *Environmental Science and Technology.* doi:10.1021/es203941n.

Girish, B. E., S. Neeraj, and J. G. Suryawanshi. 2017. "Investigations on Premixed Charge Compression Ignition (PCCI) Engines: A Review." In *Fluid Mechanics Fluid,* 1455–62. doi:10.1007/978-81-322-2743-4_139.

Goldsborough, S. Scott, Simone Hochgreb, Guillaume Vanhove, Margaret S. Wooldridge, Henry J. Curran, and Chih Jen Sung. 2017. "Advances in Rapid Compression Machine Studies of Low- and Intermediate-Temperature Autoignition Phenomena." *Progress in Energy and Combustion Science.* doi:10.1016/j.pecs.2017.05.002.

Gumus, M., Sayin, C., and Canakci, M. 2012. "The impact of fuel injection pressure on the exhaust emissions of a direct injection diesel engine fueled with biodiesel–diesel fuel blends". *Fuel.* 95. 486–494. https://doi.org/10.1016/J.FUEL.2011.11.020

Han, S., B. Choongsik, and S. B. Choi. 2012. "Effects of Operating Parameters on Mode Transition between Low Temperature Combustion and Conventional Combustion in a Light Duty Diesel Engine." *International Journal of Engine Research* 14 (3): 231–46. doi:10.1177/1468087412454249.

Han, Xiaoye, Prasad Divekar, Graham Reader, Ming Zheng, and Jimi Tjong. 2015. "Active Injection Control for Enabling Clean Combustion in Ethanol-Diesel Dual-Fuel Mode." *SAE International Journal of Engines.* doi:10.4271/2015-01-0858.

Hanson, Reed, Scott Curran, Robert Wagner, and Rolf Reitz. 2013. "Effects of Biofuel Blends on RCCI Combustion in a Light-Duty, Multi-Cylinder Diesel Engine." *SAE International Journal of Engines.* doi:10.4271/2013-01-1653.

Hanson, Reed M., Sage L. Kokjohn, Derek A. Splitter, and Rolf D. Reitz. 2010. "An Experimental Investigation of Fuel Reactivity Controlled PCCI Combustion in a Heavy-Duty Engine." In *SAE Technical Papers.* doi:10.4271/2010-01-0864.

Harada, Akira, Naoki Shimazaki, Satoru Sasaki, Takeshi Miyamoto, Hisashi Akagawa, and Kinji Tsujimura. 1998. "The Effects of Mixture Formation on Premixed Lean Diesel Combustion Engine." In *SAE Technical Papers.* doi:10.4271/980533.

Harari, Praveen Adappa. 2017. "Comprehensive Review on Enabling Reactivity Controlled Compression Ignition (RCCI) in Diesel Engines." *Integrated Research Advances* 5 (1): 2018.

Hazar, Hanbey, and Hüseyin Aydin. 2010. "Performance and Emission Evaluation of a CI Engine Fueled with Preheated Raw Rapeseed Oil (RRO)-Diesel Blends." *Applied Energy.* doi:10.1016/j.apenergy.2009.05.021.

Hellier, Paul, Nicos Ladommatos, and Talal Yusaf. 2015. "The Influence of Straight Vegetable Oil Fatty Acid Composition on Compression Ignition Combustion and Emissions." *Fuel.* doi:10.1016/j.fuel.2014.11.021.

Herfatmanesh, Mohammad Reza, Pin Lu, Mohammadreza Anbari Attar, and Hua Zhao. 2013. "Experimental Investigation into the Effects of Two-Stage Injection on Fuel Injection Quantity, Combustion and Emissions in a High-Speed Optical Common Rail Diesel Engine." *Fuel.* doi:10.1016/j.fuel.2013.01.013.

Hiroyasu, H., T. Kadota, and M. Arai. 1980. "Supplementary Comments: Fuel Spray Characterization in Diesel Engines." In *Combustion Modeling in Reciprocating Engines.* doi:10.1007/978-1-4899-5298-1_12.

Hoang, Anh Tuan, Van Vang Le, Van Viet Pham, and Boi Chau Tham. 2019. "An Investigation of Deposit Formation in the Injector, Spray Characteristics, and Performance of a Diesel Engine Fueled with Preheated Vegetable Oil and Diesel Fuel." *Energy Sources, Part A: Recovery, Utilization and Environmental Effects.* doi:10.1080/15567036.2019.1582731.

Hoekman, S. Kent, and Curtis Robbins. 2012. "Review of the Effects of Biodiesel on NOx Emissions." *Fuel Processing Technology*. doi:10.1016/j.fuproc.2011.12.036.

Horibe, N., S. Harada, T. Ishiyama, and M. Shioji. 2009. "Improvement of Premixed Charge Compression Ignition-Based Combustion by Two-Stage Injection." *International Journal of Engine Research*. doi:10.1243/14680874JER02709.

Hotta, Yoshihiro, Minaji Inayoshi, Kiyomi Nakakita, Kiyoshi Fujiwara, and Ichiro Sakata. 2005. "Achieving Lower Exhaust Emissions and Better Performance in an HSDI Diesel Engine with Multiple Injection." In *SAE Technical Papers*. doi:10.4271/2005-01-0928.

Hountalas, D. T., G. C. Mavropoulos, and T. C. Zannis. 2007. "Comparative Evaluation of EGR, Intake Water Injection and Fuel/Water Emulsion as NOx Reduction Techniques for Heavy Duty Diesel Engines." In *SAE Technical Papers*. doi:10.4271/2007-01-0120.

Huang, Haozhong, Wenwen Teng, Qingsheng Liu, Chengzhong Zhou, Qingxin Wang, and Xueqiang Wang. 2016. "Combustion Performance and Emission Characteristics of a Diesel Engine under Low-Temperature Combustion of Pine Oil–Diesel Blends." *Energy Conversion and Management*. doi:10.1016/j.enconman.2016.09.090.

Hwang, Joonsik, Donghui Qi, Yongjin Jung, and Choongsik Bae. 2014. "Effect of Injection Parameters on the Combustion and Emission Characteristics in a Common-Rail Direct Injection Diesel Engine Fueled with Waste Cooking Oil Biodiesel." *Renewable Energy* 63: 9–17. doi:10.1016/j.renene.2013.08.051.

Ilango, T., and S. Natarajan. 2014. "Effect of Compression Ratio on Partially Premixed Charge Compression Ignition Engine Fuelled with Methanol Diesel Blends-an Experimental Investigation." *International Journal of Mechanical and Production Engineering* 2 (8): 2320–492.

Imtenan, S., M. Varman, H. H. Masjuki, M. A. Kalam, H. Sajjad, M. I. Arbab, and I. M. Rizwanul Fattah. 2014. "Impact of Low Temperature Combustion Attaining Strategies on Diesel Engine Emissions for Diesel and Biodiesels: A Review." *Energy Conversion and Management* 80 (x): 329–56. doi:10.1016/j.enconman.2014.01.020.

Isaac Joshua Ramesh Lalvani, J., M. Parthasarathy, B. Dhinesh, and K. Annamalai. 2016. "Pooled Effect of Injection Pressure and Turbulence Inducer Piston on Performance, Combustion, and Emission Characteristics of a DI Diesel Engine Powered with Biodiesel Blend." *Ecotoxicology and Environmental Safety* 134: 336–43. doi:10.1016/j.ecoenv.2015.08.020.

Jaeger, William K., and Thorsten M. Egelkraut. 2011. "Biofuel Economics in a Setting of Multiple Objectives and Unintended Consequences." *Renewable and Sustainable Energy Reviews*. doi:10.1016/j.rser.2011.07.118.

Jaichandar, S., and K. Annamalai. 2012. "Effects of Open Combustion Chamber Geometries on the Performance of Pongamia Biodiesel in a Di Diesel Engine." *Fuel*. doi:10.1016/j.fuel.2012.04.004.

Jaliliantabar, Farzad, Barat Ghobadian, Antonio Paolo Carlucci, Gholamhassan Najafi, Rizalman Mamat, Antonio Ficarella, Luciano Strafella, Angelo Santino, and Stefania De Domenico. 2020. "A Comprehensive Study on the Effect of Pilot Injection, EGR Rate, IMEP and Biodiesel Characteristics on a CRDI Diesel Engine." *Energy*. doi:10.1016/j.energy.2019.116860.

Jayswal, Abhishek, Xiang Li, Anand Zanwar, Helen H. Lou, and Yinlun Huang. 2011. "A Sustainability Root Cause Analysis Methodology and Its Application." *Computers and Chemical Engineering*. doi:10.1016/j.compchemeng.2011.05.004.

Jia, Ming, Maozhao Xie, Hong Liu, Wei Haur Lam, and Tianyou Wang. 2011. "Numerical Simulation of Cavitation in the Conical-Spray Nozzle for Diesel Premixed Charge Compression Ignition Engines." *Fuel*. doi:10.1016/j.fuel.2011.04.017.

Jia, Ming, Maozhao Xie, Tianyou Wang, and Zhijun Peng. 2011. "The Effect of Injection Timing and Intake Valve Close Timing on Performance and Emissions of Diesel PCCI Engine with a Full Engine Cycle CFD Simulation." *Applied Energy*. doi:10.1016/j.apenergy.2011.03.024.

Jia, Zhiqin, and Ingemar Denbratt. 2015. "Experimental Investigation of Natural Gas-Diesel Dual-Fuel RCCI in a Heavy-Duty Engine." *SAE International Journal of Engines.* doi:10.4271/2015-01-0838.

Kaddatz, John, Michael Andrie, Rolf Reitz, and Sage Kokjohn. 2012. "Light-Duty Reactivity Controlled Compression Ignition Combustion Using a Cetane Improver." In *SAE Technical Papers.* doi:10.4271/2012-01-1110.

Kakaee, Amir Hasan, Pourya Rahnama, and Amin Paykani. 2015. "Influence of Fuel Composition on Combustion and Emissions Characteristics of Natural Gas/Diesel RCCI Engine." *Journal of Natural Gas Science and Engineering.* doi:10.1016/j.jngse.2015.04.020.

Kanda, Tomohiro, Takazo Hakozaki, Tatsuya Uchimoto, Jyunichi Hatano, Naoto Kitayama, and Hiroshi Sono. 2005. "PCCI Operation with Early Injection of Conventional Diesel Fuel." In *SAE Technical Papers.* doi:10.4271/2005-01-0378.

Kass, Michael D., Samuel A. Lewis, Matthew M. Swartz, Shean P. Huff, Doh Won Lee, Robert M. Wagner, and John M. E. Storey. 2009. "Utilizing Water Emulsification to Reduce NOx and Particulate Emissions Associated with Biodiesel." *Transactions of the ASABE* 52 (1): 5–13.

Khandal, S. V., N. R. Banapurmath, V. N. Gaitonde, and S. S. Hiremath. 2017. "Paradigm Shift from Mechanical Direct Injection Diesel Engines to Advanced Injection Strategies of Diesel Homogeneous Charge Compression Ignition (HCCI) Engines – A Comprehensive Review." *Renewable and Sustainable Energy Reviews.* doi:10.1016/j.rser.2016.11.058.

Klein-Marcuschamer, Daniel, Piotr Oleskowicz-Popiel, Blake A. Simmons, and Harvey W. Blanch. 2010. "Technoeconomic Analysis of Biofuels: A Wiki-Based Platform for Lignocellulosic Biorefineries." *Biomass and Bioenergy.* doi:10.1016/j.biombioe.2010.07.033.

Kocher, Lyle, Daniel Van Alstine, Mark Magee, and Greg Shaver. 2013. "A Nonlinear Model-Based Controller for Premixed Charge Compression Ignition Combustion Timing in Diesel Engines." In *Proceedings of the American Control Conference.* doi:10.1109/acc.2013.6580519.

Kohli, Kirtika, Ravindra Prajapati, Samir K. Maity, and Brajendra K. Sharma. 2019. "Hydrocracking of Heavy Crude/Residues with Waste Plastic." *Journal of Analytical and Applied Pyrolysis.* doi:10.1016/j.jaap.2019.03.013.

Kokjohn, Sage L., and Rolf D. Reitz. 2013. "Reactivity Controlled Compression Ignition and Conventional Diesel Combustion: A Comparison of Methods to Meet Light-Duty NOx and Fuel Economy Targets." *International Journal of Engine Research.* doi:10.1177/1468087413476032.

Kokjohn, Sage L., Mark P. B. Musculus, and Rolf D. Reitz. 2015. "Evaluating Temperature and Fuel Stratification for Heat-Release Rate Control in a Reactivity-Controlled Compression-Ignition Engine Using Optical Diagnostics and Chemical Kinetics Modeling." *Combustion and Flame.* doi:10.1016/j.combustflame.2015.04.009.

Kokjohn, Sage, Rolf D. Reitz, Derek Splitter, and Mark Musculus. 2012. "Investigation of Fuel Reactivity Stratification for Controlling PCI Heat-Release Rates Using High-Speed Chemiluminescence Imaging and Fuel Tracer Fluorescence." *SAE International Journal of Engines.* doi:10.4271/2012-01-0375.

Komninos, N. P., and C. D. Rakopoulos. 2016. "Heat Transfer in Hcci Phenomenological Simulation Models: A Review." *Applied Energy.* doi:10.1016/j.apenergy.2016.08.061.

Ladommatos, N., S. Abdelhalim, and H. Zhao. 2000. "The Effects of Exhaust Gas Recirculation on Diesel Combustion and Emissions." *International Journal of Engine Research.* doi:10.1243/1468087001545290.

Laha, Priyanka, and Basab Chakraborty. 2017. "Energy Model – A Tool for Preventing Energy Dysfunction." *Renewable and Sustainable Energy Reviews.* doi:10.1016/j.rser.2017.01.106.

Lee, Roland Arthur, and Jean-Michel Lavoie. 2013. "From First- to Third-Generation Biofuels: Challenges of Producing a Commodity from a Biomass of Increasing Complexity." *Animal Frontiers.* doi:10.2527/af.2013-0010.

Li, J., W. M. Yang, H. An, and D. Zhao. 2015. "Effects of Fuel Ratio and Injection Timing on Gasoline/Biodiesel Fueled RCCI Engine: A Modeling Study." *Applied Energy.* doi:10.1016/j.apenergy.2015.05.114.

Li, J., W. M. Yang, H. An, and S. K. Chou. 2015. "Modeling on Blend Gasoline/Diesel Fuel Combustion in a Direct Injection Diesel Engine." *Applied Energy.* doi:10.1016/j.apenergy.2014.08.105.

Li, J., W. M. Yang, H. An, D. Z. Zhou, W. B. Yu, J. X. Wang, and L. Li. 2015. "Numerical Investigation on the Effect of Reactivity Gradient in an RCCI Engine Fueled with Gasoline and Diesel." *Energy Conversion and Management.* doi:10.1016/j.enconman.2014.12.071.

Li, Jing, Wen Ming Yang, Thong Ngee Goh, Hui An, and Amin Maghbouli. 2014. "Study on RCCI (Reactivity Controlled Compression Ignition) Engine by Means of Statistical Experimental Design." *Energy.* doi:10.1016/j.energy.2014.10.071.

Li, Jing, Wenming Yang, and Dezhi Zhou. 2017. "Review on the Management of RCCI Engines." *Renewable and Sustainable Energy Reviews* 69 (November 2016): 65–79. doi:10.1016/j.rser.2016.11.159.

Li, T., R. Moriwaki, H. Ogawa, R. Kakizaki, and M. Murase. 2012. "Dependence of Premixed Low-Temperature Diesel Combustion on Fuel Ignitability and Volatility." *International Journal of Engine Research.* doi:10.1177/1468087411422852.

Lif, Anna, and Krister Holmberg. 2006. "Water-in-Diesel Emulsions and Related Systems." *Advances in Colloid and Interface Science.* doi:10.1016/j.cis.2006.05.004.

Liu, Haifeng, Shuaiying Ma, Zhong Zhang, Zunqing Zheng, and Mingfa Yao. 2015. "Study of the Control Strategies on Soot Reduction under Early-Injection Conditions on a Diesel Engine." *Fuel.* doi:10.1016/j.fuel.2014.09.011.

Liu, Haifeng, Xin Wang, Zunqing Zheng, Jingbo Gu, Hu Wang, and Mingfa Yao. 2014. "Experimental and Simulation Investigation of the Combustion Characteristics and Emissions Using N-Butanol/Biodiesel Dual-Fuel Injection on a Diesel Engine." *Energy.* doi:10.1016/j.energy.2014.07.041.

Ma, Shuaiying, Zunqing Zheng, Haifeng Liu, Quanchang Zhang, and Mingfa Yao. 2013. "Experimental Investigation of the Effects of Diesel Injection Strategy on Gasoline/ Diesel Dual-Fuel Combustion." *Applied Energy.* doi:10.1016/j.apenergy.2013.04.012.

Martin, Jonathan, Chenxi Sun, Andre Boehman, and Jacqueline O'Connor. 2016. "Experimental Study of Post Injection Scheduling for Soot Reduction in a Light-Duty Turbodiesel Engine." In *SAE Technical Papers.* doi:10.4271/2016-01-0726.

Matsumoto, S., and W. W. Kang. 1989. "Formation and Applications of Multiple Emulsions." *Journal of Dispersion Science and Technology.* doi:10.1080/01932698908943184.

Maurya, Rakesh Kumar, and Mohit Raj Saxena. 2018. "Characterization of Ringing Intensity in a Hydrogen-Fueled HCCI Engine." *International Journal of Hydrogen Energy.* doi:10.1016/j.ijhydene.2018.03.194.

Mehregan, Mina, and Mohammad Moghiman. 2018. "Effects of Nano-Additives on Pollutants Emission and Engine Performance in a Urea-SCR Equipped Diesel Engine Fueled with Blended-Biodiesel." *Fuel.* doi:10.1016/j.fuel.2018.02.172.

Mohamed Shameer, P., K. Ramesh, R. Sakthivel, and R. Purnachandran. 2017. "Effects of Fuel Injection Parameters on Emission Characteristics of Diesel Engines Operating on Various Biodiesel: A Review." *Renewable and Sustainable Energy Reviews* 67 (3): 1267–81. doi:10.1016/j.rser.2016.09.117.

Mohan, Balaji, Wenming Yang, and Siaw Kiang Chou. 2013. "Fuel Injection Strategies for Performance Improvement and Emissions Reduction in Compression Ignition Engines – A Review." *Renewable and Sustainable Energy Reviews* 28 (x): 664–76. doi:10.1016/j.rser.2013.08.051.

Montgomery, D. T., and R. D. Reitz. 1996. "Six-Mode Cycle Evaluation of the Effect of EGR and Multiple Injections on Particulate and NOx Emissions from a D. I. Diesel Engine." In *SAE Technical Papers*. doi:10.4271/960316.

Naik, S. N., Vaibhav V. Goud, Prasant K. Rout, and Ajay K. Dalai. 2010. "Production of First and Second Generation Biofuels: A Comprehensive Review." *Renewable and Sustainable Energy Reviews*. doi:10.1016/j.rser.2009.10.003.

Nanthagopal, K., B. Ashok, Raghuram Srivatsava Garnepudi, Kavalipurapu Raghu Tarun, and B. Dhinesh. 2019. "Investigation on Diethyl Ether as an Additive with Calophyllum Inophyllum Biodiesel for CI Engine Application." *Energy Conversion and Management*. doi:10.1016/j.enconman.2018.10.064.

Nazha, M. A. A., H. Rajakaruna, and S. A. Wagstaff. 2001. "The Use of Emulsion, Water Induction and EGR for Controlling Diesel Engine Emissions." In *SAE Technical Papers*. doi:10.4271/2001-01-1941.

Ndayishimiye, Pascal, and Mohand Tazerout. 2011. "Use of Palm Oil-Based Biofuel in the Internal Combustion Engines: Performance and Emissions Characteristics." *Energy*. doi:10.1016/j.energy.2010.12.046.

Nieman, Derek E., Adam B. Dempsey, and Rolf D. Reitz. 2012. "Heavy-Duty RCCI Operation Using Natural Gas and Diesel." *SAE International Journal of Engines*. doi:10.4271/2012-01-0379.

Pandey, Sunil Kumar, S. R. Sarma Akella, and R. V. Ravikrishna. 2018. "Novel Fuel Injection Strategies for PCCI Operation of a Heavy-Duty Turbocharged Diesel Engine." *Applied Thermal Engineering*. doi:10.1016/j.applthermaleng.2018.08.001.

Park, Cheolwoong, Sanghoon Kook, and Choongsik Bae. 2004. "Effects of Multiple Injections in a HSDI Diesel Engine Equipped with Common Rail Injection System." In *SAE Technical Papers*. doi:10.4271/2004-01-0127.

Park, J. W., K. Y. Huh, and K. H. Park. 2000. "Experimental Study on the Combustion Characteristics of Emulsified Diesel in a Rapid Compression and Expansion Machine." *Proceedings of the Institution of Mechanical Engineers, Part D: Journal of Automobile Engineering*. doi:10.1243/0954407001527862.

Parks, James E., Vitaly Prikhodko, John M. E. Storey, Teresa L. Barone, Samuel A. Lewis, Michael D. Kass, and Shean P. Huff. 2010. "Emissions from Premixed Charge Compression Ignition (PCCI) Combustion and Affect on Emission Control Devices." *Catalysis Today*. doi:10.1016/j.cattod.2010.02.053.

Parrish, Scott E., R. J. Zink, Y. Sivathanu, and J. Lim. 2010. "Spray Patternation of a Multi-Hole Injector Utilizing Planar Line-of-Sight Extinction Tomography." *ILASS Americas, 22nd Annual Conference on Liquid Atomization and Spray Systems*, November 2015.

Petersen, Ben R., Isaac W. Ekoto, and Paul C. Miles. 2010. "An Investigation into the Effects of Fuel Properties and Engine Load on Uhc and Co Emissions from a Light-Duty Optical Diesel Engine Operating in a Partially Premixed Combustion Regime." In *SAE Technical Papers*. doi:10.4271/2010-01-1470.

Pierpont, D. A., and R. D. Reitz. 1995. "Effects of Injection Pressure and Nozzle Geometry on D. I. Diesel Emissions and Performance." In *SAE Technical Papers*. doi:10.4271/950604.

Puhan, S., Jegan, R., Balasubbramanian, K., and Nagarajan, G. 2009. "Effect of injection pressure on performance, emission and combustion characteristics of high linolenic linseed oil methyl ester in a DI diesel engine". *Renewable Energy*. 34. 1227–1233. https://doi.org/10.1016/j.renene.2008.10.001

Purushothaman, K., and Nagarajan, G. 2009. "Effect of injection pressure on heat release rate and emissions in CI engine using orange skin powder diesel solution". *Energy Conversion and Management*. 50. 962–969. https://doi.org/10.1016/j.enconman.2008.12.030

Qi, D. H., H. Chen, R. D. Matthews, and Y. Z. H. Bian. 2010. "Combustion and Emission Characteristics of Ethanol-Biodiesel-Water Micro-Emulsions Used in a Direct Injection Compression Ignition Engine." *Fuel*. doi:10.1016/j.fuel.2009.06.029.

Ramadhas, A. S., S. Jayaraj, and C. Muraleedharan. 2005. "Characterization and Effect of Using Rubber Seed Oil as Fuel in the Compression Ignition Engines." *Renewable Energy*. doi:10.1016/j.renene.2004.07.002.

Ramalingam, Senthil, Silambarasan Rajendran, and Pranesh Ganesan. 2017. "Assessment of Engine Operating Parameters on Working Characteristics of a Diesel Engine Fueled with 20% Proportion of Biodiesel Diesel Blend." *Energy*. doi:10.1016/j. energy.2017.09.134.

Reitz, Rolf D., and Ganesh Duraisamy. 2015. "Review of High Efficiency and Clean Reactivity Controlled Compression Ignition (RCCI) Combustion in Internal Combustion Engines." *Progress in Energy and Combustion Science*. doi:10.1016/j.pecs.2014.05.003.

Sadhik Basha, J., and R. B. Anand. 2011. "Role of Nanoadditive Blended Biodiesel Emulsion Fuel on the Working Characteristics of a Diesel Engine." *Journal of Renewable and Sustainable Energy*. doi:10.1063/1.3575169.

Saleh, H. E. 2009. "Effect of Exhaust Gas Recirculation on Diesel Engine Nitrogen Oxide Reduction Operating with Jojoba Methyl Ester." *Renewable Energy*. doi:10.1016/j. renene.2009.03.024.

Saroj, Ray. 2013. *Analysis of Combustion and Emission Characteristics of a Diesel Engine Fueled with Alternative Fuel by Thermodynamic Modeling*, Thesis, 1–37.

Schleussner, Carl Friedrich, Joeri Rogelj, Michiel Schaeffer, Tabea Lissner, Rachel Licker, Erich M. Fischer, Reto Knutti, Anders Levermann, Katja Frieler, and William Hare. 2016. "Science and Policy Characteristics of the Paris Agreement Temperature Goal." *Nature Climate Change*. doi:10.1038/nclimate3096.

Sims, Ralph E. H., Warren Mabee, Jack N. Saddler, and Michael Taylor. 2010. "An Overview of Second Generation Biofuel Technologies." *Bioresource Technology* 101 (6): 1570–80. doi:10.1016/j.biortech.2009.11.046.

Singh, Akhilendra Pratap, and Avinash Kumar Agarwal. 2018. "Low-Temperature Combustion: An Advanced Technology for Internal Combustion Engines." In *Advances in Internal Combustion Engine Research*. doi:10.1007/978-981-10-7575-9_2.

Singh, Pranil J., Jagjit Khurma, and Anirudh Singh. 2010. "Preparation, Characterisation, Engine Performance and Emission Characteristics of Coconut Oil Based Hybrid Fuels." *Renewable Energy*. doi:10.1016/j.renene.2010.02.007.

Song, H., B. T. Tompkins, J. A. Bittle, and T. J. Jacobs. 2012. "Comparisons of NO Emissions and Soot Concentrations from Biodiesel-Fuelled Diesel Engine." *Fuel*. doi:10.1016/j. fuel.2012.01.004.

Splitter, Derek, Reed Hanson, Sage Kokjohn, and Rolf Reitz. 2011. "Reactivity Controlled Compression Ignition (RCCI) Heavy-Duty Engine Operation at Mid-and High-Loads with Conventional and Alternative Fuels." In *SAE 2011 World Congress and Exhibition*. doi:10.4271/2011-01-0363.

Splitter, Derek, Reed Hanson, Sage Kokjohn, Martin Wissink, and Rolf Reitz. 2011. "Injection Effects in Low Load RCCI Dual-Fuel Combustion." In *SAE Technical Papers*. doi:10.4271/2011-24-0047.

Splitter, Derek, Rolf Reitz, and Reed Hanson. 2010. "High Efficiency, Low Emissions RCCI Combustion by Use of a Fuel Additive." In *SAE Technical Papers*. doi:10.4271/2010-01-2167.

Stanglmaier, Rudolf H., and Charles E. Roberts. 1999. "Homogeneous Charge Compression Ignition (HCCI): Benefits, Compromises, and Future Engine Applications." In *SAE Technical Papers*. doi:10.4271/1999-01-3682.

Stanglmaier, Rudolf H., Jianwen Li, and Ronald D. Matthews. 1999. "The Effect of In-Cylinder Wall Wetting Location on the HC Emissions from SI Engines." In *SAE Technical Papers*. doi:10.4271/1999-01-0502.

Tat, Mustafa Ertunc, and J. H. Van Gerpen. 2003. *Measurement of Biodiesel Speed of Sound and Its Impact on Injection Timing*. National Renewable Energy Laboratory, NREL/SR-510-31462. doi:NREL/SR-510-31462.

Tauzia, Xavier, Alain Maiboom, and Samiur Rahman Shah. 2010. "Experimental Study of Inlet Manifold Water Injection on Combustion and Emissions of an Automotive Direct Injection Diesel Engine." *Energy.* doi:10.1016/j.energy.2010.05.007.

Tesfa, B., R. Mishra, F. Gu, and A. D. Ball. 2012. "Water Injection Effects on the Performance and Emission Characteristics of a CI Engine Operating with Biodiesel." *Renewable Energy.* doi:10.1016/j.renene.2011.06.035.

Thangaraja, J., and C. Kannan. 2016. "Effect of Exhaust Gas Recirculation on Advanced Diesel Combustion and Alternate Fuels – A Review." *Applied Energy.* doi:10.1016/j.apenergy.2016.07.096.

Thiyagarajan, S., Ankit Sonthalia, V. Edwin Geo, B. Ashok, K. Nanthagopal, V. Karthickeyan, and B. Dhinesh. 2019. "Effect of Electromagnet-Based Fuel-Reforming System on High-Viscous and Low-Viscous Biofuel Fueled in Heavy-Duty CI Engine." *Journal of Thermal Analysis and Calorimetry.* doi:10.1007/s10973-019-08123-w.

Tsolakis, A., A. Megaritis, M. L. Wyszynski, and K. Theinnoi. 2007. "Engine Performance and Emissions of a Diesel Engine Operating on Diesel-RME (Rapeseed Methyl Ester) Blends with EGR (Exhaust Gas Recirculation)." *Energy.* doi:10.1016/j.energy.2007.05.016.

Vallinayagam, R., S. Vedharaj, W. M. Yang, P. S. Lee, K. J. E. Chua, and S. K. Chou. 2013. "Combustion Performance and Emission Characteristics Study of Pine Oil in a Diesel Engine." *Energy.* doi:10.1016/j.energy.2013.05.061.

Vanegas, A., H. Won, C. Felsch, M. Gauding, and N. Peters. 2008. "Experimental Investigation of the Effect of Multiple Injections on Pollutant Formation in a Common-Rail Di Diesel Engine." In *SAE Technical Papers.* doi:10.4271/2008-01-1191.

Vedharaj, S., R. Vallinayagam, W. M. Yang, S. K. Chou, K. J. E. Chua, and P. S. Lee. 2013. "Experimental Investigation of Kapok (Ceiba Pentandra) Oil Biodiesel as an Alternate Fuel for Diesel Engine." *Energy Conversion and Management.* doi:10.1016/j.enconman.2013.08.042.

Venkataramana, P. 2013. *Homogeneous Charge Compression Ignition.* International Journal of Engineering Research & Technology (IJERT).

Verma, Gaurav, Hemant Sharma, Sukrat S. Thipse, and Avinash Kumar Agarwal. 2016. "Spark Assisted Premixed Charge Compression Ignition Engine Prototype Development." *Fuel Processing Technology.* doi:10.1016/j.fuproc.2016.07.006.

Vigneswaran, R., K. Annamalai, B. Dhinesh, and R. Krishnamoorthy. 2018. "Experimental Investigation of Unmodified Diesel Engine Performance, Combustion and Emission with Multipurpose Additive along with Water-in-Diesel Emulsion Fuel." *Energy Conversion and Management.* doi:10.1016/j.enconman.2018.07.039.

Vinod Babu, V. B. M., M. M. K. Madhu Murthy, and G. Amba Prasad Rao. 2017. "Butanol and Pentanol: The Promising Biofuels for CI Engines – A Review." *Renewable and Sustainable Energy Reviews.* doi:10.1016/j.rser.2017.05.038.

Walker, N. Ryan, Martin L. Wissink, Dan A. DelVescovo, and Rolf D. Reitz. 2014. "Natural Gas for High Load Dual-Fuel Reactivity Controlled Compression Ignition (RCCI) in Heavy-Duty Engines." In *ASME 2014 Internal Combustion Engine Division Fall Technical Conference,* ICEF 2014. doi:10.1115/1.4030110.

Wang, Hu, Adam B. Dempsey, Mingfa Yao, Ming Jia, and Rolf D. Reitz. 2014. "Kinetic and Numerical Study on the Effects of Di-Tert -Butyl Peroxide Additive on the Reactivity of Methanol and Ethanol." *Energy and Fuels.* doi:10.1021/ef500867p.

Wiemann, Sebastian, Robert Hegner, Burak Atakan, Christof Schulz, and Sebastian A. Kaiser. 2018. "Combined Production of Power and Syngas in an Internal Combustion Engine – Experiments and Simulations in SI and HCCI Mode." *Fuel.* doi:10.1016/j.fuel.2017.11.002.

Wu, Binyang, Qiang Zhan, Xiaoyang Yu, Guijun Lv, Xiaokun Nie, and Shuai Liu. 2017. "Effects of Miller Cycle and Variable Geometry Turbocharger on Combustion and Emissions in Steady and Transient Cold Process." *Applied Thermal Engineering.* doi:10.1016/j.applthermaleng.2017.02.074.

Xing-Cai, Lü, Yang Jian-Guang, Zhang Wu-Gao, and Huang Zhen. 2004. "Effect of Cetane Number Improver on Heat Release Rate and Emissions of High Speed Diesel Engine Fueled with Ethanol-Diesel Blend Fuel." *Fuel.* doi:10.1016/j.fuel.2004.05.003.

Yamane, K., A. Ueta, and Y. Shimamoto. 2001. "Influence of Physical and Chemical Properties of Biodiesel Fuels on Injection, Combustin and Exhaust Emission Characteristics in a Direct Injection Compression Ignition Engine." *International Journal of Engine Research.* doi:10.1243/1468087011545460.

Yang, Binbin, Mingfa Yao, Wai K. Cheng, Yu Li, Zunqing Zheng, and Shanju Li. 2014. "Experimental and Numerical Study on Different Dual-Fuel Combustion Modes Fuelled with Gasoline and Diesel." *Applied Energy.* doi:10.1016/j.apenergy.2013.07.034.

Yao, Mingfa, Hu Wang, Zunqing Zheng, and Yan Yue. 2009. "Experimental Study of Multiple Injections and Coupling Effects of Multi-Injection and EGR in a HD Diesel Engine." In *SAE Technical Papers.* doi:10.4271/2009-01-2807.

Yao, Mingfa, Zhaolei Zheng, and Haifeng Liu. 2009. "Progress and Recent Trends in Homogeneous Charge Compression Ignition (HCCI) Engines." *Progress in Energy and Combustion Science.* doi:10.1016/j.pecs.2009.05.001.

Yap, D., A. Megaritis, and M. L. Wyszynski. 2004. "An Investigation into Bioethanol Homogeneous Charge Compression Ignition (HCCI) Engine Operation with Residual Gas Trapping." *Energy and Fuels.* doi:10.1021/ef0400215.

Ye, Peng, and André L. Boehman. 2012. "An Investigation of the Impact of Injection Strategy and Biodiesel on Engine NO x and Particulate Matter Emissions with a Common-Rail Turbocharged Di Diesel Engine." *Fuel.* doi:10.1016/j.fuel.2012.02.021.

Ying, Wang, He Li, Zhou Longbao, and Li Wei. 2010. "Effects of DME Pilot Quantity on the Performance of a DME PCCI-DI Engine." *Energy Conversion and Management.* doi:10.1016/j.enconman.2009.10.023.

Yoon, Seung Hyun, Hyun Kyu Suh, and Chang Sik Lee. 2009. "Effect of Spray and EGR Rate on the Combustion and Emission Characteristics of Biodiesel Fuel in a Compression Ignition Engine." *Energy and Fuels.* doi:10.1021/ef800949a.

Zhang, F., H. F. Liu, J. Yu, and M. Yao. 2016. "Direct Numerical Simulation of N-Heptane/Air Auto-Ignition with Thermal and Charge Stratifications under Partially-Premixed Charge Compression Ignition (PCCI) Engine Related Conditions." *Applied Thermal Engineering.* doi:10.1016/j.applthermaleng.2016.05.100.

Zheng, Ming, Graham T. Reader, and J. Gary Hawley. 2004. "Diesel Engine Exhaust Gas Recirculation – A Review on Advanced and Novel Concepts." *Energy Conversion and Management.* doi:10.1016/S0196-8904(03)00194-8.

# 7 Solid Waste Management

## 7.1 INTRODUCTION

Population growth, increase in economy, and development in industries and cities have resulted in a rise in solid waste generation. Solid waste management (SWM) is much required because wastes are a serious problem. Safe disposal of wastes is correlated with environmental factors, public health, and aesthetic issues. The wastes created by humans due to domestic and industrial activities have adverse effects towards the environment. Better waste management systems could be developed by understanding the various resources for waste generation and how society deals environmental conditions related to wastes. Commonly, solid wastes are categorized as material which has no worth to its owner anymore and is disposed of. The majority of solid washes are composed of household wastes (kitchen wastes and garden cuttings), metals, papers, plastics, and glass. Solid waste management involves managing solid waste production, storage, waste collection, transition transportation for processing, and waste disposal. The technique requires finding ways that best address safety, economic, engineering, and environmental concerns while remaining sensitive to people's attitudes (Muralikrishna and Manickam 2017).

The generation of enormous amounts of greenhouse gases is one of the major issues stemming from the rise in solid wastes, which will in turn create global warming. Many prominent issues related to their management are pollution of water and soil, development of foul smell, and transmission of diseases. Even after numerous attempts at handling, most of the waste ends up in a landfill. The problems with landfills include the need for a lot of space, money, and energy to sun the system properly. In brief, solid waste management is a global challenge and the world takes both an economic and environmental toll. Therefore, there is a need for efficient waste management and waste generation reduction (Awasthi et al. 2019). Sustainable management of solid waste needs civic knowledge and involvement, specific laws and regulations at the level of local councils and policy, and practical solutions.

Different research for the efficient handling of solid wastes has been conducted in various parts of the world. Research in Brazil has demonstrated that effective sanitation (including drinking water, drainage, and SWM) can be accomplished by public involvement together with private companies and local people (Dias et al. 2018). Further research in São Paulo, Brazil, showed that socioeconomic factors affect the generation of municipal solid waste (MSW) and that social dimensions and inequalities must therefore be remembered while preparing for MSW management (Vieira, Morais, and Matheus 2018). A study in Saguenay, Quebec (Canada), has stressed that knowledge on the quantity and nature of residual household waste created by residents can help to prepare its management effectively (Guérin et al. 2018). Examination of household solid waste (HSW) in Japan's cities of Okayama and Otsu has shown that the production and composition of HSW is determined by individual consumption expenses. Greener lifestyles, intense recycling, product accountability, root isolation, and composting should be practiced to handle HSW better (Gu et al. 2017).

Landfilling, composting, and incineration have been the commonly employed ways of treating MSW in China, resulting in air emissions. A study on air pollution and greenhouse gas emissions associated with MSW management in China stressed that limiting, processing, waste recycling, advanced disposal techniques, efficient laws/regulations, public knowledge, and education are strategies for managing air pollution associated with MSW clearance (Tian et al. 2013). Contemporary literature stresses that we will exercise the four R's—Reduce, Reuse, Recycle, and Recover—for efficient management of solid wastes. Many studies on this subject emphasize the development of a circular economy during the handling of MSWs. Wastes can be treated as a tool, as they may yield several economic advantages if they are properly disposed of. Examination of three case studies, including San Francisco's Zero Waste Strategy, Flanders' Sustainable Materials Management Policy, and Japan's Sound Material Cycle Culture Project, shows that the current change to see waste as a tool for the development of a circular economy is in the initial phase and more improvement in policy policies, preparation, and action is required (Silva et al. 2017).

This chapter considers the various types of waste generated and how they are segregated/characterized for storage and collection purposes. It also gives detail about the various types of recycling and recovery of materials from the wastes. The waste-created environmental pollution and health risks for humans are discussed to illustrate the impact of the solid waste. The chapter tells about various technologies for the handling of solid waste and future policies for waste management which can improve management techniques with respect to the growth of the population and economy.

## 7.2   WASTE QUANTITIES AND CHARACTERIZATION

In solid waste management, it is important to have the knowledge of waste quantities and their characterization. Characterization of wastes is mainly done based on the generator source, property, and geological positions. Wastes are classified based on their forms. Based on their decomposition, wastes are of two types: biodegradable and non-biodegradable wastes. Biodegradable wastes are organic wastes from food and other kitchen leftovers which can turned into usable product with the action of microbes. Non-biodegradable wastes are mainly hard plastics, which can be processed into reusable products or end up in landfills.

Based on their effect, other form of wastes are classified as hazardous and non-hazardous. Hazardous wastes are those which present chemical, physical, radioactive, and infectious hazards after their usage; these wastes require safety handling and should be disposed of by experts in accordance with specific regulations. Wastes are mostly distributed into recyclable and non-recyclable depending on their recycling viability.

Biowaste and combustible and non-combustibles components are typical of solid waste. For the management of solid waste, determination of caloric value, organic fractions, moisture content, and composition of chemicals in various fractions of wastes should be done.

Some of the wastes produced widely will be listed below. MSW is significant waste created mostly from companies and households. The collection of these wastes is mainly the responsibility of municipalities, local bodies, and private corporations. The major concentration of this is food waste, which is heterogeneous waste. The main constituents of the MSW are plastics, glass, leather, metals, paper, and wood. It also includes toxic wastes such as discharged batteries having mercury (Castaldi 2014). Waste is generated during construction and demolition (CD) through construction, renovation, and destruction of houses, highways, and bridges. The main constituents of this waste are concrete, metals, plastic, cement, woods, and bricks, which are inert materials; most of its fractions can be recycled. Hazardous wastes of CD are asbestos and particulate matters. Quantification of solid waste is necessary for handling and CD waste is quantified with various techniques that have been explained in many papers (Wu et al. 2014).

Technology developments have led to production and consumption of more electronic devices such as televisions, computers, mobile phones, tablets, notebooks, and digital cameras. Waste of such items is called electronic waste (e-waste). Increasing competition, improved availability, social needs, and shorter periods of usage of these goods have culminated in a dramatic rise in e-waste production. Due to the presence of heavy metals, e-waste has significantly raised toxicity, leading to environmental pollution. In the current situation, e-waste recycling is primary, and the different portions such as glass, plastic, and metals (including rare earth metals and minor metals) may be recycled (Tansel 2017).

Medical wastes come from health care centers, hospitals, and clinics. The aging population, increases in disease due to lifestyle, low-cost medical treatment, and the usage of disposable items have to lead to the rise in medical wastes. Some specific constituents of these wastes are cotton, syringes, medicine, gloves, bandages, pathologic waste, paper, plastic, placentas, and chemicals. Fewer medical wastes have to be incinerated compulsorily due to stringent rules, and some other materials which have biological contaminants should be disposed of with proper treatment. Medical wastes from health care centers are commonly categorized as infectious and noninfectious wastes. In hospitals, bags of different colors are used to collect the wastes. If bags are yellow, this indicates that the waste they contain needs greater attention, and they are used to gather contagious, pathologic, and sharp objects, while black bags are used for common wastes and paper. The generation of medical wastes differs across different wards of hospitals. Medical care wastes demonstrate variation in the characteristics of its constituents, like calorific value, moisture content, and bulk density (Diaz et al. 2008).

Solid wastes have sources which are divided into five main categories. Animal wastes constitute a major portion of solid waste, which is 42.7% on a weight basis; 31.3% is mining waste; 15.7% is agricultural materials; 7.2% is household, municipal, and commercial; and 3.1% is industrial wastes. On a tonnage basis, animal waste is measured to be about 1500 million metric tons per year. The distribution of solid wastes in different cities differs considerably. Waste generation varies according to time and place. Commonly, solid wastes have organic wastes as its major concentration rather than other materials. In Asian countries, the total amount of organic matter in solid waste from major cities ranges from 50% to 70% (Muralikrishna and Manickam 2017).

## 7.3   STORAGE AND COLLECTION OF SOLID WASTES

In municipal solid management, collection and transportation (CT) of solid waste places significant costs on the entire system, with fuel usage, social exasperation, and air emissions throughout the network. Around the world, various techniques are followed and proposed to have economically viable, eco-friendly, socially accessible, sustainable, and technologically upgraded techniques for collection and transportation (Yadav and Karmakar 2020). In waste management, the major cost of consumption is accounted for collection of waste and transportation (Yadav et al. 2016). In developing countries like India, 70–80% of cost is spent for CT, whereas it is less in developed countries like Sweden, where it is 5–75% of the total management cost (Sonesson 2000; Ghose, Dikshit, and Sharma 2006). Unstructured recycling and human workers are mostly used in developing countries (Amponsah and Salhi 2004) for CT while developed countries uses mechanical collection facilities for the separation of MSW (Rodrigues, Martinho, and Pires 2016; Mohsenizadeh, Tural, and Kentel 2020). All countries practice CT technique in diverse manners: Indian municipalities follow a standardized manual prescribed by the Ministry of Urban Development (MoUD 2016), while Danish municipalities follow CT methods which must be decentralized in compliance with the ordinances laid by local urban authorities (Christensen 2010). There are various techniques which are followed, most commonly door to door, curbside/alley, pumping at designated place, backyard, and the pneumatic technique (Yadav and Karmakar 2020).

In door-to-door techniques, producers of the waste give their waste at the doorstep to the waste collectors (Yadav et al. 2016). It is an easy technique for citizens, being economical and less of a nuisance. In order to avoid these wastes from becoming mixed, vehicles with various chambers are required to collect different wastes at the needed time. The curbside/alley technique does not require any human to give the wastes to the collector, and it also eliminates the time taken for CT (Zbib and Wøhlk 2019). Wastes are kept in containers nearer to their homes and are collected daily by waste collection vehicles. But when there is a delay in collection of wastes, it can lead to foul smell, fleas, and nuisance due to stray animals (Hermelin et al. 2019).

Dumping at designated places is another technique in which the wastes are dumped at a common place to be collected periodically (Yadav et al. 2017). It is done for places where there is no easy vehicle accessibility. This technique is commonly followed in low-income and developing countries (Batool and Ch 2009; Parrot, Sotamenou, and Dia 2009). It is less expensive, and uses less human labor, but it creates nuisance, bad aesthetics, and problems with fleas and animals. In the backyard technique, garbage collectors go directly to the collection point with a tote barrel and shift it to the vehicle (Lella, Mandla, and Zhu 2017). It does not require any collection bins on roads. It requires skilled labor and it is expensive, so it is mostly used in developed countries with lower population density.

CT with automobiles obtrudes major adverse externalities in metropolitan centers such as air emissions and manual labor costs (Miller, Spertus, and Kamga 2014). To avoid these effects, the pneumatic collection and transport technique is followed in some urban areas. Wastes are collected with the pneumatic technique and transferred by pipes with pressurized air (Bi et al. 2019). It requires professional human power, large initial capital investment, and highly planned regional conglomeration, rendering it deal for advanced economies. If the system is built effectively, it can reduce the greenhouse gas emissions and limit air and noise pollution; furthermore, it requires more energy, which creates issues including pipe blockages due to excessive waste (Miller, Spertus, and Kamga 2014; Mangialardi et al. 2016).

## 7.4  FACILITIES FOR MATERIALS RECOVERY AND RECYCLING

Human actions generate voluminous waste materials due to the various necessities of their lifestyle (Kelly and He 2014; Mubarok et al. 2017). The waste generated by humans limits the capacity of natural recycling systems and it must be managed with respect to their impact on the environment and people's health (Jones et al. 2013). Often, nutrient-rich wastes, organic matter, and resources are not effectively handled and processed properly for recycling and reuse. On one hand, material recovery and recycling can build sustainable livelihoods, and on the other, they can reduce wastage, support the green economy, and enhance overall environmental health and cost recovery (Banu et al. 2018).

The key factors in solid waste management are the solid waste separation process, recycling and material recovery processes for different types of solid waste, and the regulatory basis for SW sustainability management. Waste management requires the attention of every individual towards how waste materials can be reused and recycled. Recycling of resources helps in attaining positive social, economic, and environmental benefits. In addition, recycling helps to conserve natural resources, reduce energy usage, shift human labor, and create job opportunities with the benefits of curbing waste and minimizing contamination of the environment. Solid waste recycling helps in transforming unnecessary waste into usable materials, converting hazardous wastes into non-hazardous products, and encouraging waste reuse. Recycling of solid waste has major importance in terms of energy conservation, recovery of resources, and reduction in environmental risk.

Waste separation and recovery of materials are important as they not only reduce waste accumulation but also allow money to be generated. Waste accumulation can be reduced by receiving household wastes separately as paper, scrap metals, plastic, etc., in a collective manner. Likewise, recovery of biogas from landfills and thermal energy from incineration plants would help to eliminate waste production and aid in the effective utilization of materials. Adequate recovery of valuable metals from e-waste will delay the decline of natural resources, improving and promoting the usage of urban mines. Utilization of livestock and manure, farm waste, household sewage, and other organic solid wastes through aerobic compositing and anaerobic digestion and recycling of the organic products and nutrients, etc. are a few effective techniques to understand for resource utilization and recovery systems (Awasthi et al. 2019).

Food wastes (like kitchen vegetables wastes) are produced everyday continuously in ever-higher quantities all around the world; their sustainable disposal poses an increasing environmental issue. Different value-added products can be generated through valorization from food wastes such as chemicals and fuels (Donato et al. 2014; Venkata Mohan et al. 2018). A few examples of biofuels produced are bio-diesel, biomethanol, bioethanol, biobutanol, and biohydrogen. Similarly, the various chemicals obtained from wastes include waste esters, organic acids, pigments, bio-surfactants, sugars, and biopolymers (Saratale, Jung, and Oh 2016; Ong et al. 2018).

Currently, solid waste disposal and management is rising as a major problem due to the adverse effects of solid waste on people and nature's health (Sharholy et al. 2007). Technologies and processes have been developed for the utilization of the solid wastes in the developments of biofuels and valuable chemicals/products (Kumar and Samadder 2017). Various products, such as antioxidants, metals, flavo-noids, vitamins, polyphenols, syngas, fibers, and carotenoids, have been recovered from municipal and industrial wastes by using different techniques and technologies created for the production of biological products (Sun et al. 2018). Gas fermentation is a modern paradigm which appears to be viable solution to address some waste management problems (Arafat and Jijakli 2013; Ramachandriya et al. 2016). Syngas produced from fermentation of waste may be used a possible source of carbon and energy resources for fuel production and different chemicals, including petrochemi-cal products (Mohan, Varjani, and Pandey 2018).

Phosphorus is important to all forms of life. All over the world, phosphorus resources are being depleted. Therefore, phosphorus recovery from phosphorus-bearing waste has considerable importance (Petzet et al. 2011). Recovery of phos-phorus from solid industrial waste has great impact and it should be done on an industrial scale, which can improve the economies of waste-processing industries (Kleemann et al. 2017). Phosphorus-rich incinerated sewage sludge ash (ISSA) was generated as a by-product in sewage treatment plants with dewatered sewage sludge in the incineration process (Mattenberger et al. 2008). The generated ISSA phos-phorus is economical and able to be recovered in an eco-friendly way, which leads to better sewage sludge management and greater phosphorus reserves (Guedes et al. 2014). The chemical activity occurring during various extraction processes for phosphorus and metal(loid)s in the recovery of phosphorus, and also the phosphorus precipitation behavior in leachate for the manufacture of phosphorus as fertilizer, are both of great importance (Cyr, Coutand, and Clastres 2007).

Electronic waste or e-waste commonly refers to discarded electrical and elec-tronic devices (Ghosh et al. 2015). High-valve metals are present in greater amounts in e-wastes; for example, the concentration of gold in e-wastes is 10 times greater than that in gold ores (Xin et al. 2012). A strategic analysis of metal recovery technology techniques from e-waste requires an improvement to generate low-toxicity, finan-cially feasible, and eco-friendly sustainable methods (Kim et al. 2011). Numerous methods exist for e-waste metal recovery and they are being extensively studied for future applications (Chagnes and Pospiech 2013).

Agricultural biomass is an abundant source and is a major part of potential future energy strategies, as it lowers emissions of greenhouse gases and can meet long-term energy demands (Hiloidhari, Das, and Baruah 2014). Chemical and energy production from green agri-biomass offers an effective substitute for petroleum refining, and it can lead to stable growth in the community (Maurya et al. 2018). Due to the growth of the agri-based bioeconomy, there has been an increase in crop production throughout the world (Elmekawy et al. 2014). Though most of these wastes have high nutrient value and are rich in energy content and organic compounds, they are not used wisely for material recovery and recycling for product development (Michelin et al. 2018). Lignocellulosic wastes can be taken for valorization, for producing value-added products which have economic importance, and for promoting the growth of new techniques for creating income by using phosphorus within novel technology sectors (Costa et al. 2017). Technologies like gasification can be used for the conversion of biomass into energy, which produces power or fuel for automobiles and power plants (Bartela, Kotowicz, and Dubiel-Jurgaś 2018). Recently, interest in agri-biomass is growing, with various research being conducted for creating technologies and products which can recover energy and generate commercial income with recovered products (Yu and Tsang 2017).

Metals are being recovered through bioleaching and biohydrometallurgy; these are techniques that are of great importance for sustainable development in metallurgy and mineral processing, especially for phosphorus, and will lead to an improved environment and economy (Johnson 2014). They are considered as feasible methods for the recovery and recycling of basic, costly, and critical metals from minimal-grade ores, mine waste, and screenings (Fashola, Ngole-Jeme, and Babalola 2016). In this, they use microbes for their potential role in recovery of metals. Biotechnological studies have been made on microbes with respect to three different metal resistance mechanisms, which have made a pathway for leaching metal, metal recovery from their ores, and heavy metal treatments (Mehrotra and Sreekrishnan 2017). Research has to be done to develop and explore the challenges and opportunities posed by the use of micro-organisms as biosorbents, mining-related pollution, regulation, and the futuristic development of industrial-scale techniques (Feng, Yang, and Wang 2015).

## 7.5  HEALTH AND SAFETY RISKS

Wastes which are collected and dumped present various hazards to human health and many environmental issues. The chemical elements present in the wastes can reach or directly contact human beings and cause problems.

Around 10% of household waste is constituted of plastics, and the generation of plastic wastes is rapidly increasing every year due to overpopulation (Verma et al. 2016). In a 2013 report on plastic waste management, India's Central Pollution Control Board stated that plastic litter and careless disposal of non-biodegradable wastes lead to more environmental issues (Bhawan and Nagar 2013). Plastics that are commonly used in our day-to-day lives, which are termed "everywhere plastics," most often contain phthalates and bisphenol-A or BPA. This chemical, present in various products, can easily mix with food and water and then enter into the human body. Some research studies show that phthalates and BPA are linked with many growth problems and hormonal imbalances. Babies and young children are found to be most exposed to these plastics during their early growth period, and they pose a high risk to their health. Many experimental studies have been done on BPA, and it has been found to be a neurotoxin and endocrine disruptor that is very harmful to humans. Researchers found that exposure to BPA will lead to many health concerns like metabolic diseases, fertility disorders, and even cognitive impacts in adolescents (Rochester 2013). BPA and phthalate exposure have been studied and a correlation has been found with steroid hormone level alteration in adults (Sathyanarayana et al. 2017) and sometimes even in babies (Araki et al. 2014). It also affects the female reproductive system, causing damage to the uterus. Therefore, it is important to monitor the levels of BPA and phthalates in various products and wastes (Watkins et al. 2017).

Electronic wastes containing the discarded elements of electronic devices have both hazardous and non-hazardous components. Elements like cadmium, arsenic, lead, and mercury, along with the plastics present in these wastes, are hazardous to human health. Persons involved in e-waste recycling and recovery are susceptible to cancer (Trivedi et al. 2020).

Silver is used as a disinfectant since it is bactericidal; it is widely used for medical purposes and nanotechnology research work. It also poses problems in medical research and treatment due to its wide range of possible effects on bacterial resistance (Chopra 2007). Chlorine and ammonium substances are used as antimicrobial agents in pathology centers, clinics, landfill areas, etc. It is recommended not use large amounts in closed environments due to its toxic fumes.

When pesticides are not stored properly, there is a chance of leakage, and rainwater can get in contact during the rainy reason, which can lead to the poisoning of groundwater. The toxic groundwater can affect the food chain and can lead to hazardous DNA mutations leading to multigenerational genetic effects in humans.

Normally, mercury appears in a liquid state and is of a shining silver color. It can be easily transported by wind, and it settles at the bottom of water bodies, where it contacts bacteria; this process forms methyl mercury, which can enter into the food chain. Although mercury is very beneficial in certain ways, it is toxic or dangerous in some forms and may be assimilated through the dermal pathway. The mercury effect can damage the vascular system, digestive system, and nervous system with additional hepatic and immune system failure. It can even lead to deformity of neurological organs and problems during pregnancy (Secretariat of the Basel Convention and WHO 2004).

The manufacture of diagnostic medical equipment uses metal mercury, and safe disposal of this equipment is quite challenging. Mercury used in the batteries of medical devices also requires safe disposal. Governments like those of the United States and the European Union have largely stopped the usage of mercury-powered batteries due to their severe effects, but some countries are using them to a certain extent. Medical care equipment is slowly being replaced with mercury-free alternatives. Mercury containments in untreated waste water/effluents that are directly released into water bodies contribute around 5% of the overall mercury pollution to the ecosystem, damaging the marine ecosystem. When it comes in contact with the marine ecosystem, it easily gets into the food chain. As per the reports of the US Environmental Protection Agency and the EU, half of the mercury contaminations are due to dental practices and medical incinerations (Fisher 2003).

Health care wastes should be handled with care, and proper training should be given to the staff involved in the waste management. Clinical wastes are more prone to be infectious and can present immediate health-related problems. Ineffective waste management can lead to many immunodeficiency diseases and viral infections. Sometimes contagious disease can spread unknowingly to the mass population. If body fluid samples, syringes, and patient waste are not handled or disposed safely, then other patients, staff, and family members will be prone to contagions. It was found that lower effectiveness of drugs and disinfectants to microbial action may be due to improper waste handling (Gańczak, Milona, and Szych 2006). Expired medications and old chemicals are dumped as wastes. The dumped wastes can enter into the human body through inhalation, injection, and epidermal contact. Sometimes, these wastes react with flammable, corrosive, or otherwise reactive chemicals and become even more toxic for humans. Workers who have contact with medical care wastes are reported to have infections according to blood and other fluid samples, and it is considered as an occupational threat. In 2009, it was found that the reuse of syringes from infected persons was the cause for hepatitis B spread in Gujarat (Harhay et al. 2009).

## 7.6   ENVIRONMENTAL POLLUTION

While solid waste is being disposed, it produces significant amounts of greenhouse gases, which are emitted into the environment and have great impact on climatic change. Solid waste is created at every individual processing stage, i.e., waste collection, transport, and separation, and it gets deposited in landfills, recycled, incinerated, or subjected to a waste-to-energy conversion process. In solid waste management, organic wastes are broken down in aerobic and anaerobic digestion processes, which produce methane ($CH_4$) and carbon dioxide ($CO_2$). The breakdown process is done mainly with the help of bacteria at the sites of landfill and bacterial biomass. In addition to the bacterial activity, the organic wastes are controlled by other physical factors like temperature, moisture, pH conditions, etc. In addition, wastes dumped at landfills produce landfill gases, which are comprised of $CO_2$ 40-50% of $CH_4$ and 40% to 60% of $CO_2$ by volume (Verma, Kaur, and Tripathi 2019).

These are two categories of landfill site: open dump sites and managed landfills. The two types of landfill site produce considerable amounts of harmful gas as their emissions. $CH_4$, which is generated at uncontrollable rates from landfills, poses serious issues to the surrounding local environment and it will have great impact on the local biodiversity; in addition, it normally produces foul odors. If the amount of $CH_4$ generation increases from 5% to 15% to the air, then it can create fire hazards at the landfill/dump sites. Among the greenhouse gases, one of the harmful is methane, which is produced by anthropogenic activity; $CH_4$ emissions to the global atmosphere are mainly contributed by solid wastes from landfill sites and have impact on global climate change. Every nation faces a greater challenge in reducing the anthropogenic activity, which is the main factor for greenhouse gas emissions (Gómez et al. 2009). For sustainable development in waste management, the landfills' high rate of methane gas emission production requires immediate action to produce environmental and social benefits and to reduce the negative impact on public health and communities (Papageorgiou, Barton, and Karagiannidis 2009).

Landfills are said to be biochemical reactors which produce environmental pollutants in large quantities, including landfill gases, dust, odor gas, and soluble or insoluble products (Torno et al. 2011). Research done over various landfill sites from different areas has shown that the foul smells are caused by ammonia, dimethyl disulfide, dimethyl sulfide, acetic acid, xylene, toluene, styrene, n-butyl aldehyde, n-butanone, methanol, and acetone (Fang et al. 2012). The main mechanisms for the generation and diffusion of these pollutants from landfill sites are dependent on their morphology, structure, composition, age, and the surrounding atmospheric conditions (Nolasco et al. 2008). Pollutants that are formed due to the solid wastes in the landfill sites last up to three decades or more, even after the sites are closed according to the report of Ritzkowshi et al. (Ritzkowski, Heyer, and Stegmann 2006).

Air pollution is due in large part to the emissions of landfill gases, which consist of methane (50–60%), carbon dioxide (over 40%), and remaining non-methane volatile gases (El-Fadel, Findikakis, and Leckie 1997). Since they are major contributors to environmental effects and are of toxicological significance, emissions of landfill gases have been an increasing issue for the climate. Greenhouse gases and chlorofluorocarbon can damage the ozone layer (Newman et al. 2009), substances like benzene, toluene, etc., have mutation effects and can cause heart-related issues (Durmusoglu, Taspinar, and Karademir 2010). One study found that methane is 21 times greater than carbon dioxide in creating in global warming problems (IPCC 1996). Around the world, 18% of the methane emissions from the waste field is due to anthropogenic activity (Bogner et al. 2008).

Landfills in the United States are considered as the second primary cause of methane due to anthropogenic activity; they account for 23% of the overall anthropogenic methane emitted. In 2010, approximately 5135 Gg of methane emissions were estimated in the United States (US Environmental Protection Agency 2012). In 2006, the second highest anthropogenic methane emissions in Europe were generated from landfills; the amount was estimated at around 3373 Gg of methane emissions due to the waste disposal (MacCarthy et al. 2008). The methane emissions from landfills of China in 1994 were accounted to be 2.03 Tg, with overall concentration of 26.3% to the overall methane gas generated from waste disposal and treatment (Beijing 2004). However, the greenhouse effects created by solid waste management have to be reduced, so it is important to collect and recycle the gases from landfills and avoid emission-creating products (Tian et al. 2013).

Air pollution is mainly caused by odor generated during the decomposition of wastes. Microbes are the major factor in the emission of foul gases, due to their reaction with organic matter. The anaerobic process in the compost of wastes produces odorous gases which are harmful to the environment and human health (Sánchez et al. 2015). These problems are a major concern in developing nations, and most of the plants have been closed in the past few years (Chung 2007). If food wastes are composted, then there should be action taken to ameliorate the release of odorous gases in processing centers. The concentrations of emitted gases are mainly depends on the pH, oxygen level, temperature, and composition of the substances.

In solid waste management, incineration is mainly done to reduce waste, and it requires equipment for processing. During the incineration phase, wastes are significantly reduced, but certain toxins present in the waste or formed during the reaction will escape from pollution controlling equipment and diffuse into the environment. These contaminants can travel through the various media of the environment such as land, water, plants, and phytoplankton. Due to these, human health is greatly affected directly or indirectly through various media like drinking water and infected food products, and also through ingestion and absorption of toxins through skin (Tian et al. 2012). During the incineration of solid waste, four categories of hazardous air pollutants are released into the environment. They are toxic heavy metals (such as nickel, chromium, lead, mercury, stibium, etc.), particulate matters, acid gaseous emissions (such as oxides of nitrogen, sulfur dioxide, hydrochloric acid, etc.), and incomplete combustion pollutants like carbon monoxide, polychlorinated dibenzo-p-dioxins, and dibenzofurans (PCDD/Fs), etc.

Mercury is mostly present is discarded batteries and fluorescent lamps that get their final deposition as solid wastes; during the incineration process, this mercury will get released into the environment. According to studies, globally 8% of the anthropogenic mercury emissions are accounted to be from disposed waste and incineration (Pirrone et al. 2010). PCDD/F compounds have chemical properties similar to those of the unburned products over flow combustion parameters and are commonly referred to as dioxins and furans. In incineration processes, the most toxic substances formed are PCDD/Fs. Several works have shown that most of the PCDD/F emissions are from residues, and 90% of them enter as fly ash into the environment (Zhang, Hai, and Cheng 2012). While the total emissions of PCDD/Fs are not high in volume, their severe effects cannot be neglected. According to the US Environmental Protection Agency, they are hazardous to people, causing cancer, several reproductive disorders, immune and system failure, and other human disorders (Fries and Paustenbach 1990). Yuan and Li investigated the generation of PCDD/Fs during the incineration process and suggested that the emissions of PCDD/Fs can be solved with emissions of flue gases (Yuan and Li 2012).

## 7.7 DIFFERENT TECHNOLOGIES FOR SOLID WASTE MANAGEMENT

Waste treatment management is done using various technologies, among which landfills are the most basic and common technology. The other technologies are compositing and recycling process. All these technologies are widely used for the treatment of wastes, and recently more attention has been given to technologies for converting waste to energy, among which are various thermochemical and biological technologies. These conversion technologies are not more complex than existing techniques, but they have certain advantages like volume reduction, elimination of toxic substances, and recovery of energy (Rajendran et al. 2019).

Landfills are the most common way to dispose of solid waste on land without causing any damage or threat to local health or safety, by using the concepts of engineering to restrict the waste to the smallest possible area, to minimize it to the lowest practical volume, and to cover it with a layer of soil at the end of every day's operation or at other regular intervals as necessary. Therefore, this process basically consists of spreading the material, which is accompanied by its gradual compaction with the least exposed region to the lowest possible volume and then fully covering with soil. Since the area is exposed is small, the required amount of soil is also less. Covering with soil or inorganic matter will help the waste to be free from flies and rodents, and it also conserves the heat from the decomposition of waste, thus helping to exterminate fly larva and other pathogenic species. Landfilling is most commonly categorized into three types based on site conditions: they are the ramp, area, and trench methods (Rao, Sultana, and Kota 2017). Landfilling involves high management costs, and it has resulted in many idle sites with serious pollution emissions (Sahariah et al. 2015).

Composting is a process in which the wastes are decomposed with the help of microbes into rich organic manure (Suthar, Mutiyar, and Singh 2012). It is traditionally a common process done mainly in rural areas (Narayana 2009). Composting is classified into three types: traditional pit composting, traditional heap composting, and modern large-scale composting. In traditional pit composting, the most common aerobic and anaerobic processes are used, and heap composting is done using aeration techniques. Modern large-scale composting types include in-vessel composting, windrow composting, and effective microorganism-based quick composting (Pergola et al. 2018). The windrow composting technique is used for the treatment of wastes at large scales. In the aerated turning windrow composting technique, the wastes are distributed through long stretches followed by periodic turning by mechanical aeration activity. Bulking substances are combined with solid wastes in another form of windrow composting called static pile composting. Since there is no mechanical turning involved in the process, the air passages in the layers of waste are formed as gaps and crevices because of the inclusion of bulking substances (Misra, Roy, and Hiraoka 2003). Windrow composting is widely used in Ecuador to recycle urban solid waste (organic/biodegradable) to minimize the waste treatment costs and also for the culturing of vegetables (Gavilanes-Terán et al. 2016). In some Canadian provinces (Saskatchewan, Alberta, and British Columbia), 200 drum composting units were installed for a population of 34.9 million (Zhang et al. 2013). More than 2000 species of microbes (e.g., species of *Rhodobacter*, *Streptomyces*, *Lactobacillus*, etc.) are capable of reducing the lignocellulosic wastes and accelerating the composting cycle (Aminah et al. 2016). Urban development and increasing consumption habits have increased the level of waste generated in the urban areas of the world. It is estimated that 40% of the gross wastes deposited in landfills consists of biodegradable products (Manios 2004). According to one study (Narayana 2009), about 70–80% of waste produced by Asian cities including Singapore, Mumbai, and Tokyo is biodegradable.

Basically, earthworm-mediated composting or vermicomposting is the widely followed technique all over the world because of its rapid decomposition rate of waste substances, its substantial enhancement of the finished product, and its eco-friendly sustainability (Suthar, Mutiyar, and Singh 2012). Vermicompost is earthworm's excreta and it is humus rich. Certain earthworm species (*Perionyx excavates, Metaphire posthuma, Ponthoscolex corethrurus, Lampito mauritii, L. terrestris, Lumbricus rubellus, Eudrillus euginae*, and *Eisenia fetida*) have proved to detoxify successfully and balance a broad group of compound domestic and industrial wastes (Bhattacharya and Kim 2016).

Waste treatment does not stop with the reduction of waste volume; it can also be utilized for energy recovery. Due to the rise in the global population and the demand for energy, the world is trying to find more sources for the production of energy. The International Energy Agency reported that the demand for energy supply is growing rapidly and that this is a major concern in developing countries. According to the report, it is predicted that there will be decline in oil and gas demand by the end of 2030 and the waste-to-energy production will rise by another 54%(International Energy Agency 2018). Energy generation from waste is clean and green. It can eliminate the emission of carbon dioxide while burning fossil fuels and methane emissions from landfills; this technique is globally accepted and adopted (Scarlat, Fahl, and Dallemand 2019). Two-thirds of solid wastes is composed of biodegradable substances which can be used for capturing $CO_2$-neutral energy, thereby lowering the reliance on fossil fuels (Christensen et al. 2015).

Solid wastes have high energy content, and their latent energy can be used for waste management and conversion technologies for heat, oil or gas, into electricity (Das et al. 2019). The two main waste-to-energy technologies are thermochemical conversion and biochemical conversion. In thermochemical conversion, the energy is produced from waste material by thermal decomposition. The main types of this technology are incineration, gasification, pyrolysis, and carbonization.

Incineration is technology in which wastes are combusted and converted into heat energy, flue gas, and ash (Pan et al. 2015). The process occurs at a temperature of 800–1000°C with a supply of sufficient oxygen in which thermal breakdown and deposition of wastes are done. The heat energy recovered from the combustion process can be utilized for heat exchangers or power plants. The flue gases from the process should be cleaned and pollutant-free gaseous should be released into atmosphere. The ash from the process is due to the inorganic waste and they are released as solid pollutants with the flue gases. Incineration will decrease the solid waste volume up to 80–85% of the original waste, and it results in a significant reduction in the volume for disposal (Pham et al. 2015). In fact, incineration can be used for the treatment of medical wastes and other hazardous wastes where high temperature is required to destroy the microbes and toxins (Rodriguez-Garcia et al. 2014). Potential pollution correlated with waste incineration is a big problem for incinerators. The various emissions from incineration are hydrogen fluoride, hydrochloric acid, ammonia, methane, furan, polychlorinated dibenzodioxine, fluorine, chlorine, sulfur, SO, $NO_x$, $CO_2$, CO, and N (Tabasová et al. 2012). It is cost intensive and requires skilled labor and specialized handling equipment (Abramov et al. 2018). Emissions can be controlled by altering the composition of fuel, the moisture content particle size, and design of the incinerator (Karademir, Bakoglu, and Ayberk 2003).

Pyrolysis is a combustion technique in which organic compounds are broken down in the absence of oxygen at high temperature to produce various biofuels such as bio-oil, biochar, and syngas. The syngas is composed of CO and $H_2$ at 85%, and the remainder consists of $CO_2$ and $CH_4$. The pyrolysis is done at temperatures of 400–1200°C, and the bio-oil which is produced has a high calorific value of 17 MJ/kg, which can be added to petroleum diesel. The product yield of pyrolysis is mainly dependent on their process parameters such as temperature and residence time for the production of biochar (Tripathi, Sahu, and Ganesan 2016). The various advantages of the pyrolysis are as follows: fuel gas production, ability to add the bio-oil to diesel to make a blend mixture, less pollution emission compared to incineration, and the compact nature of the pyrolysis. The major disadvantages of this technique are the formation of hazardous gases such as polyacrylonitriles and hydrogen cyanides, greater oil viscosity, high moisture content, and lower total energy output; this technique also requires pre-treatment of the solid wastes for production of bio-oil, which incurs more cost, and furthermore, treatment is required during oil storage due to its instability (Czajczyńska et al. 2017).

Gasification is another thermal process in which wastes are treated like incineration. The main difference between incineration and gasification lies in their products. In incineration, high heat energy is produced, whereas in gasification the product is combustible gases. Thermal decomposition of wastes is done under partial oxidization in the temperature range of 800–900°C, which converts solid wastes into combustible and non-combustible gases. The gas produced has low calorific value and it can used for direct burning or as a fuel for gas turbines and gas engines. The gas or syngas produced is a mixture of $CO_2$, $CO$, and $H_2O$; the remaining by-product is called vitreous slag. The produced gas can be used for the production of alcohols (such as methanol). Solid waste gasification is done on in a series of sequential endothermic and exothermic phases in relation to the reactant and final products (Rollinson 2018). The advantage of gasification technology is the production of syngas; the vitreous slag by-product can be used in road construction as a filling substance, and chemicals such as sulfuric acid can be produced from the sulfur, which is in impure form. The limitations of this technology are the high cost, high capital investment, high coal requirement, and production of hazardous organic polyhalogenated compounds (Das et al. 2019). The gasification technique uses a variety of reactor beds such as rotary kilns, moving grate furnaces, vertical shafts, entrained beds, fluidized beds, all determined based on their applications (Pham et al. 2015).

Hydrothermal carbonization is another thermal solid waste conversion technique in which the waste is converted into carbonaceous residues (hydrochar) under a water-based setup with a temperature range from 180°C to 350°C. The advantages are the low energy required to initiate the setup, the exothermic nature, and the ease of operation with high-moisture-content waste products. The disadvantages are cost and the increase in water toxicity (Mihajlović et al. 2018).

Biochemical conversion is a technique in which wastes are converted with the assistance of microbes and enzymes. The two types of biochemical conversion are anaerobic digestion and fermentation. Solid wastes which have high moisture content and putrescible fractions are preferred for these processes due to their microbial activity nature.

Anaerobic digestion is a sequential biological process where various microbes in the absence of oxygen break down biodegradable materials. The feedstocks mostly chosen for anaerobic digestion are wet organic wastes such as agricultural wastes, animal slurries, and food wastes. The final product of the digester is biogas, which contains $CH_4$ 60% and $CO_2$ 40%, which can be combusted for heat energy and generation of electrical energy or upgradation of the biogas to automotive fuels and natural gas. The digestate which is collected as a by-product has high levels of rich nutrients and it can be used a soil fertilizer. Anaerobic digestion is eco-friendly with various benefits like renewable energy production, the prospect of recycling nutrients, and reduction in waste volume (Xu et al. 2018). Anaerobic digestion plants have received attention in many countries in recent years. In Germany, there are about 9000 plants for biogas production and in 2013 Europe had about 14,572 biogas plants (Torrijos 2016). It was found that the methane content was rich if the biogas plant was provided with a mixture of food waste and wastewater sludge, and sequentially an increase in food waste up to 35% not only increased the methane yield but also improved the mechanism of methane formation (Koch, Helmreich, and Drewes 2015). The major drawbacks of anaerobic digestion is its longer duration: the digestion process can take 20–40 days for the microbial reaction. Hydrolysis, acidogenesis, acetogenesis, and methanogenesis are the four major biological processes happening during anaerobic digestion (Zhen et al. 2017).

Fermentation techniques are used to generate bioethanol or biohydrogen from wet organic wastes includes sewage sludge, agricultural wastes, and food wastes. Certain wastes need hydrolysis or saccharification for the conversion of carbohydrates into sugars before the start of fermentation (Rajendran et al. 2019).

In biohydrogen production, dark hydrogen fermentation is done in which the sugars are converted into hydrogen and vaporous fatty acids with the help of hydrogen-generating bacteria (like *Enterobacter* and *Clostridium*). Since hydrogen is found to have a high energy density per unit mass, it can be used as a substitute for liquid fuels; in fact, it is has energy content 2.8 times greater the hydrocarbon fuels, which makes it a versatile energy source (Chandrasekhar, Lee, and Lee 2015). The combustion of hydrogen fuel produces water as the final product, and thus it is considered as an option for reducing greenhouse gas emissions. Production of hydrogen by the dark hydrogen fermentation technique is quite in its early stages, and the research has been limited to the pilot scale so far. The estimated hydrogen yield is 4 mol/mol of glucose by dark fermentation (Lin et al. 2016). Process parameters like pH levels, evaporative fatty acids, and accumulated hydrogen of the bacterial strain are the main causes for the low hydrogen yield (Ren et al. 2016).

As for the production of bioethanol, conversion of sugars into bioethanol is done in the fermentation process. Compared to biohydrogen production, the production of ethanol is very old and well developed. In this, microbes like *Candida shehatae*, *Pachysolen tannophilus*, *Zymomonas mobilis*, *Escherichia coli*, and *Saccharomyces cerevisiae* are used for the conversion of glucose to ethanol. The industrial-scale production of ethanol is done by process such as sugar fermentation, distillation, dehydration, and denaturation (if needed). Pretreatment of wastes and enzymatic hydrolysis are done for cellulosic wastes before bioethanol production in order to improve the glucose concentration level. Ethanol as fuel can be used in automotive fuel and it has the potential to replace petrol in the global market. However, the rate for bioethanol production is high compared to hydrocarbon fuels, and research is ongoing to find low-cost enzymes and a suitable selection of microorganisms which have the ability to produce inhibitory compounds (Pham et al. 2015).

## 7.8   FUTURE SOLID WASTE MANAGEMENT POLICY

Based on the research analysis of Xiao et al. (2020), with the well-developed integrated system dynamic model, three outcomes are suggested as follows: prudent social policy, increased biochemical treatment capacity, and significant usage of residues. Firstly, the integration of the economic and demographic policies to MSW management should be more prudent. According to their results, economic and demographic policies were used for the determination of the MSW production per capita and the overall quantity of MSW generated. It is apparent that sluggish economic policy would greatly lower the overall quantity of MSW generation and their end treatment. Decreasing the gross domestic product to 1% would result in lowering the total amount of MSW generation to 13.33% and the overall volume of MSW landfill demand to 2.44% by 2035. According to their report, MSW generation per capita will increase if a strict population policy is implemented. So before framing policies for solid waste management, economic and demographic policies should be taken in account to have control over the solid waste generation and end treatment demand.

Secondly, to address the rising amounts of household food waste, biochemical treatment centers must be built. If the MSW were segregated effectively, then there will be a sustained increase in household food waste and the required biochemical plants for the treatment of the waste will be insufficient. According to their policy implementation results, the amount of household waste in 2020 will be around 6.33 Mt in Shanghai, which will require half again as many biochemical treatment plants as currently exist. So policymakers are urged to give more attention in finding out the amount of food waste generation and to construct enough biochemical treatment plants for the required volume. This could avoid the wastes being returned to conventional and undesirable treatment methods.

Thirdly, it is important to reduce the residues coming from both biochemical treatment and incineration plants to avoid the increase in landfill dumping. The end products of the treatment plants are ultimately dumped in landfill sites, and the increase in food waste treatment in biochemical treatment plants and incineration of more fluid waste will result in a greater amount of residual wastes; this in turn will increase the landfill pollution and area consumption. In Europe, merely 40–60% of the final ashes from the incineration process are used in road construction or related work. So the policy framework should consider the usage of residuals from waste treatment plants to decrease the waste volume and increase the economy (Xiao et al. 2020).

Bioeconomy is considered as the production of biological renewable resources and the conversion of these sources into products like bioproducts, food, animal fodder, and bioenergy (European Commission 2012). Bioeconomy has played a major role in solving global climatic change, rising demand for food and materials, and restricted natural resources (Eickhout 2012). For sustainable criteria of bioeconomy, bioenergy, and biofuel products, it is important to select the proper biomass for sustainable energy production with minimum greenhouse emissions. A roundtable on sustainable bioeconomy is a Swiss-based organization, they expanded their reach by consider biomaterials and bioenergy as part of its research. There is an immediate need for innovation to produce more effective technology, which can combine better with the development of policies. Furthermore, research is needed in the future for the suggestions of various sustainable criteria for various products and to analyze bioeconomy plans in various states and regions. It is important to study the growth of the circular bioeconomy and its mutual impacts over time, experiencing the synergies and restrictions needed to achieve a circular bioeconomy (Garud and Gehman 2012). It is important to recognize the need for a significant modeling framework for life cycle assessment (LCA), life cycle costing (LCC), and social life assessment (s-LCA) in the creation of future policy choices. While little information on attribution methods was available, LCA was less structured (De Menna et al. 2018). LCC becomes much more important when considering its standard microeconomic aspects, which puts focus on expense rather than on long-term economic gain.

As both economic LCC and societal LCC are conducted with LCA, subsequent strategies that incorporate different steps of the supply chain and cost bearers as well as potential external considerations on certain structures will become more applicable. Further studies should therefore be undertaken to develop a comprehensive theoretical structure for LCA-LCC, which should incorporate guidelines on the extent of the food waste management system, the economic impact, the external factors required, and a pragmatic approach for multifunctionality and product substitution (Lam et al. 2018). The absence of a consistent method also results in a large degree of variability and uncertainty in food waste studies. This incomparability impedes the systematization of case studies and the analysis of the cross-study results. Food waste studies have a variety of attributional scopes. It is proposed that a variety of typologies can be invented and that the differences between situations such as prevention, valorization, and management be centered. Cut-off levels based on cost modeling and alternatives should be emphasized in a comprehensive analytical framework. Policymakers are also recommended to develop proposals on the uniformity of expense typologies, measurement objectives, and performance parameters (Mak et al. 2020).

## 7.9 CONCLUSION

Solid waste management is global issue and it needs to be improved technically. The chapter discussed the variety of wastes generated and techniques needed for their disposal. Proper management planning and execution requires a multi-dimensional approach, and the various steps of waste collection, separation, storage, recycling, and treatment should be made in effective ways that will avoid environmental pollution and risks to human safety. The individual's contribution is very important in waste management, and awareness should be created to reduce waste and littering. Technologies should be improved for the disposal and effective recovery of energy from wastes, and this process should meet the global demand with respect to the increasing population and waste production. Policymakers should take account of various strategies and implications to have sustainable solid waste management. Finally, for unavoidable wastes, various techniques should be developed for their safe disposal in order to reduce their environmental impact.

# REFERENCES

Abramov, Sergey, Jing He, Dominik Wimmer, Marie Louise Lemloh, E. Marie Muehe, Benjamin Gann, Ellen Roehm, et al. 2018. "Heavy Metal Mobility and Valuable Contents of Processed Municipal Solid Waste Incineration Residues from Southwestern Germany." *Waste Management*. doi:10.1016/j.wasman.2018.08.010.

Aminah, Siti, Ab Muttalib, Sharifah Norkhadijah, Syed Ismail, and Sarva Mangala Praveena. 2016. "Application of Effective Microorganism (EM) in Food Waste Composting: A Review." *Asia Pacific Environmental and Occupational Health Journal* 2 (April): 37–47.

Amponsah, S. K., and S. Salhi. 2004. "The Investigation of a Class of Capacitated Arc Routing Problems: The Collection of Garbage in Developing Countries." *Waste Management*. doi:10.1016/j.wasman.2004.01.008.

Arafat, Hassan A., and Kenan Jijakli. 2013. "Modeling and Comparative Assessment of Municipal Solid Waste Gasification for Energy Production." *Waste Management*. doi:10.1016/j.wasman.2013.04.008.

Araki, Atsuko, Takahiko Mitsui, Chihiro Miyashita, Tamie Nakajima, Hisao Naito, Sachiko Ito, Seiko Sasaki, et al. 2014. "Association between Maternal Exposure to Di(2-Ethylhexyl) Phthalate and Reproductive Hormone Levels in Fetal Blood: The Hokkaido Study on Environment and Children's Health." *PLoS One*. doi:10.1371/journal.pone.0109039.

Awasthi, Mukesh Kumar, Junchao Zhao, Parimala Gnana Soundari, Sumit Kumar, Hongyu Chen, Sanjeev Kumar Awasthi, Yumin Duan, Tao Liu, Ashok Pandey, and Zengqiang Zhang. 2019. "Sustainable Management of Solid Waste." In *Sustainable Resource Recovery and Zero Waste Approaches*, 79–99. doi:10.1016/B978-0-444-64200-4.00006-2.

Banu, J. Rajesh, R. Yukesh Kannah, M. Dinesh Kumar, M. Gunasekaran, Periyasamy Sivagurunathan, Jeong Hoon Park, and Gopalakrishnan Kumar. 2018. "Recent Advances on Biogranules Formation in Dark Hydrogen Fermentation System: Mechanism of Formation and Microbial Characteristics." *Bioresource Technology*. doi:10.1016/j.biortech.2018.07.034.

Bartela, Łukasz, Janusz Kotowicz, and Klaudia Dubiel-Jurgaś. 2018. "Investment Risk for Biomass Integrated Gasification Combined Heat and Power Unit with an Internal Combustion Engine and a Stirling Engine." *Energy*. doi:10.1016/j.energy.2018.02.152.

Batool, Syeda Adila, and Muhammad Nawaz Ch. 2009. "Municipal Solid Waste Management in Lahore City District, Pakistan." *Waste Management*. doi:10.1016/j.wasman.2008.12.016.

Beijing. 2004. *The People's Republic of China Initial National Communication on Climate Change Executive Summary*. The United Nations Framework Convention on Climate Change.

Bhattacharya, Satya Sundar, and Ki Hyun Kim. 2016. "Utilization of Coal Ash: Is Vermitechnology a Sustainable Avenue?" *Renewable and Sustainable Energy Reviews* 58. Elsevier: 1376–86. doi:10.1016/j.rser.2015.12.345.

Bhawan, Parivesh, and East Arjun Nagar. 2013. *Overview of Plastic Waste Management*. Central Pollution Control Board, 2. doi:10.1016/j.aquaculture.2006.08.054.

Bi, Haijun, Huabing Zhu, Lei Zu, Shuanghua He, Yong Gao, and Song Gao. 2019. "Pneumatic Separation and Recycling of Anode and Cathode Materials from Spent Lithium Iron Phosphate Batteries." *Waste Management and Research*. doi:10.1177/0734242X18823939.

Bogner, Jean, Riitta Pipatti, Seiji Hashimoto, Cristobal Diaz, Katarina Mareckova, Luis Diaz, Peter Kjeldsen, et al. 2008. "Mitigation of Global Greenhouse Gas Emissions from Waste: Conclusions and Strategies from the Intergovernmental Panel on Climate Change (IPCC) Fourth Assessment Report. Working Group III (Mitigation)." *Waste Management and Research*. doi:10.1177/0734242X07088433.

Castaldi, Marco J. 2014. "Perspectives on Sustainable Waste Management." *Annual Review of Chemical and Biomolecular Engineering* 5 (1): 547–62. doi:10.1146/annurev-chembioeng-060713-040306.

Chagnes, Alexandre, and Beata Pospiech. 2013. "A Brief Review on Hydrometallurgical Technologies for Recycling Spent Lithium-Ion Batteries." *Journal of Chemical Technology and Biotechnology.* doi:10.1002/jctb.4053.

Chandrasekhar, Kuppam, Yong Jik Lee, and Dong Woo Lee. 2015. "Biohydrogen Production: Strategies to Improve Process Efficiency through Microbial Routes." *International Journal of Molecular Sciences.* doi:10.3390/ijms16048266.

Chopra, Ian. 2007. "The Increasing Use of Silver-Based Products as Antimicrobial Agents: A Useful Development or a Cause for Concern?" *Journal of Antimicrobial Chemotherapy.* doi:10.1093/jac/dkm006.

Christensen, Thomas H. 2010. *Solid Waste Technology & Management.* doi:10.1002/9780470666883.

Christensen, Thomas Højlund, Anders Damgaard, and Thomas Fruergaard Astrup. 2015. "Waste to Energy the Carbon Perspective." *Waste Management World*, January–February: 24–28.

Chung, Ying Chien. 2007. "Evaluation of Gas Removal and Bacterial Community Diversity in a Biofilter Developed to Treat Composting Exhaust Gases." *Journal of Hazardous Materials.* doi:10.1016/j.jhazmat.2006.10.045.

Costa, Carlos E., Aloia Romaní, Joana T. Cunha, Björn Johansson, and Lucília Domingues. 2017. "Integrated Approach for Selecting Efficient Saccharomyces Cerevisiae for Industrial Lignocellulosic Fermentations: Importance of Yeast Chassis Linked to Process Conditions." *Bioresource Technology.* doi:10.1016/j.biortech.2016.12.016.

Cyr, Martin, Marie Coutand, and Pierre Clastres. 2007. "Technological and Environmental Behavior of Sewage Sludge Ash (SSA) in Cement-Based Materials." *Cement and Concrete Research.* doi:10.1016/j.cemconres.2007.04.003.

Czajczyńska, D., L. Anguilano, H. Ghazal, R. Krzyżyńska, A. J. Reynolds, N. Spencer, and H. Jouhara. 2017. "Potential of Pyrolysis Processes in the Waste Management Sector." *Thermal Science and Engineering Progress.* doi:10.1016/j.tsep.2017.06.003.

Das, Subhasish, S. H. Lee, Pawan Kumar, Ki Hyun Kim, Sang Soo Lee, and Satya Sundar Bhattacharya. 2019. "Solid Waste Management: Scope and the Challenge of Sustainability." *Journal of Cleaner Production.* doi:10.1016/j.jclepro.2019.04.323.

Dias, Cintia Mara Miranda, Luiz P. Rosa, Jose M. A. Gomez, and Alexandre D'avignon. 2018. "Achieving the Sustainable Development Goal 06 in Brazil: The Universal Access to Sanitation as a Possible Mission." *Anais Da Academia Brasileira de Ciencias.* doi:10.1590/0001-3765201820170590.

Diaz, L. F., L. L. Eggerth, Sh Enkhtsetseg, and G. M. Savage. 2008. "Characteristics of Healthcare Wastes." *Waste Management.* doi:10.1016/j.wasman.2007.04.010.

Donato, Paola Di, Ilaria Finore, Gianluca Anzelmo, Licia Lama, Barbara Nicolaus, and Annarita Poli. 2014. "Biomass and Biopolymer Production Using Vegetable Wastes as Cheap Substrates for Extremophiles." *Chemical Engineering Transactions.* doi:10.3303/CET1438028.

Durmusoglu, Ertan, Fatih Taspinar, and Aykan Karademir. 2010. "Health Risk Assessment of BTEX Emissions in the Landfill Environment." *Journal of Hazardous Materials.* doi:10.1016/j.jhazmat.2009.11.117.

Eickhout, By Bas. 2012. "A Strategy for a Bio-Based Economy." *Green New Deal Series* 9: 1–52.

El-Fadel, Mutasem, Angelos N. Findikakis, and James O. Leckie. 1997. "Environmental Impacts of Solid Waste Landfilling." *Journal of Environmental Management.* doi:10.1006/jema.1995.0131.

Elmekawy, Ahmed, Ludo Diels, Lorenzo Bertin, Heleen De Wever, and Deepak Pant. 2014. "Potential Biovalorization Techniques for Olive Mill Biorefinery Wastewater." *Biofuels, Bioproducts and Biorefining.* doi:10.1002/bbb.1450.

European Commission. 2012. *European Strategy: Innovating for Sustainable Growth: A Bioeconomy for Europe*. Official Journal of the European Union.

Fang, Jing Jing, Na Yang, Dan Yan Cen, Li Ming Shao, and Pin Jing He. 2012. "Odor Compounds from Different Sources of Landfill: Characterization and Source Identification." *Waste Management*. doi:10.1016/j.wasman.2012.02.013.

Fashola, Muibat Omotola, Veronica Mpode Ngole-Jeme, and Olubukola Oluranti Babalola. 2016. "Heavy Metal Pollution from Gold Mines: Environmental Effects and Bacterial Strategies for Resistance." *International Journal of Environmental Research and Public Health*. doi:10.3390/ijerph13111047.

Feng, Shoushuai, Hailin Yang, and Wu Wang. 2015. "Microbial Community Succession Mechanism Coupling with Adaptive Evolution of Adsorption Performance in Chalcopyrite Bioleaching." *Bioresource Technology*. doi:10.1016/j.biortech.2015.04.122.

Fisher, John. 2003. *Elemental Mercury and Inorganic Mercury Compounds: Human Health Aspects*. World Health Organization.

Fries, George F., and Dennis J. Paustenbach. 1990. "Evaluation of Potential Transmission of 2, 3, 7, 8-Tetrachlorodibenzo-p-Dioxin- Contaminated Incinerator Emissions to Humans via Foods." *Journal of Toxicology and Environmental Health*. doi:10.1080/15287399009531369.

Gańczak, Maria, Marta Milona, and Zbigniew Szych. 2006. "Nurses and Occupational Exposures to Bloodborne Viruses in Poland." *Infection Control & Hospital Epidemiology*. doi:10.1086/500333.

Garud, Raghu, and Joel Gehman. 2012. "Metatheoretical Perspectives on Sustainability Journeys: Evolutionary, Relational and Durational." *Research Policy*. doi:10.1016/j. respol.2011.07.009.

Gavilanes-Terán, Irene, Janneth Jara-Samaniego, Julio Idrovo-Novillo, Ma Angeles Bustamante, Raúl Moral, and Concepción Paredes. 2016. "Windrow Composting as Horticultural Waste Management Strategy – A Case Study in Ecuador." *Waste Management*. doi:10.1016/j.wasman.2015.11.026.

Ghose, M. K., A. K. Dikshit, and S. K. Sharma. 2006. "A GIS Based Transportation Model for Solid Waste Disposal – A Case Study on Asansol Municipality." *Waste Management*. doi:10.1016/j.wasman.2005.09.022.

Ghosh, B., M. K. Ghosh, P. Parhi, P. S. Mukherjee, and B. K. Mishra. 2015. "Waste Printed Circuit Boards Recycling: An Extensive Assessment of Current Status." *Journal of Cleaner Production*. doi:10.1016/j.jclepro.2015.02.024.

Gómez, Guadalupe, Montserrat Meneses, Lourdes Ballinas, and Francesc Castells. 2009. "Seasonal Characterization of Municipal Solid Waste (MSW) in the City of Chihuahua, Mexico." *Waste Management*. doi:10.1016/j.wasman.2009.02.006.

Gu, Binxian, Takeshi Fujiwara, Renfu Jia, Ruiyang Duan, and Aijun Gu. 2017. "Methodological Aspects of Modeling Household Solid Waste Generation in Japan: Evidence from Okayama and Otsu Cities." *Waste Management and Research*. doi:10.1 177/0734242X17738338.

Guedes, Paula, Nazaré Couto, Lisbeth M. Ottosen, and Alexandra B. Ribeiro. 2014. "Phosphorus Recovery from Sewage Sludge Ash through an Electrodialytic Process." *Waste Management*. doi:10.1016/j.wasman.2014.02.021.

Guérin, Julie Élize, Maxime Charles Paré, Sylvain Lavoie, and Nancy Bourgeois. 2018. "The Importance of Characterizing Residual Household Waste at the Local Level: A Case Study of Saguenay, Quebec (Canada)." *Waste Management*. doi:10.1016/j. wasman.2018.04.019.

Harhay, Michael O., Scott D. Halpern, Jason S. Harhay, and Piero L. Olliaro. 2009. "Health Care Waste Management: A Neglected and Growing Public Health Problem Worldwide." *Tropical Medicine and International Health*. doi:10.1111/j.1365-3156.2009.02386.x.

Hermelin, Danny, Judith Madeleine Kubitza, Dvir Shabtay, Nimrod Talmon, and Gerhard J. Woeginger. 2019. "Scheduling Two Agents on a Single Machine: A Parameterized Analysis of NP-Hard Problems." *Omega (United Kingdom).* doi:10.1016/j. omega.2018.08.001.

Hiloidhari, Moonmoon, Dhiman Das, and D. C. Baruah. 2014. "Bioenergy Potential from Crop Residue Biomass in India." *Renewable and Sustainable Energy Reviews.* doi:10.1016/j.rser.2014.01.025.

International Energy Agency. 2018. *2018 World Energy Outlook: Executive Summary.* OECD/IEA.

IPCC. 1996. "Methane Emissions from Rice Cultivation: Flooded Rice Fields." *Revised 1996 IPCC Guidelines for National Greenhouse Gas Inventories: Reference Manual.* IPCC.

Johnson, D. Barrie. 2014. "Biomining-Biotechnologies for Extracting and Recovering Metals from Ores and Waste Materials." *Current Opinion in Biotechnology.* doi:10.1016/j. copbio.2014.04.008.

Jones, Peter Tom, Daneel Geysen, Yves Tielemans, Steven Van Passel, Yiannis Pontikes, Bart Blanpain, Mieke Quaghebeur, and Nanne Hoekstra. 2013. "Enhanced Landfill Mining in View of Multiple Resource Recovery: A Critical Review." *Journal of Cleaner Production.* doi:10.1016/j.jclepro.2012.05.021.

Karademir, Aykan, M. Bakoglu, and Savaş Ayberk. 2003. "PCDD/F Removal Efficiencies of Electrostatic Precipitator and Wet Scrubbers in IZAYDAS Hazardous Waste Incinerator." *Fresenius Environmental Bulletin* 12 (10): 12288–332.

Kelly, Patrick T., and Zhen He. 2014. "Nutrients Removal and Recovery in Bioelectrochemical Systems: A Review." *Bioresource Technology.* doi:10.1016/j.biortech.2013.12.046.

Kim, Eun Young, Min Seuk Kim, Jae Chun Lee, and B. D. Pandey. 2011. "Selective Recovery of Gold from Waste Mobile Phone PCBs by Hydrometallurgical Process." *Journal of Hazardous Materials.* doi:10.1016/j.jhazmat.2011.10.034.

Kleemann, Rosanna, Jonathan Chenoweth, Roland Clift, Stephen Morse, Pete Pearce, and Devendra Saroj. 2017. "Comparison of Phosphorus Recovery from Incinerated Sewage Sludge Ash (ISSA) and Pyrolysed Sewage Sludge Char (PSSC)." *Waste Management.* doi:10.1016/j.wasman.2016.10.055.

Koch, Konrad, Brigitte Helmreich, and Jörg E. Drewes. 2015. "Co-Digestion of Food Waste in Municipal Wastewater Treatment Plants: Effect of Different Mixtures on Methane Yield and Hydrolysis Rate Constant." *Applied Energy.* doi:10.1016/j.apenergy.2014.10.025.

Kumar, Atul, and S. R. Samadder. 2017. "A Review on Technological Options of Waste to Energy for Effective Management of Municipal Solid Waste." *Waste Management.* doi:10.1016/j.wasman.2017.08.046.

Lam, Chor Man, Iris K. M. Yu, Francisco Medel, Daniel C. W. Tsang, Shu Chien Hsu, and Chi Sun Poon. 2018. "Life-Cycle Cost-Benefit Analysis on Sustainable Food Waste Management: The Case of Hong Kong International Airport." *Journal of Cleaner Production.* doi:10.1016/j.jclepro.2018.03.160.

Lella, Jaydeep, Venkata Ravibabu Mandla, and Xuan Zhu. 2017. "Solid Waste Collection/ Transport Optimization and Vegetation Land Cover Estimation Using Geographic Information System (GIS): A Case Study of a Proposed Smart-City." *Sustainable Cities and Society* 35 (April). Elsevier: 336–49. doi:10.1016/j.scs.2017.08.023.

Lin, Richen, Jun Cheng, Lingkan Ding, Wenlu Song, Min Liu, Junhu Zhou, and Kefa Cen. 2016. "Enhanced Dark Hydrogen Fermentation by Addition of Ferric Oxide Nanoparticles Using Enterobacter Aerogenes." *Bioresource Technology.* doi:10.1016/j. biortech.2016.02.009.

MacCarthy, Joanna, Jenny Thomas, Sarah Choudrie, Neil Passant, Glen Thistlethwaite, Tim Murrells, John Watterson, et al. 2008. *UK Greenhouse Gas Inventory, 1990 to 2008: Annual Report for Submission under the Framework Convention on Climate Change.*

Mak, Tiffany M. W., Xinni Xiong, Daniel C. W. Tsang, Iris K. M. Yu, and Chi Sun Poon. 2020. "Sustainable Food Waste Management towards Circular Bioeconomy: Policy Review, Limitations and Opportunities." *Bioresource Technology*. doi:10.1016/j. biortech.2019.122497.

Mangialardi, Giovanna, Gianluca Trullo, Francesco Valerio, and Angelo Corallo. 2016. "Sustainability of a Pneumatic Refuse System in the Metropolitan Area: A Case Study in Southern Apulia Region." *Procedia – Social and Behavioral Sciences* 223: 799–804.

Manios, T. 2004. "The Composting Potential of Different Organic Solid Wastes: Experience from the Island of Crete." *Environment International*. doi:10.1016/ S0160-4120(03)00119-3.

Mattenberger, H., G. Fraissler, T. Brunner, P. Herk, L. Hermann, and I. Obernberger. 2008. "Sewage Sludge Ash to Phosphorus Fertiliser: Variables Influencing Heavy Metal Removal during Thermochemical Treatment." *Waste Management*. doi:10.1016/j. wasman.2008.01.005.

Maurya, Rakesh Kumar, Amit R. Patel, Prabir Sarkar, Harpreet Singh, and Himanshu Tyagi. 2018. "Biomass, Its Potential and Applications." In *Biorefining of Biomass to Biofuels*, 25–52. Springer.

Mehrotra, Akanksha, and T. R. Sreekrishnan. 2017. "Heavy Metal Bioleaching and Sludge Stabilization in a Single-Stage Reactor Using Indigenous Acidophilic Heterotrophs." *Environmental Technology (United Kingdom)*. doi:10.1080/09593330.2016.1275821.

Menna, Fabio De, Jana Dietershagen, Marion Loubiere, and Matteo Vittuari. 2018. "Life Cycle Costing of Food Waste: A Review of Methodological Approaches." *Waste Management*. doi:10.1016/j.wasman.2017.12.032.

Michelin, Michele, Héctor A. Ruiz, Maria de Lourdes T. M. Polizeli, and José A. Teixeira. 2018. "Multi-Step Approach to Add Value to Corncob: Production of Biomass-Degrading Enzymes, Lignin and Fermentable Sugars." *Bioresource Technology*. doi:10.1016/j.biortech.2017.09.128.

Mihajlović, Marija, Jelena Petrović, Snežana Maletić, Marijana Kragulj Isakovski, Mirjana Stojanović, Zorica Lopičić, and Snežana Trifunović. 2018. "Hydrothermal Carbonization of Miscanthus × giganteus: Structural and Fuel Properties of Hydrochars and Organic Profile with the Ecotoxicological Assessment of the Liquid Phase." *Energy Conversion and Management*. doi:10.1016/j.enconman.2018.01.003.

Miller, Benjamin, Juliette Spertus, and Camille Kamga. 2014. "Costs and Benefits of Pneumatic Collection in Three Specific New York City Cases." *Waste Management*. doi:10.1016/j.wasman.2014.06.008.

Misra, R. V., R. N. Roy, and H. Hiraoka. 2003. "Composting Process and Techniques." *On-Farm Composting Methods*. doi:10.1017/CBO9781107415324.004.

Mohan, S. Venkata, Sunita Varjani, and Ashok Pandey. 2018. *Biomass, Biofuels, Biochemicals: Microbial Electrochemical Technology: Sustainable Platform for Fuels, Chemicals and Remediation*. doi:10.1016/C2017-0-00856-X.

Mohsenizadeh, Melika, Mustafa Kemal Tural, and Elçin Kentel. 2020. "Municipal Solid Waste Management with Cost Minimization and Emission Control Objectives: A Case Study of Ankara." *Sustainable Cities and Society*. doi:10.1016/j.scs.2019.101807.

MoUD. 2016. *Municipal Solid Waste Management Manual*. Ministry of Urban Development, Government of India.

Mubarok, M. Z., R. Winarko, S. K. Chaerun, I. N. Rizki, and Z. T. Ichlas. 2017. "Improving Gold Recovery from Refractory Gold Ores through Biooxidation Using Iron-Sulfur-Oxidizing/Sulfur-Oxidizing Mixotrophic Bacteria." *Hydrometallurgy* 168. Elsevier B. V.: 69–75. doi:10.1016/j.hydromet.2016.10.018.

Muralikrishna, Iyyanki, and Valli Manickam. 2017. "Solid Waste Management." In *Environmental Management*. doi:10.1016/B978-0-12-811989-1.00016-6.

Narayana, Tapan. 2009. "Municipal Solid Waste Management in India: From Waste Disposal to Recovery of Resources?" *Waste Management.* doi:10.1016/j.wasman.2008.06.038.

Newman, P. A., L. D. Oman, A. R. Douglass, E. L. Fleming, S. M. Frith, M. M. Hurwitz, S. R. Kawa, et al. 2009. "What Would Have Happened to the Ozone Layer If Chlorofluorocarbons (CFCs) Had Not Been Regulated?" *Atmospheric Chemistry and Physics.* doi:10.5194/acp-9-2113-2009.

Nolasco, Dácil, R. Noemí Lima, Pedro A. Hernández, and Nemesio M. Pérez. 2008. "Non-Controlled Biogenic Emissions to the Atmosphere from Lazareto Landfill, Tenerife, Canary Islands." *Environmental Science and Pollution Research.* doi:10.1065/espr2007.02.392.

Ong, Khai Lun, Guneet Kaur, Nattha Pensupa, Kristiadi Uisan, and Carol Sze Ki Lin. 2018. "Trends in Food Waste Valorization for the Production of Chemicals, Materials and Fuels: Case Study South and Southeast Asia." *Bioresource Technology.* doi:10.1016/j.biortech.2017.06.076.

Pan, Shu Yuan, Michael Alex Du, I. Te Huang, I. Hung Liu, E. E. Chang, and Pen Chi Chiang. 2015. "Strategies on Implementation of Waste-to-Energy (WTE) Supply Chain for Circular Economy System: A Review." *Journal of Cleaner Production.* doi:10.1016/j.jclepro.2015.06.124.

Papageorgiou, A., J. R. Barton, and A. Karagiannidis. 2009. "Assessment of the Greenhouse Effect Impact of Technologies Used for Energy Recovery from Municipal Waste: A Case for England." *Journal of Environmental Management.* doi:10.1016/j.jenvman.2009.04.012.

Parrot, Laurent, Joel Sotamenou, and Bernadette Kamgnia Dia. 2009. "Municipal Solid Waste Management in Africa: Strategies and Livelihoods in Yaoundé, Cameroon." *Waste Management.* doi:10.1016/j.wasman.2008.05.005.

Pergola, Maria, Alessandro Persiani, Assunta Maria Palese, Vincenzo Di Meo, Vittoria Pastore, Carmine D'Adamo, and Giuseppe Celano. 2018. "Composting: The Way for a Sustainable Agriculture." *Applied Soil Ecology.* doi:10.1016/j.apsoil.2017.10.016.

Petzet, S., B. Peplinski, S. Y. Bodkhe, and P. Cornel. 2011. "Recovery of Phosphorus and Aluminium from Sewage Sludge Ash by a New Wet Chemical Elution Process (SESAL-Phos-Recovery Process)." *Water Science and Technology.* doi:10.2166/wst.2011.682.

Pham, Thi Phuong Thuy, Rajni Kaushik, Ganesh K. Parshetti, Russell Mahmood, and Rajasekhar Balasubramanian. 2015. "Food Waste-to-Energy Conversion Technologies: Current Status and Future Directions." *Waste Management.* doi:10.1016/j.wasman.2014.12.004.

Pirrone, N., S. Cinnirella, X. Feng, R. B. Finkelman, H. R. Friedli, J. Leaner, R. Mason, et al. 2010. "Global Mercury Emissions to the Atmosphere from Anthropogenic and Natural Sources." *Atmospheric Chemistry and Physics.* doi:10.5194/acp-10-5951-2010.

Rajendran, Karthik, Richen Lin, David M. Wall, and Jerry D. Murphy. 2019. *Influential Aspects in Waste Management Practices. Sustainable Resource Recovery and Zero Waste Approaches.* Elsevier B. V. doi:10.1016/b978-0-444-64200-4.00005-0.

Ramachandriya, D. Karthikeyan, Dimple K. Kundiyana, Ashokkumar M. Sharma, Ajay Kumar, Hasan K. Atiyeh, Raymond L. Huhnke, and Mark R. Wilkins. 2016. "Critical Factors Affecting the Integration of Biomass Gasification and Syngas Fermentation Technology." *AIMS Bioengineering.* doi:10.3934/bioeng.2016.2.188.

Rao, M. N., Razia Sultana, and Sri Harsha Kota. 2017. "Chapter 2 – Municipal Solid Waste BT." In *Solid and Hazardous Waste Management.* doi:https://doi.org/10.1016/B978-0-12-809734-2.00002-X.

Ren, Nan Qi, Lei Zhao, Chuan Chen, Wan Qian Guo, and Guang Li Cao. 2016. "A Review on Bioconversion of Lignocellulosic Biomass to H2: Key Challenges and New Insights." *Bioresource Technology.* doi:10.1016/j.biortech.2016.03.124.

Ritzkowski, M., K. U. Heyer, and R. Stegmann. 2006. "Fundamental Processes and Implications during in Situ Aeration of Old Landfills." *Waste Management*. doi:10.1016/j. wasman.2005.11.009.

Rochester, Johanna R. 2013. "Bisphenol A and Human Health: A Review of the Literature." *Reproductive Toxicology*. doi:10.1016/j.reprotox.2013.08.008.

Rodrigues, Susana, Graça Martinho, and Ana Pires. 2016. "Waste Collection Systems. Part A: A Taxonomy." *Journal of Cleaner Production*. doi:10.1016/j. jclepro.2015.09.143.

Rodriguez-Garcia, G., N. Frison, J. R. Vázquez-Padín, A. Hospido, J. M. Garrido, F. Fatone, D. Bolzonella, M. T. Moreira, and G. Feijoo. 2014. "Life Cycle Assessment of Nutrient Removal Technologies for the Treatment of Anaerobic Digestion Supernatant and Its Integration in a Wastewater Treatment Plant." *Science of the Total Environment*. doi:10.1016/j.scitotenv.2014.05.077.

Rollinson, Andrew N. 2018. "Fire, Explosion and Chemical Toxicity Hazards of Gasification Energy from Waste." *Journal of Loss Prevention in the Process Industries*. doi:10.1016/j. jlp.2018.04.010.

Sahariah, Banashree, Linee Goswami, Imran U. Farooqui, Prashanta Raul, Pradip Bhattacharyya, and Satya Sundar Bhattacharya. 2015. "Solubility, Hydrogeochemical Impact, and Health Assessment of Toxic Metals in Municipal Wastes of Two Differently Populated Cities." *Journal of Geochemical Exploration*. doi:10.1016/j. gexplo.2015.06.003.

Sánchez, Antoni, Adriana Artola, Xavier Font, Teresa Gea, Raquel Barrena, David Gabriel, Miguel Ángel Sánchez-Monedero, Asunción Roig, María Luz Cayuela, and Claudio Mondini. 2015. "Greenhouse Gas Emissions from Organic Waste Composting." *Environmental Chemistry Letters*. doi:10.1007/s10311-015-0507-5.

Saratale, Ganesh D., Moo Young Jung, and Min Kyu Oh. 2016. "Reutilization of Green Liquor Chemicals for Pretreatment of Whole Rice Waste Biomass and Its Application to 2,3-Butanediol Production." *Bioresource Technology*. doi:10.1016/j. biortech.2016.01.028.

Sathyanarayana, Sheela, Samantha Butts, Christina Wang, Emily Barrett, Ruby Nguyen, Stephen M. Schwartz, Wren Haaland, and Shanna H. Swan. 2017. "Early Prenatal Phthalate Exposure, Sex Steroid Hormones, and Birth Outcomes." *Journal of Clinical Endocrinology and Metabolism*. doi:10.1210/jc.2016-3837.

Scarlat, Nicolae, Fernando Fahl, and Jean François Dallemand. 2019. "Status and Opportunities for Energy Recovery from Municipal Solid Waste in Europe." *Waste and Biomass Valorization*. doi:10.1007/s12649-018-0297-7.

Secretariat of the Basel Convention, and WHO. 2004. *Preparation of National Healthcare Waste Management Plans in Sub-Saharan Countries: Guidance Manual*. http://www .who.int/water_sanitation_health/medicalwaste/en/guidancemanual.pdf.

Sharholy, Mufeed, Kafeel Ahmad, R. C. Vaishya, and R. D. Gupta. 2007. "Municipal Solid Waste Characteristics and Management in Allahabad, India." *Waste Management*. doi:10.1016/j.wasman.2006.03.001.

Silva, Angie, Michele Rosano, Laura Stocker, and Leen Gorissen. 2017. "From Waste to Sustainable Materials Management: Three Case Studies of the Transition Journey." *Waste Management*. doi:10.1016/j.wasman.2016.11.038.

Sonesson, Ulf. 2000. "Modelling of Waste Collection – A General Approach to Calculate Fuel Consumption and Time." *Waste Management and Research*. doi:10.1034/j.1399-3070.2000.00099.x.

Sun, Lu, Minoru Fujii, Tomohiro Tasaki, Huijuan Dong, and Satoshi Ohnishi. 2018. "Improving Waste to Energy Rate by Promoting an Integrated Municipal Solid-Waste Management System." *Resources, Conservation and Recycling*. doi:10.1016/j. resconrec.2018.05.005.

Suthar, Surindra, Pravin K. Mutiyar, and Sushma Singh. 2012. "Vermicomposting of Milk Processing Industry Sludge Spiked with Plant Wastes." *Bioresource Technology.* doi:10.1016/j.biortech.2012.03.101.

Tabasová, Andrea, Jiří Kropáč, Vít Kermes, Andreja Nemet, and Petr Stehlík. 2012. "Waste-to-Energy Technologies: Impact on Environment." *Energy* 44 (1): 146–55. doi:10.1016/j.energy.2012.01.014.

Tansel, Berrin. 2017. "From Electronic Consumer Products to E-Wastes: Global Outlook, Waste Quantities, Recycling Challenges." *Environment International.* doi:10.1016/j.envint.2016.10.002.

Tian, Hezhong, Jiajia Gao, Jiming Hao, Long Lu, Chuanyong Zhu, and Peipei Qiu. 2013. "Atmospheric Pollution Problems and Control Proposals Associated with Solid Waste Management in China: A Review." *Journal of Hazardous Materials.* doi:10.1016/j.jhazmat.2013.02.013.

Tian, Hezhong, Jiajia Gao, Long Lu, Dan Zhao, Ke Cheng, and Peipei Qiu. 2012. "Temporal Trends and Spatial Variation Characteristics of Hazardous Air Pollutant Emission Inventory from Municipal Solid Waste Incineration in China." *Environmental Science and Technology.* doi:10.1021/es302343s.

Torno, Susana, Javier Toraño, Mario Menendez, Malcolm Gent, and Cristina Allende. 2011. "Prediction of Particulate Air Pollution from a Landfill Site Using CFD and LIDAR Techniques." *Environmental Fluid Mechanics.* doi:10.1007/s10652-010-9187-7.

Torrijos, Michel. 2016. "State of Development of Biogas Production in Europe." *Procedia Environmental Sciences.* doi:10.1016/j.proenv.2016.07.043.

Tripathi, Manoj, J. N. Sahu, and P. Gancsan. 2016. "Effect of Process Parameters on Production of Biochar from Biomass Waste through Pyrolysis: A Review." *Renewable and Sustainable Energy Reviews.* doi:10.1016/j.rser.2015.10.122.

Trivedi, Mala, Manish Mathur, Parul Johri, Aditi Singh, and Rajesh K. Tiwari. 2020. "Waste Management: A Paradigm Shift." In *Environmental Concerns and Sustainable Development,* 337–63. Singapore: Springer Singapore. doi:10.1007/978-981-13-6358-0_14.

US Environmental Protection Agency. 2012. "Trends in Greenhouse Gas Emissions." *US Greenhouse Inventory Report.*

Venkata Mohan, S., P. Chiranjeevi, Shikha Dahiya, and A. Naresh Kumar. 2018. "Waste Derived Bioeconomy in India: A Perspective." *New Biotechnology.* doi:10.1016/j.nbt.2017.06.006.

Verma, Nemit, Manpreet Kaur, and A. K. Tripathi. 2019. "Greenhouse Gas Emissions from Municipal Solid Waste Management Practice." In *Environmental Concerns and Sustainable Development: Volume 2: Biodiversity, Soil and Waste Management,* 399–408. doi:10.1007/978-981-13-6358-0_17.

Verma, Rinku, K. S. Vinoda, M. Papireddy, and A. N. S. Gowda. 2016. "Toxic Pollutants from Plastic Waste – A Review." *Procedia Environmental Sciences.* doi:10.1016/j.proenv.2016.07.069.

Vieira, Victor H. Argentino de Morais, and Dácio R. Matheus. 2018. "The Impact of Socioeconomic Factors on Municipal Solid Waste Generation in São Paulo, Brazil." *Waste Management and Research.* doi:10.1177/0734242X17744039.

Watkins, Deborah J., Brisa N. Sánchez, Martha Maria Téllez-Rojo, Joyce M. Lee, Adriana Mercado-García, Clara Blank-Goldenberg, Karen E. Peterson, and John D. Meeker. 2017. "Phthalate and Bisphenol A Exposure during in Utero Windows of Susceptibility in Relation to Reproductive Hormones and Pubertal Development in Girls." *Environmental Research.* doi:10.1016/j.envres.2017.07.051.

Wu, Zezhou, Ann T. W. Yu, Liyin Shen, and Guiwen Liu. 2014. "Quantifying Construction and Demolition Waste: An Analytical Review." *Waste Management.* doi:10.1016/j.wasman.2014.05.010.

Xiao, Shijiang, Huijuan Dong, Yong Geng, Xu Tian, Chang Liu, and Haifeng Li. 2020. "Policy Impacts on Municipal Solid Waste Management in Shanghai: A System Dynamics Model Analysis." *Journal of Cleaner Production* 262. Elsevier Ltd: 121366. doi:10.1016/j.jclepro.2020.121366.

Xin, Baoping, Wenfeng Jiang, Hina Aslam, Kai Zhang, Changhao Liu, Renqing Wang, and Yutao Wang. 2012. "Bioleaching of Zinc and Manganese from Spent Zn–Mn Batteries and Mechanism Exploration." *Bioresource Technology*. doi:10.1016/j.biortech.2011.12.013.

Xu, Fuqing, Yangyang Li, Xumeng Ge, Liangcheng Yang, and Yebo Li. 2018. "Anaerobic Digestion of Food Waste – Challenges and Opportunities." *Bioresource Technology*. doi:10.1016/j.biortech.2017.09.020.

Yadav, V., S. Karmakar, A. K. Dikshit, and S. Vanjari. 2016. "Transfer Stations Siting in India: A Feasibility Demonstration." *Waste Management* 47 (2): 1–4.

Yadav, Vinay, A. K. Bhurjee, Subhankar Karmakar, and A. K. Dikshit. 2017. "A Facility Location Model for Municipal Solid Waste Management System under Uncertain Environment." *Science of the Total Environment*. doi:10.1016/j.scitotenv.2017.02.207.

Yadav, Vinay, and Subhankar Karmakar. 2020. "Sustainable Collection and Transportation of Municipal Solid Waste in Urban Centers." *Sustainable Cities and Society* 53. Elsevier B. V.: 101937. doi:10.1016/j.scs.2019.101937.

Yadav, Vinay, Subhankar Karmakar, A. K. Dikshit, and Shivkumar Vanjari. 2016. "A Feasibility Study for the Locations of Waste Transfer Stations in Urban Centers: A Case Study on the City of Nashik, India." *Journal of Cleaner Production*. doi:10.1016/j.jclepro.2016.03.017.

Yu, Iris K. M., and Daniel C. W. Tsang. 2017. "Conversion of Biomass to Hydroxymethylfurfural: A Review of Catalytic Systems and Underlying Mechanisms." *Bioresource Technology*. doi:10.1016/j.biortech.2017.04.026.

Yuan, Li-Jun, and Yu-Chun Li. 2012. "Comparison of Technologies for Controlling Dioxin in Flue Gas from Sintering Process and Waste Incineration." *China Metallurgy* 2: 2012.

Zbib, Hani, and Sanne Wøhlk. 2019. "A Comparison of the Transport Requirements of Different Curbside Waste Collection Systems in Denmark." *Waste Management*. doi:10.1016/j.wasman.2019.01.037.

Zhang, Dr. Baiyu, Dr. Leonard Lye, Khoshrooz Kazemi, and Weiyun Lin. 2013. *Development of Advanced Composting Technologies for Municipal Organic Waste Treatment in Small Communities in Newfoundland and Labrador*. www.mun.ca/harriscentre/reports/arf/2011/11-12-WMARF-Final-Zhang.pdf.

Zhang, Gang, Jing Hai, and Jiang Cheng. 2012. "Characterization and Mass Balance of Dioxin from a Large-Scale Municipal Solid Waste Incinerator in China." *Waste Management*. doi:10.1016/j.wasman.2012.01.024.

Zhen, Guangyin, Xueqin Lu, Hiroyuki Kato, Youcai Zhao, and Yu You Li. 2017. "Overview of Pretreatment Strategies for Enhancing Sewage Sludge Disintegration and Subsequent Anaerobic Digestion: Current Advances, Full-Scale Application and Future Perspectives." *Renewable and Sustainable Energy Reviews*. doi:10.1016/j.rser.2016.11.187.

# 8 Assessment of Physicochemical Properties and Analytical Characterization of Lignocellulosic Biomass

## 8.1 INTRODUCTION

The rising global population, industrial expansion, and rapid increases in economic growth and urbanization have led to a gradual increase in the worldwide energy demand, which has resulted in the gradual depletion of fossil fuels (Patel, Zhang, and Kumar 2016). A most promising and alternative source for bioenergy production is lignocellulosic energy, which is one of the fastest-growing renewable energy resources (Saba et al. 2015). The main composition of the lignocellulosic biomass, which is also referred to as plant dry matter, consists of cellulose, hemicellulose, and lignin (Carrier et al. 2011). The agriculture, forest, and industrial sectors are the main contributors to the availability of lignocellulosic biomass feedstock used for energy purposes (Mabee, McFarlane, and Saddler 2011). The most promising feedstocks are from agricultural waste and forest residues, as they are abundantly available and are relatively low in cost. Adverse impacts on the environment come from land degradation and desertification due to traditional practice of burning lignocellulosic biomass for the purposes of cooking and heating (Sansaniwal, Rosen, and Tyagi 2017).

The processes included in the conversion of the entire biomass into the required biofuel include the usage of lignocellulosic biomass with process such as logistics, pre-treatment, and its transformation. Collection, handling, storage, and transportation are included in this logistics process (Chundawat et al. 2010). Drying along with grinding and sieving the biomass are involved in the pre-treatment process. Feeding, conversion, intermediate process separation, and advancement along with the collection of the products are included in this conversion process. The essential reference data for the design along with the application of this process are the physicochemical characteristics of the lignocellulosic biomass (Dutta et al. 2012).

The availability of the lignocellulosic biomass feedstock for energy purposes is mainly from agriculture, forest, industries, and other sectors in large scale. The agriculture sector mostly provides lignocellulosic energy products such as crop residues, oil, sugar, and starch crops such as herbaceous crops and crop straw, for example, switchgrass, miscanthus, reed, corn stalk, wheat straw, rice straw, cotton stalk, rapeseed, sugarcane, and corn (Limayem and Ricke 2012). The forestry sector provides particular forest by-products such as plantations with short rotations, for example, eucalyptus, poplar, willow, branches of timber blocks, barks, wood chips collected from trees, and firewood. The industrial sector provides residues from agro-industrial processes, including lignocellulosic substances such as rice husk, sugarcane bagasse, and corncob (Goyal, Seal, and Saxena 2008). Other sectors provide wood industry residues and lignocellulosic waste such as waste wood from industrial processes; sawmills generated sawdust and residues of lignocellulosic material which are found in country parks and botanical gardens, such as prunings and grass (Perlack 2005).

To eradicate moisture and contamination, increasing ratios of H/C and other energy contents are used, and this helps to improve overall fuel properties of the biomass. The effective utilization of various physical techniques such as thermal treatment at a warm temperature, i.e. torrefaction, and chemical process with acidic solutions or alkali metals and grinding are undertaken (Kan, Strezov, and Evans 2016). Many studies concentrate on the properties of the biomass. Xu et al. (2013) talked about the progress and prospects for the application and also had a look over the studies of the composition of biomass and their structure using the infrared methods. Lin et al. (2015) stated that the liquid products from the pyrolysis of biomass mainly depends on the bond in the middle of the composition of the biomass. The simple configuration of three types of forest biomass, namely spruce, pine, and birch, was studied. A review of production and composition of miscanthus biomass for usage in bioenergy was given by Arnoult and Brancourt-Hulmel (2015). A summary of the hemicellulose, cellulose, and lignin substances of many biomasses such as softwood, hardwood, and residues from agricultural processes along with grasses was given by Isikgor and Becer (2015). The basic composition of biomass, which included H, O, C, K, Si, Al, N, Ca, S, Mg, C, P, Fe, and Mn, their comparative benefits, and causes of biomass and coal composition were examined, along with the disadvantages that counterbalance the advantages of biomass for biofuel (Vassilev, Vassileva, and Vassilev 2015). The technological and other barriers were compensated for by the economic, environmental, and social benefits. The physicochemical characteristics of the analytical characterization methods, lignocellulosic biomass, and the current developments in biomass conversion were the main focus of the review by Cai et al. (2017).

## 8.2  LIGNOCELLULOSIC BIOMASS FEEDSTOCKS AVAILABLE FOR ENERGY PURPOSES

Any renewable organic matter that is generated by plants through a photosynthesis process is called biomass. The presence of combustible organic materials like carbon, hydrogen oxygen, and a negligible amount of sulfur and nitrogen constitutes biomass (Pu et al. 2016). Biomass contributes 10% of the world's energy supply. India is the world's 7th largest country with 328 million hectares and is widely endowed with renewable energy sources. India generates approximately 450 million metric tons of biomass per year, of which about 200 million tons are surplus and provide 32% of all primary energy consumption in the nation (Demirbas, Balat, and Balat 2009).

## 8.2.1 AGRICULTURE

Agricultural residues are obtained from the lignocellulosic energy crops, crop residues, and sugar and starch energy crops. Some examples are herbaceous crops including switchgrass, miscanthus, reeds, rice straw, corn stalks, wheat straw, rapeseed, and sugarcane. These are dried in the atmosphere for thermochemical conversion (Sims et al. 2006).

## 8.2.2 FOREST

Dedicated forestry and forestry bioproducts constitute forest residues. Such bioproducts include short-rotation plants like willow poplar and eucalyptus and wood chips from branches obtained during thinning (Sims et al. 2006).

## 8.2.3 INDUSTRY

Lignocellulosic agro residues can also be obtained from industries and industrial wood materials. Some examples for industrial waste include sawdust from sawmills, sugarcane bagasse from the sugar industry, industrial waste wood, corn cob, and rice husk (Sims et al. 2006).

## 8.3 CHOICE OF PRE-TREATMENT BASED ON BIOMASS TYPES

Cellulose, hemicellulose, and lignin are the three main compositions of the lignocellulosic biomass. Rigid lignin and hemicellulose coverings surrounded by cellulose microfibrils restrict the hydrolysis susceptibility of cellulose, which is a major component (Chen and Chen 2014). Hence, to liberate the cellulose from the lignin hemicellulose seal and to reduce the cellulose crystallinity, it is necessary to provide an effective pre-treatment; such pre-treatments include acid hydrolysis, steam explosion (SE), ammonia fiber expansion (AFEX), wet alkaline oxidation, and hot water pre-treatment (Haghighi Mood et al. 2013). Pre-treatment of biomass minimize the degradation products formation which inhibits hydrolysis and hence fermentation should be done by an ideal pre-treatment along with the reduction of lignocellulosic recalcitrance. Worldwide, intensive research into pre-treatment methods is ongoing. Still, not all are developed enough to be used in large-scale process applications (Taherzadeh and Karimi 2008).

## 8.3.1 ACID/ALKALI TREATMENT

Due to multiple advantages of alkali and acidic pre-treatment, it has been a promising process in biomass solubilization; its advantages include simplicity of the equipment, ease of operation, high methane conversion efficiency, and low cost. Acids such as HCl, $H_2SO_4$, $H_3PO_4$, and $HNO_3$ are used to perform acidic hydrolysis, whereas alkaline solutions such as NaOH, KOH, $Ca(OH)_2$, $Mg(OH)_2$, CaO, and ammonia are used in the alkali pre-treatment (Den et al. 2018). Thus, the process can be carried out under ambient temperature as the adding of acid or base removes the need for higher temperature. The change in the substrate kinds and properties (which are studied due to their distinct affinity for organic compounds) can affect the acidic or alkali pre-treatment. Acidic and alkali pre-treatment are mainly used for lignocellulosic biomass and solubilization of lignin and hemicellulose, respectively. The main catalysts used in alkali pre-treatment are sodium hydroxide and ammonia, and it is observed that hardwood and agricultural residues show great resistance to alkali pre-treatment whereas softwood does get highly affected due to high lignin content (Singh, Suhag, and Dhaka 2015).

### 8.3.2 Ammonia Fiber Expansion

When compared to other pre-treatment processes, the most promising pre-treatment process for the lignocellulose biomass is AFEX (Eggeman and Elander 2005). During pre-treatment of biomass, no liquid stream will be formed since it is a dry-to-dry process, making it low cost compared to other pre-treatment processes, which include dilute-acid and steam explosion. Different kinds of cellulosic biomasses such as corn stover (Laureano-Perez et al. 2009), switchgrass (Alizadeh et al. 2005), miscanthus, and rice straw are highly converted by AFEX pre-treatment (Wolosker et al. 2008). Several experiments were performed in parallel with 22-mL stainless steel reactors according to the method previously mentioned by Balan et al. (2009) for quickly optimizing the AFEX pre-treatment conditions. To remove the residual ammonia, the AFEX–pre-treated biomass was placed under a fume hood throughout the night, and then it was stored at 4°C until use.

### 8.3.3 Steam Explosion

The process used for the production of hardboard, also known as steam explosion, was developed in 1925, when the inventor patented an apparatus which consists of a minimum of three stream guns operated in a continuous manner; as one is discharged, the remaining two are heated (Datar et al. 2007). The waste stream can also be utilized for preheating the woodchips with inlet and outlet valves which are specially designed to allow a completely sealed apparatus and to promote a fibrate action. This technology was patented in the early 1980s for the production of sugars by pre-treatment of aspen woodchips. Similar patents have been made on steam cooking pre-treatment by some inventors who concentrated on the improvement of the ruminant digestibility of lignocellulosic and non-lignocellulosic biomass (Duque et al. 2016). In order to produce sugar, the inventor's patent looked forward to increasing the cellulose accessibility of the residues of hardwoods and agriculture to microorganisms and enzymes. He proposed three improvements in steam cooking, including a formula to determine the minimum time in pre-treatment for gaining maximum cellulose digestion, along with venting of volatiles before the explosion and addition of catalyst in order to maximize the hydrolysis of glucose and xylose by two-step pre-treatment; fractionating the lignocellulosic biomass and removal of lignin were focused on by other steam explosion processes (Ropars et al. 1992).

### 8.3.3.1 Variables Affecting Steam Explosion Pre-Treatment

Temperature, moisture, particle size, and residual time are the main factors influencing SE pre-treatment. The efficiency of pre-treatment is highly influenced by the composition of biomass, which has already been proven. Softwoods have high recalcitrance to SE, as it has lower content in the acetyl group of their hemicelluloses, which is used to catalyze the biomass autohydrolysis process (Martín-Davison et al. 2015).

### 8.3.3.2 Moisture and Particle Size

The effectiveness of heat transfer and steam consumption has a strong relationship with the moisture and particle size of biomass feedstock. Small biomass particles are preferred. Bulky chips may lead to surface overcooking, leading to degraded product formation, and incomplete hydrolysis takes place, whereas small biomass particles have faster heat transfer in the reactor (Brownell, Yu, and Saddler 1986). The essential components of SE operation are temperature and residence time. Increasing the temperature leads to higher recovery of xylose and cellulose, whereas there is no influence of temperature on the recovery of total free glucose after SE of poplar at 200 and 220°C (Pereira Ramos 2003).

## 8.4 PHYSICOCHEMICAL PROPERTIES OF LIGNOCELLULOSIC BIOMASS FOR ENGINEERING APPLICATIONS

The most highly available renewable resource in the globe for biofuel generation is lignocellulosic biomass. Dependence on biofuel extracted from lignocellulosic biomass can lead to less reliance on fossil fuels and can help significantly in climate change mitigation. The analytical characterization methods and physicochemical properties will give suitable biomass conversion techniques. The main objective of this chapter is to present a complete study of the physical and chemical properties of biomass, which include grind ability, flowability, particle size, density, thermal properties, moisture sorption, calorific value, and elemental composition. It also focuses on offering systematic information to users by describing techniques and physicochemical properties of biomass for converting it and for the biofuel's application (Fatma et al. 2018)

### 8.4.1 Density

The density of the biomass or any substance is defined as mass per unit volume. The density of biomass is classified as particle density, and bulk concentration is mentioned below.

### 8.4.1.1 Particle Density

Excluding the pore space volume for a group of biomass particles, the particle density is defined as the ratio of the mass of all particles to the volume occupied by the particles.

$$\rho p = mp * Vp \qquad (8.1)$$

where ρp is the density of the particle, mp is the mass of a particle, and Vp is the true volume of the particle. The pressure difference can be used to determine the true volume with a known pressurized gas quantity flowing into a reference volume cell contains biomass matter (Xue and Fox 2015).

$$Vp = Vc - VR\left(P_1 P_2 - 1\right) \qquad (8.2)$$

Where $V_c$ is the sample cell volume, $V_R$ is the reference volume, $P_1$ is the pressure after pressurizing the reference volume, and $P_2$ is the pressure after including $V_c$. The biomass particle density will be an important factor considered in a computational fluid dynamics simulation (Himmel et al. 2007).

### 8.4.1.2  Bulk Density

The biomass particle mass upon the total volume of biomass particle is known as bulk density, which includes the volume of pore space in and around the biomass matter. This is a physicochemical property which is used in planning the biomass system of logistics, transport, and handling, which depends on the size and shape of biomass matter, the moisture content, the density of particles, and the characteristics of the surface. According to the ASTM standard E873-82, the bulk density measurement of biomass is performed (ASTM E873-82 2006). A regular box (300 mm width × 300 mm depth × 600 mm height) is used to receive the poured biomass sample, and the remaining materials are removed by scraping across the top of the container with a straight edge. The bulk density can be found from the net mass of the sample (Rabier et al. 2006).

Extreme variations in bulk densities such as from 15–201 kg m$^{-3}$ for cereal grain straws to 280–480 kg m$^{-3}$ for hardwood biomass chips are shown by different biomass samples, which are low compared to of coal's bulk density, of about 901 kg m$^{-3}$ (Jose and Bhaskar 2015). The bulk density is proportional to the fill tightness degree; the loose fill had a bulk density of 50 to 265 kg m$^{-3}$ and packed fill had 69 to 326 kg m$^{-3}$ for switchgrass (after tapping). Porosity, defined as the bulk samples' pore spaces, is shown by the following formula:

$$\varepsilon 0 = 1 - \rho b \rho p \qquad (8.3)$$

where $\varepsilon_0$ is the bulk biomass sponginess, $\rho_b$ is the bulk density, and $\rho_p$ is the density of the particle (Lam et al. 2007).

### 8.4.2  Flowability

The property of flowability plays a major role in handling, storage, and transportation of biomass. The feeding subsystem of a biomass conversion system is usually bridging the biomass particles (Ueki et al. 2011). Flowability is a means to measure the flow of biomass from one place to another. The parameters to describe biomass flowability are the cohesion coefficient, the angle of repose, the flow index, and the compressibility index. The steepest angle where a large quantity of biomass particles remains constant without slumping, known as angle of repose, ranges from 0° to 90° (Rouse 2014). The methods used for measuring the angle of repose are fixed funnel, titling box, and revolving cylinder. This method shown here is the method of static funnel represented in the ASTM standard C144 (Rosillo-Calle 2012). A cone is formed from a funnel, which is used to pour biomass matter slowly into it. The poring should be stopped when the biomass matter has accumulated and reaches a defined elevation or base of determined width. The formula for finding the angle of repose is as follows:

$$\phi = \tan{-1}\left(HR\right) \tag{8.4}$$

where the angle of repose is $\phi$, and $H$ and $R$ are the piling cone's height and radius, respectively (Mani 2015).

The various types of flowability of biomass are low, medium, cohesive, high, and very cohesive flowing. Mani has clearly shown that the flowability results in different angles of repose. When the angle of repose is less than 55°, the flow index is FI<1 and the flowability grade is very cohesive. When it is between 45° and 55°, then FI lies between 1 and 2 and the flowability grade is cohesive. When it is between 38° and 45°, then FI lies between 2 and 4 and the flowability grade is low flowing. When it is between 30° and 38°, then FI lies between 4 and 10 and the flowability grade is medium flowing. When it is less than 30°, then FI is higher than 10 and the flowability grade is high flowing (Szalay, Kelemen, and Pintye-Hódi 2015). The bulk density of biomass increases under consolidated pressure and the compressibility index of biomass is determined. At any given consolidation pressure, this can be determined by the following equation:

$$Cb = 1 - \rho bi \rho bf \tag{8.5}$$

where $C_b$, $\rho_{bi}$, and $\rho_{bf}$ are the compressibility, initial bulk density before consolidation, and final bulk density, respectively (Jenike 1964). The flow behavior of biomass material can be quantified using a shear tester. This method is done as per the ASTM standard D6128-16. The unconstrained yield stress ($\sigma_c$) and the main associating stress ($\sigma_1$) are noted. The slope of the linear fit of $\sigma_c$ versus $\sigma_1$ plot is used to get the flow function for biomass particles. Flow index is the transpose of the flow function (Conshohocken 2001).

## 8.4.3 PARTICLE SIZE

The fluidization and mixing, the surface area for mass and heat transfer, and the flow behavior of biomass particles are all affected by the size and shape of biomass feedstock particles (Vidal et al. 2011). Hence, the efficiency of various conversions and the energy input requirements can be seen for feedstock particles of varying shapes and sizes (van Loo and Koppejan 2012). Several difficulties arise due to irregularity in the shape of lignocellulosic biomass such as elemental quantities of width, length, and thickness. The handling, storage, and processing facilities require accurate knowledge of biomass particle size and shape. The most important parameter for the particles is the aspect ratio (AR), which affects the transfer of heat in the feedstock thermal conversion (Gera et al. 2002). It is shown as AR = $bl$ where AR, $b$, and $l$ are aspect ratio, width (i.e., the smallest space between of 2 parallel lines at a tangent to the outline projected by the particle), and length (i.e., the biggest space between two parallel lines at right angles to the width tangents), respectively. The two main particle size description methods are analysis of sieving and imaging particle analysis (Shastri et al. 2014).

### 8.4.4 MOISTURE SORPTION

Another important factor for biomass harvest, handling, transport, and storage is the moisture sorption. In order to preserve the biomass feedstock quality, suitable ventilation and storing operations are necessary (Lin et al. 2016). The characterization of the water sorption behavior of biomass is done with equilibrium moisture content (EMC), which is described as the moisture content of a substance in equilibrium in a specified atmosphere with regard to temperature and comparative humidity. It depends on the specific surface area and the composition microstructure of the biomass (Kaur 2009). The equilibrium moisture content is determined using various static and dynamic methods. From the saturation of different concentrations of salt solution, several different relative humidities are obtained with respect to saturated and dried-out air (Mann et al. 2015).

### 8.4.5 THERMAL PROPERTIES

The thermochemical conversion characteristics of biomass are greatly influenced by the thermal properties of biomass. The important properties of thermal are specific heat and thermal conductivity, which are explained here.

#### 8.4.5.1 Thermal Conductivity

The heat conduction in and around the fiber is changed with respect to the biomass material when it is heated during a thermochemical conversion process, which affects their behavior. Masonet al. used a custom-built test apparatus which is a part of a method that gives their thermal conductivities (Mason et al. 2016). It was observed that wood materials have high thermal conductivity in the range of 0.080 to 0.419 W m$^{-1}$ K$^{-1}$ and residues from agriculture have a very low thermal conductivity of around 0.05 W m$^{-1}$ K$^{-1}$. Being an anisotropic material, the thermal conductivity of biomass is dependent on the porosity, density, moisture, heating direction, and temperature (Dahlquist 2013). The correlations interrelating the thermal conductivities, density and moisture content were given as follows:

$$Keff = \left\{ sg\left(0.2 + 0.004Md\right) + 0.0238Md > 40\% \right.$$

$$\left. sg\left(0.2 + 0.0055Md\right) + 0.0238Md < 40\% \right\}$$

$$(8.6)$$

where $K_{eff}$, $M_d$, and $s_g$ are the thermal conductivity (W m$^{-1}$ K$^{-1}$), specific gravity, and moisture percentage of biomass, respectively (Pandey 2008).

### 8.4.5.2 Specific Heat

One more important thermal property of biomass is the specific heat, which is used to indicate the capacity of the heat of the material, which in turn is required for calculations regarding thermodynamics of the material depending on the moisture content in the biomass and its temperature. Ranging from 0°C to 106°C, the expression for the specific heat of dried wood is given by the following:

$$Cp\theta = 0.266 + 0.001116\theta \qquad (8.7)$$

where $\theta$ and $Cp\theta$ are temperature and specific heat at the temperature $\theta$ (in°C and J g$^{-1}$ K$^{-1}$, respectively). The relationship between the specific heat of residual samples from dried agricultural products withparticle sizes of less than 200 μm and temperature is given below:

$$CpT = 5.340T - 299 \qquad (8.8)$$

where $T$ and $CpT$ are the temperature (314 K $\leq T \leq$ 355 K) and specific heat at the temperature $T$ (in K and J kg$^{-1}$ K$^{-1}$, respectively). The following expression shows the impact on specific heat by moisture content.

$$Cp = MARB * Cw + (1 - Mwet)Cp\theta \qquad (8.9)$$

where MARB and Cw are the moisture content as received and water-specific heat, respectively.

## 8.5   CHEMICAL PROPERTIES

The chemical properties of biomass fuels can be classified as proximate analysis, energy content, thermogravimetric analysis, ultimate analysis, and composition analysis.

### 8.5.1   PROXIMATE ANALYSIS

Proximate analysis can predict the amount of moisture, ash, volatile matter, and fixed carbon of the lignocellulosic biomass sample. Moisture content characterizes the number of water molecules present in the biomass sample, expressed in weight percentage. Biomass has a major influence not only on producing a harvest but also on the transportation, storage, drying, and resultant product (Karunanithy, Muthukumarappan, and Donepudi 2013). Moisture sorption is the external moisture present in the biomass (outside the cell walls) which is above the equilibrium moisture content. But the cell walls absorb inherent moisture. In the thermochemical conversion process, the evaporation of moisture plays a major role in the efficiency of the system (Basu 2013). The ash content is the solid residue left after complete combustion of a particular biomass sample. The primary elements of biomass ash are oxides of silica, magnesium, aluminum oxide, titanium, sodium, and potassium.

The chemical analysis report of the biomass sample is helpful to predict the char deposited in the thermochemical conversion reactors and boilers (Vassilev et al. 2013). The volatile matter is expressed as the condensable vapor and permanent gases released from biomass when it is combusted at the inert atmosphere. In the pyrolysis process, increased bio-oil production with a high-volatility biomass sample was reported. The yield of bio-oil depends upon the heating condition, retention time, and heating rate (Chouhan and Sarma 2013). For determination of the volatile matter, the biomass sample is heated in a tubular furnace at 925°C for 7 minutes at inert atmosphere. The weight loss gives the percentage of volatile matter. Fixed carbon is the solid combustible residue that remains after the volatile matter is expelled from the biomass sample.

$$\text{Fixed carbon} = 100 - \text{moisture} + \text{ash} + \text{volatile matter}.$$

As per the ASTM standard E1756-08, the biomass moisture content is estimated (ASTM E1756 2002). According to the standard, a measured sample of biomass is kept at a temperature of 105°C in a tubular furnace or muffle furnace from 3 hours to 72 hours and weighed afterwards cooling. To ensure complete evaporation of water content from the biomass, the process is carried out until the weight is constant.

The ash quantity can be determined according to ASTM standard E1755-01 in the furnace at a temperature of 575°C for 3 h in a normal atmosphere (ASTM 2015). The biomass sample is cooled before being weighed. After that, the biomass sample once again heated for 1 h at 575°C until the biomass sample changes by an amount of less than 0.3 mg from the earlier weighed sample. According to ASTM standard E872-82 for standard volatile matter content of the biomass, a sample is measured. The biomass sample is placed in a crucible, which is then placed in a furnace at 950°C for 7 min at inert atmosphere so that the biomass will not get fully combusted. Then the crucible is chilled and placed in a desiccator, and then the sample is weighed.

### 8.5.2 Ultimate Analysis

The ultimate analysis is an elemental composition analysis used to determine the carbon, nitrogen, hydrogen, sulfur content and oxygen content of the biomass sample. The first four elements are measured directly in the analyzer, while the oxygen content is usually determined by difference (Abdullah, Sulaiman, and Taib 2013). The ultimate analysis is performed in the elemental analyzer, with biomass samples combusted in the controlled environment and the succeeding release of gaseous yields measured. The results are expressed in the dry basis or ash dry basis (Golden, Reed, and Das 1988). In the combustion process, C, H, N, and S are converted into $CO_2$, $H_2O$, $NO_x$, and $SO_2$, respectively. The combustion products are cleared out of the process through a high concentration of copper, which will convert $O_2$ and $NO_x$ to nitrogen. Therefore, the ultimate analysis is determined by the following process.

The carbon content of the biomass sample will be around 45%, and that of coal will be around 60%. The oxygen content of the biomass sample is higher than that in coal (De Jong and Van Ommen 2014). Moreover, the hydrogen content will be around 6%. Sulfur and nitrogen are neglected for calculating the CHO index due to negligible in amount present in the biomass.the CHO index to predict the oxidation state of organic carbons in organic material as expressed below:

$$CHO\ index = 2[O] - [H] \div [C] \tag{8.10}$$

where [O], [H], and [C] are the mole fractions of oxygen, hydrogen, and carbon, respectively. A higher CHO index expresses more oxidized compound, while a lower CHO index expresses less oxidized compound (Vassilev, Vassileva, and Vassilev 2015).

### 8.5.3 Energy Content

The energy content of the biomass sample is the quantity of energy deposited in the biomass and is usually expressed as the heat of combustion, where complete combustion takes place with oxygen under standard conditions ("Biomass as Energy Source: Resources, Systems and Applications" 2013).

Heating or calorific value is usually measured as the energy content of particular biomass. There are two values of heat: the first is higher heating value (HHV) and the second is lower heating value (LHV). The higher heating value is the amount of heat existing in biomass, including latent heat of vaporization. The lower heating value excludes the latent heat of vaporization (Sheng and Azevedo 2005). According to ASTM standard D5865-13 the higher calorific value of the biomass sample is determined. Inside a bomb calorimeter, a small biomass sample is placed in an oxygen environment, and the heat released from the combustion is measured. The produced steam from hydrogen contained in the biomass sample is also measured when the product of combustion is condensed (Ohliger, Förster, and Kneer 2013). LHV can be calculated from HHV of biomass according to equations 8.10 and 8.11 (Panoutsou et al. 2013).

$$HHVARB = HHVDB(1 - MARB) \tag{8.11}$$

$$LHVARB = HHVDB(1 - MARB) - 2.447MARB - 22.023ECH \tag{8.12}$$

$$DB(1 - ECH, DB) \tag{8.13}$$

The lower heating value has an impact on the moisture content of the biomass sample. The higher heating value of the biomass sample on a dry basis can be calculated from the presence of H, C, N, S, O, and ash in biomass based on the following empirical equation derived by the International Energy Agency (van Loo and Koppejan 2012):

$$HHVDB = 0.3491ECC, DB + 1.1783ECH.DB$$

$$+ 0.1005ECS, DB - 0.0151ECN, DB \qquad (8.14)$$

$$- 0.1034ECO, DB - 0.0211ADB$$

The heating value of biomass sample varies according to soil condition and climatic condition. While there is no fixed value for woody biomass, it typically ranges from 18 to 22 kJ/mol (Williams et al. 2016). For herbaceous biomasses, the range is from 15 to 19.5 kJ/mol.

### 8.5.4 COMPOSITIONAL ANALYSIS

The chemical composition of lignocellulosic biomass consists of cellulose, hemicellulose, and lignin. Cellulose is composed of simple polymers with glucopyranose linked via glycosidic bonds (Wu et al. 2015). Hemicellulose is comprised of short branched sugar molecules. Lignin constitutes a phenolic polymer linked with large molecular structures (Steyermark 1984). These chemical compositions vary depending on the type of lignocellulosic biomass. The performance of cellulose, lignin, and hemicellulose varies in biological and thermochemical conversion processes (Isikgor and Becer 2015). For example, lignin content interferes with cellulosic bioethanol production during biological conversion. The economics and conversion yield of the process depend on the accuracy of compositional analysis (Sluiter et al. 2010).

The ASTM standard E1757-01 is followed for compositional analysis for the preparation of biomass samples. The specification of the standard outlines a technique for sample preparation in terms of moisture content and specific particle size (Hames et al. 2004). The first step is to dry the sample at 45°C for 30 h and mill the sample so that it passes through a 2-mm screen; then the milled sample is sieved and a mesh fraction of –20/80 is chosen for further analysis. There are three method types: 1. near-infrared spectroscopy (NIRS) analysis, 2. sulfuric acid analysis, 3. kinetic analysis (Cai et al. 2013).

The most commonly used method is two-step sulfuric acid hydrolysis. A research report reviewed the sulfuric acid hydrolysis method for biomass samples, including its uncertainties (Templeton et al. 2010). This procedure has been used for nearly 100 years with modifications for different objects and conditions. The most updated procedure for determining lignin and structural carbohydrates in biomass has been published by the National Renewable Energy Laboratory (NREL) (Sluiter et al. 2010). This NREL procedure is for measuring carbohydrates and lignin content of biomass samples without extractives. This procedure employs two-step hydrolysis for quantifying the biomass. This method starts with the ethanol extraction unit setup. It can accurately predict the polysaccharide sugars and lignin content. After the extraction, strong sulfuric acid hydrolysis at normal condition followed by dilute sulfuric acid hydrolysis at a higher temperature is used for the breakdown of monomers (Sluiter et al. 2004).

Ultraviolet spectroscopy is used to quantify the acid-soluble lignin, whereas klason lignin can be gravimetrically estimated from acid-insoluble residue. We can achieve reliable results from the sulfuric acid hydrolysis method about biomass composition, but it is time consuming, high cost, and labor intensive and requires the removal of extractives. The special features of the NIRS technique are low cost less complicated procedure. Hazardous chemicals are strictly prohibited in the analysis, and the procedure is fast and precise (Savy and Piccolo 2014). Different infrared spectrums can be used for various functional groups in the structural components of biomass. Moreover, while it won't give a compositional analysis, it will identify chemical bonds present in the biomasses.

Carrier et al. (2011) reported on a kinetic model for predicting the biomass composition, but it can estimate only cellulose and hemicellulose content from a particular biomass sample by using thermogravimetric analysis. Cai et al. (2013) reported that activation energy models are adopted to find the chemical composition of biomass. Biomass is analyzed in a thermogravimetric analyzer under an inert atmosphere. The pyrolysis kinetic behavior of biomass is described by three activation energy models. The biochemical conversion process uses lignin and cellulose content as a key factor for processing the biomass. The overall cellulose degradation in the biochemical process is greater than the lignin content. Because lignin is rigid, it won't easily degrade. The forest residues and agricultural residues are suitable for the thermochemical conversion process. Sugar crops are suitable for bioethanol fermentation (Maurya, Singla, and Negi, 2015).

## 8.6 AN ASSESSMENT OF THE SUSTAINABILITY OF LIGNOCELLULOSIC BIOMASS FOR BIOREFINING

A development of new bioenergy organization is crucial for assessing the environment and public benefits of biofuel generation. The sustainability objectives must be followed in the policies of the biofuel industry (Ewing and Msangi 2009). Such an evaluation committee is set up now to audit the biofuel industries in several countries. In the United States, palm oil–based biodiesel plants are audited by the roundtable sustainability standards of biofuel industries (Naylor et al. 2007). The EU issued a circular in 2009 to promote the need for renewable energy consumption to be increased by 20% before 2020 (Edgard Gnansounou 2010). The circular includes a set of sustainable norms like a 10% target of biofuel production, the importance of domestic biofuel production, and reducing greenhouse gas (GHG) emissions by 50% in 2017 and 60% in 2018 (Diao et al. 2007). Global poverty and hunger can be reduced by investigating sustainable audit results with an analysis of food security impacts. These impacts boost the economy of people in various countries.

The proper strategies should be followed for the demand and availability of feedstocks without any obstacles. Rural communities in developing countries benefit from government subsidies (Mchenry, Doepel, and de Boer 2014). Private investors can get significant profits due to the commercial interest in large-scale schemes. There should be flexibility in standards for uncertainties that exist in the use of second-generation (2G) lignocellulosic feedstock (Mchenry, Doepel, and de Boer 2014). Each year, 40 million tons of lignocellulose are generated; of that, much is thrown into landfills. Turning these wastes to power will benefit the country and won't affect the food crops. These waste materials have an impact on rural economies and increase biofuel production in domestic areas. Moreover, the use of lignocellulosic feedstocks can reduce demand with agricultural-based feedstocks. Further, the efficient utilization and conversion of biomass are challenging tasks for researchers.

The conflict between food and fuel arises due to growth in agricultural productivity. In sub-Saharan Africa, possible reasons for food and energy insecurity include the history of poverty, degraded land, and insufficient agricultural infrastructures (Nanda et al. 2014). In developing countries, increased productivity of staple crops will reduce poverty. The report suggested that around 70% of African people depend upon liquid fuels and the import of energy. So, developing domestic bioenergy generation will arouse economic benefits and employment. The government should encourage the industry by providing subsidies and incentives. Because of these, employment in local industries can be provided to the local community, and also the energy demand can be addressed (Gamser 1988).

To prevent malnourishment in nutrition-strapped regions, the availability of biomass and environment must be taken into account and new development strategies and funds must be created. In nations with abundant feedstock and domestic biofuel, generation will benefit rural peoples and investors. These biorefineries in the country will provide benefits by providing watersheds, harvesting rainwater, and improving the quality of life in the country. With use, green manure and biofertilizers won't affect the environment and groundwater contaminations. Moreover, the waste obtained from the biological conversion process can be used as a soil fertilizer. This fertilizer can be used for the feedstock land to improve the crop yield. The biochar obtained from the thermochemical conversion process is rich in metal content, like silica, magnesium, potassium, and sodium, that will increase the soil fertility. The United States has a plan to build the capacity to produce 130 billion liters of biofuel by 2020. Currently, 90% of bioethanol is produced in the United States and Brazil. EU countries like Germany, France, and Italy contributes for 60% of global biodiesel production (de Man and German 2017).

### 8.6.1 Lignocellulosic Feedstocks for Energy and Economic Sustainability

The main objectives of biofuels are promoting climate change mitigation and also offering a promising alternative to fossil fuels. As there are no shortages in the availability of renewable feedstocks, they are used in the production of biofuels (Kumar and Gayen 2011). The tremendous change in the technologies of biomass conversion and also in production of the feedstock will lead to a drastic reduction in the cost of biofuels in the future. Production of biodiesel and also bioethanol is promoted as much of the fuel (nearly 40%) used all over the globe is in liquid form. Numerous research works are being carried out for biobutanol, which may be included on the list of potential biofuels that can be launched easily in the commercial market. Biobutanol can also be considered as an innovative fuel, as it is very compatible with all engines used, has high energy density and better fuel performance than ethanol, and can be produced from a wide variety of waste biomass.

Biobutanol is capable of replacing not only bioethanol but also biodiesel, which are used in the current fuel market at approximately \$248 billion in 2020 (Green 2011). Many nations have promoted bioethanol as alternate fuel instead of gasoline, as it has an oxygen content of 35%, which helps in reduction of GHG emission as it leads to complete burning of fuel, reducing gasoline consumption by up to 32% when it is in E85 form, which has 85% and 15% of ethanol and gasoline, respectively. Many countries such as the United States, Canada, Sweden, Brazil, China, Australia, Columbia, Peru, Thailand, Paraguay, and India have already implemented many programs based on blending of ethanol and gasoline (Kim and Dale 2004).

First-generation (1G) ethanol is composed of edible parts of plants produced from agricultural products such as sugarcane, corn, and wheat. But this biofuel, which is now available on a commercial scale, has some drawbacks, including an increase in expense due to dependence on agricultural products, along with lack of feedstock selection and other problems associated with the usage of food products as fuel (Fargione et al. 2008). That there are many problems which directly or indirectly affect land usage stimulated by the cultivation of first-generation feedstock is not to be disregarded. The carbon debt can even be increased by an increase in first-generation biofuel production according to some studies, as the quantity of $CO_2$ released by direct or indirect usage of land is larger than the GHG reduction obtained by using it as an alternative for fossil fuels. First-generation ethanol derived from corn can increase GHG emission by two times over 30 years and gradually increase emission over around 170 years (Timilsina and Shrestha 2011).

The first-generation biofuel has been opposed for various reasons such as the struggle between food and fuel, the more GHG emission due to lower performance, and the effect of excessive development of agricultural areas on their surroundings, since the total area under cultivation is being increased to grow feedstock, and this affects land usage patterns (Sadeghinezhad et al. 2014). But second-generation ethanol has provided good opportunities as an alternative to fossil fuels by not producing any food-versus-fuel problems, as it is obtained from non-edible sources of plant that have originated as lignocellulosic material comprised of hemicellulose along with cellulose and lignin. The usage of 2G ethanol has drastically reduced the demand for 1G feedstock and reduced competition with farming products. 2G ethanol has less life cycle GHG emission than 1G grain ethanol (Wang, Wu, and Huo 2007).

The non-consumable remains of plants from agricultural farms, forests, and savannas are known as lignocellulosic materials, incorporating agricultural residues, remnants of energy crops, residues of wood, sawmill residues, paper mill refuse, and municipal paper waste. Some specific examples are sugarcane bagasse, corn stover, rice husk, wheat straw, switchgrass, rice hull, timothy grass, poplar wood, pinewood, willow, etc.). The lignocellulosic feedstock is available on a large scale and it has the capacity for good production of biofuels. Progress in conversion technologies can give many people access to the energy derived from residues of agriculture along with biomass from forests and crops: around 250 to 500 EJ per annum by 2050 (Coyle 2008).

## 8.6.2 Biofuels and Food Security

Bioethanol, which is derived from corn and sugarcane, along with biodiesel, derived from palm oil, rapeseed, and soya, have been the main suppliers of first-generation biofuels. This feedstock can be used for nutrition, cattle feed, and other commercial food purposes. The use of first-generation biofuels indeed shifted these crops toward the newly emerged market opportunities in biofuel, as these crops are the primary raw material for these fuels. Farmers seeking this new source of income and new employment opportunities in the agriculture and processing sectors led to a hike in the price of food around the globe (Escobar et al. 2009). These increases in price lead to a deficiency in food consumption and export to other countries. With the increase in production, biomass energy has become an effective alternative to fossil fuels but land conservation and food security have been indirectly affected. Now, approximately 14 million hectares of land, or over 1% of the world's total land, are being used for biofuel production.

The three major elements on which the Food and Agriculture Organization relies on to assess the security of food are availability, accessibility, and stability (Pingali, Raney, and Wiebe 2008). Local incomes and food prices are highly impacted by the bio-refineries, which influence these three elements. The rivalry for resources in commodities and products in the biofuel sector has highly affected the availability of food, leading to a steep hike in the cost of these food products. The ability to access food is finally determined by its price and availability. It is very hard to predict future costs, as the price for feedstock in coming years will be determined more by biofuel production feedstocks, and less by crops grown for human and cattle food (Wheeler and Von Braun 2013).

The increase in the cost of food will not alter the usage of food but will indirectly lead to malnutrition on a global level. Generation of 15 billion liters of ethanol derived from corn required 336 million metric tons of corn by the United States in 2005, which was nearly 13% of total corn production. The United States generated 26.5 billion liters of ethyl alcohol along with 1.7 billion liters of biodiesel in 2007, which were derived from corn grain and oil from soya bean, respectively. The United States produced 40% of corn and exported 60% of the corn exports at the global level. As the production rate is high, there is a high impact on the international corn price. About 92.9 million acres were used to plant corn, and 1/3 of the harvest was made used in the production of ethanol in 2007 (Soccol et al. 2010).

These have indirectly led to a hike in the cost of corn up to 73% by the end of 2010. The cost of corn was $78 per ton and $142 per ton in 2005 and 2006, respectively, which clearly shows the rise in the price of corn. There has been an increment in the amount of food from 35% to 40% in the years between 2002 and 2008. Of this increase, 70%–75% is due to the contribution by biofuel. Since there has been an increase in the ethanol derived from corn in the United States in the past decade, there has been a corresponding increase in wheat production by 30% to overcome the deficiency of food.

The historical food price inflation from 2007 to 2008 shows that the globe has started focusing on world food security (Soccol et al. 2010). Not only is there the increase in food cost; we need to consider the costs after production, such as extraction of ethanol, purification, distillation, and also fermentation waste management. Bioethanol can replace the liquid fossil fuels if and only if its total production cost is US$1 per liter. It is being assumed that upgrading to new and efficient technologies in biomass conversion and automobiles can decrease the cost to US$0.2 per liter by 2030. This can create trouble in emerging and immature countries, where importing nutrition is essential to meet their local needs. For instance, there was an average 33% import dependence in 2000 for the total cereal demand in sub-Saharan Africa along with a high level of dependence (greater than 80%) in places like Gambia, Sudan, and Zambia (Amigun, Musango, and Stafford 2011).

An FAO report suggested clearly that the food consumption of grains by emerging nations increased by 194 million tons in 2010 and 2011, but industrial production of food crops to produce ethanol increased about 250 million tons in the same time. Increases in undernourishment and starvation can occur if countries with high cultivation rates and food stability start to distribute grains to other nations in order to benefit from the high export rate and if countries with nutrition disasters are unable to import grains due to high costs. If we have a look in policies in Malaysia, the biodiesel derived from palm oil is highly promoted for making and usage, which leads to replacing forests with palm plantations. Additionally, it is noted that due to growth and change in the usage of land over the last four decades, agriculture is being decreased drastically, leading to dramatic food deficiency for future generations (Aktas et al. 2019).

FAO has clearly stated that now approximately 805 million people, which is 11.5% of the total global population, are suffering from malnutrition. About 753 million people are underfed in Africa and Asia, which is 93.5% of the total malnutrition population. However, the proportion of malnourished people has reduced from 1990 to 1992, from 1015 million to 805 million. The main reason for a hike in the price of food is the increasing market for biofuels, which also leads to high demand for feedstocks from crops. Due to energy demand, the cost of conventional fuel is going to stay high. Arable lands are increasingly being dedicated to energy production because of policies like the promotion of carbon storage in soil along with feedstock properties like fast growth, tolerance of a wide range of land and climate conditions, low agronomic requirements, and very low degradation of the lands, not like what is seen for traditional food crops. High commodity costs and incomes are enjoyed by energy crop–producing farmers and poor communities in municipal and rural areas in nutrition-importing nations (Aktas et al. 2019).

### 8.6.3 Life Cycle Assessment of Lignocellulosic Biomass and Biofuels

Life cycle assessment (LCA) is used to evaluate the socioeconomic of a product from the database of energy input and associated emissions to the power produced from bioenergy. To estimate the potential enhancement in the product chain, the ISO 14040 international standard should be followed in LCA studies, as suggested by Singh et al. (2010). LCA usually uses a cradle-to-grave approach; that is, it follows the entire life cycle of the product. It will determine the impacts on the economic ecosystem and society. LCA can be involved in all stages of biofuel production like source, conversion technologies, and the final product (Gnansounou et al. 2009).

LCA studies consist of goal and scope, system boundaries, inventory model, impact assessment, sensitive analysis, and additional parameters. To evaluate the economic and environmental impacts of any bioenergy production from biomass and to compare them with those of conventional fuel as a reference, the full cycle of bioenergy production must be taken into account (Wiloso, Heijungs, and De Snoo 2012).

## 8.7 CONCLUSIONS

Lignocellulosic biomass can be effectively converted into bioenergy, biofuels, and biochemicals by thermochemical and biochemical conversion techniques. A comprehensive understanding of biomass characterization and physicochemical properties of biomass feedstock is to be obtained to find the optimum design and operation associated with bioenergy production from biomass.

- The sieving analysis technique is an easy method to measure particle size.
- The ASTM standard E873-82 is used for measuring the bulk density of the particle. A pressurized gas flowing method is followed for particle density measurement.
- According to the ASTM standards, C144-00 and D6128-16 flowability are quantified as the angle of response and flow index, respectively.
- Either the precise thermogravimetric analysis method or the time-consuming ASTM standard approach can be used for measuring proximate analysis (e.g., moisture, ash, volatile and fixed carbon contents).
- Ultimate analysis (e.g. C, H, N, S, and O contents) of biomass can be performed using the elemental analyzer.
- The bomb calorimeter is commonly used to calculate the energy content of any particular biomass.

The sustainability of lignocellulosic biomass for biorefining, economic sustainability, biofuels, and food security life cycle assessments of biomass are discussed in this chapter.

## REFERENCES

Abdullah, Nurhayati, Fauziah Sulaiman, and Rahmad Mohd Taib. 2013. "Characterization of Banana (Musa Spp.) Plantation Wastes as a Potential Renewable Energy Source." In *AIP Conference Proceedings.* doi:10.1063/1.4803618.

Aktas, Emel, Zeynep Topaloglu, Zahir Irani, Amir Sharif, and Samsul Huda. 2019. "Food Provision to Food Security: How Can We Reduce Waste on the Supply Side?" In *Qatar Foundation Annual Research Conference Proceedings.* doi:10.5339/qfarc.2016. sshapp2342.

Alizadeh, Hasan, Farzaneh Teymouri, Thomas I. Gilbert, and Bruce E. Dale. 2005. "Pretreatment of Switchgrass by Ammonia Fiber Explosion (AFEX)." In *Applied Biochemistry and Biotechnology – Part A Enzyme Engineering and Biotechnology.* doi:10.1385/ABAB:124:1-3:1133.

Amigun, Bamikole, Josephine Kaviti Musango, and William Stafford. 2011. "Biofuels and Sustainability in Africa." *Renewable and Sustainable Energy Reviews.* doi:10.1016/j. rser.2010.10.015.

Arnoult, Stéphanie, and Maryse Brancourt-Hulmel. 2015. "A Review on Miscanthus Biomass Production and Composition for Bioenergy Use: Genotypic and Environmental Variability and Implications for Breeding." *Bioenergy Research*. doi:10.1007/s12155-014-9524-7.

ASTM. 2015. *Standard Test Method for Ash in Biomass E1755-01*. ASTM. doi:10.1520/E1755-01R07.2.

ASTM E1756. 2002. "Standard Test Method for Determination of Total Solids in Biomass." In *Annual Book of ASTM Standards*. doi:10.1520/E1756-08R15.2.

ASTM E873-82. 2006. "Standard Test Method for Bulk Density of Densified Particulate Biomass Fuels." In *Annual Book of ASTM Standards*. doi:10.1520/E0873-82R06.

Balan, Venkatesh, Bryan Bals, Shishir P. S. Chundawat, Derek Marshall, and Bruce E. Dale. 2009. "Lignocellulosic Biomass Pretreatment Using AFEX." *Methods in Molecular Biology (Clifton, N.J.)*. doi:10.1007/978-1-60761-214-8_5.

Basu, Prabir. 2013. *Biomass Gasification, Pyrolysis and Torrefaction: Practical Design and Theory*. doi:10.1016/C2011-0-07564-6.

"Biomass as Energy Source: Resources, Systems and Applications." 2013. *Choice Reviews Online*. doi:10.5860/choice.51-2104.

Brownell, H. H., E. K. C. Yu, and J. N. Saddler. 1986. "Steam-explosion Pretreatment of Wood: Effect of Chip Size, Acid, Moisture Content and Pressure Drop." *Biotechnology and Bioengineering*. doi:10.1002/bit.260280604.

Cai, Junmeng, Yifeng He, Xi Yu, Scott W. Banks, Yang Yang, Xingguang Zhang, Yang Yu, Ronghou Liu, and Anthony V. Bridgwater. 2017. "Review of Physicochemical Properties and Analytical Characterization of Lignocellulosic Biomass." *Renewable and Sustainable Energy Reviews*. doi:10.1016/j.rser.2017.03.072.

Cai, Junmeng, Weixuan Wu, Ronghou Liu, and George W. Huber. 2013. "A Distributed Activation Energy Model for the Pyrolysis of Lignocellulosic Biomass." *Green Chemistry*. doi:10.1039/c3gc36958g.

Carrier, Marion, Anne Loppinet-Serani, Dominique Denux, Jean Michel Lasnier, Frédérique Ham-Pichavant, François Cansell, and Cyril Aymonier. 2011. "Thermogravimetric Analysis as a New Method to Determine the Lignocellulosic Composition of Biomass." *Biomass and Bioenergy*. doi:10.1016/j.biombioe.2010.08.067.

Chen, Hongzhang, and Hongzhang Chen. 2014. "Chemical Composition and Structure of Natural Lignocellulose." In *Biotechnology of Lignocellulose*. doi:10.1007/978-94-007-6898-7_2.

Chouhan, A., and A. Sarma. 2013. "Critical Analysis of Process Parameters for Bio-Oil Production via Pyrolysis of Biomass: A Review." *Recent Patents on Engineering*. doi:10.2174/18722121113079990005.

Chundawat, S. P. S., V. Balan, L. Da Costa Sousa, and B. E. Dale. 2010. "Thermochemical Pretreatment of Lignocellulosic Biomass." In *Bioalcohol Production: Biochemical Conversion of Lignocellulosic Biomass*. doi:10.1533/9781845699611.1.24.

Conshohocken, West. 2001. "Standard Shear Testing Test Method for Shear Testing of Bulk Solids Using the Jenike Shear Cell 1." *Evaluation*. doi:10.1520/D6128-06.2.

Coyle, William. 2008. "The Future of Biofuels: A Global Perspective." *Journal of Rural Mental Health*. doi:10.1037/h0095952.

Dahlquist, Erik. 2013. *Technologies for Converting Biomass to Useful Energy Combustion, Gasification, Pyrolysis, Torrefaction and Fermentation*. doi:10.1201/b14561.

Datar, Rohit, Jie Huang, Pin Ching Maness, Ali Mohagheghi, Stefan Czernik, and Esteban Chornet. 2007. "Hydrogen Production from the Fermentation of Corn Stover Biomass Pretreated with a Steam-Explosion Process." *International Journal of Hydrogen Energy*. doi:10.1016/j.ijhydene.2006.09.027.

Demirbas, M. Fatih, Mustafa Balat, and Havva Balat. 2009. "Potential Contribution of Biomass to the Sustainable Energy Development." *Energy Conversion and Management*. doi:10.1016/j.enconman.2009.03.013.

Den, Walter, Virender K. Sharma, Mengshan Lee, Govind Nadadur, and Rajender S. Varma. 2018. "Lignocellulosic Biomass Transformations via Greener Oxidative Pretreatment Processes: Access to Energy and Value Added Chemicals." *Frontiers in Chemistry*. doi:10.3389/fchem.2018.00141.

Diao, Xinshen, Peter Hazell, Danielle Resnick, and James Thurlow. 2007. "The Role of Agriculture in Development: Implications for Sub-Saharan Africa." *Research Report of the International Food Policy Research Institute*. doi:10.2499/978089629161 4rr153.

Duque, A., P. Manzanares, I. Ballesteros, and M. Ballesteros. 2016. "Steam Explosion as Lignocellulosic Biomass Pretreatment." In *Biomass Fractionation Technologies for a Lignocellulosic Feedstock Based Biorefinery*. doi:10.1016/B978-0-12-802323-5.00015-3.

Dutta, Abhijit, Michael Talmadge, Jesse Hensley, Matt Worley, Doug Dudgeon, David Barton, Peter Groenendijk, et al. 2012. "Techno-Economics for Conversion of Lignocellulosic Biomass to Ethanol by Indirect Gasification and Mixed Alcohol Synthesis." *Environmental Progress and Sustainable Energy*. doi:10.1002/ep.10625.

Eggeman, Tim, and Richard T. Elander. 2005. "Process and Economic Analysis of Pretreatment Technologies." *Bioresource Technology*. doi:10.1016/j.biortech.2005.01.017.

Escobar, José C., Electo S. Lora, Osvaldo J. Venturini, Edgar E. Yáñez, Edgar F. Castillo, and Oscar Almazan. 2009. "Biofuels: Environment, Technology and Food Security." *Renewable and Sustainable Energy Reviews*. doi:10.1016/j.rser.2008.08.014.

Ewing, Mandy, and Siwa Msangi. 2009. "Biofuels Production in Developing Countries: Assessing Tradeoffs in Welfare and Food Security." *Environmental Science and Policy*. doi:10.1016/j.envsci.2008.10.002.

Fargione, Joseph, Jason Hill, David Tilman, Stephen Polasky, and Peter Hawthorne. 2008. "Land Clearing and the Biofuel Carbon Debt." *Science*. doi:10.1126/science.1152747.

Fatma, Shabih, Amir Hameed, Muhammad Noman, Temoor Ahmed, Imran Sohail, Muhammad Shahid, Mohsin Tariq, and Romana Tabassum. 2018. "Lignocellulosic Biomass: A Sustainable Bioenergy Source for Future." *Protein & Peptide Letters*. doi: 10.2174/0929866525666180122144504.

Gamser, Matthew S. 1988. "Biomass Stoves: Engineering Design, Development and Dissemination." *Research Policy*. doi:10.1016/0048-7333(88)90026-1.

Gera, D., M. P. Mathur, M. C. Freeman, and Allen Robinson. 2002. "Effect of Large Aspect Ratio of Biomass Particles on Carbon Burnout in a Utility Boiler." *Energy and Fuels*. doi:10.1021/ef0200931.

Gnansounou, E., A. Dauriat, J. Villegas, and L. Panichelli. 2009. "Life Cycle Assessment of Biofuels: Energy and Greenhouse Gas Balances." *Bioresource Technology*. doi:10.1016/j.biortech.2009.05.067.

Gnansounou, Edgard. 2010. "Production and Use of Lignocellulosic Bioethanol in Europe: Current Situation and Perspectives." *Bioresource Technology*. doi:10.1016/j.biortech.2010.02.002.

Golden, T., B. Reed, and A. Das. 1988. *Handbook of Biomass Downdraft Gasifier Engine Systems*. SERI, U.S. Department of Energy. doi:10.2172/5206099.

Goyal, H. B., Diptendu Seal, and R. C. Saxena. 2008. "Bio-Fuels from Thermochemical Conversion of Renewable Resources: A Review." *Renewable and Sustainable Energy Reviews*. doi:10.1016/j.rser.2006.07.014.

Green, Edward M. 2011. "Fermentative Production of Butanol-the Industrial Perspective." *Current Opinion in Biotechnology*. doi:10.1016/j.copbio.2011.02.004.

Haghighi Mood, Sohrab, Amir Hossein Golfeshan, Meisam Tabatabaei, Gholamreza Salehi Jouzani, Gholam Hassan Najafi, Mehdi Gholami, and Mehdi Ardjmand. 2013. "Lignocellulosic Biomass to Bioethanol, a Comprehensive Review with a Focus on Pretreatment." *Renewable and Sustainable Energy Reviews*. doi:10.1016/j.rser.2013.06.033.

Hames, Bonnie, Raymond O. Ruiz, Christopher Scarlata, Amie Sluiter, Justin Sluiter, David Templeton, and Department of Energy. 2004. "Preparation of Samples for Compositional Analysis." *Biomass Analysis Technology Team, Laboratory Analytical Procedure.*

Himmel, Michael E., Shi You Ding, David K. Johnson, William S. Adney, Mark R. Nimlos, John W. Brady, and Thomas D. Foust. 2007. "Biomass Recalcitrance: Engineering Plants and Enzymes for Biofuels Production." *Science.* doi:10.1126/science.1137016.

Isikgor, Furkan H., and C. Remzi Becer. 2015. "Lignocellulosic Biomass: A Sustainable Platform for the Production of Bio-Based Chemicals and Polymers." *Polymer Chemistry.* doi:10.1039/c5py00263j.

Jenike, A. W. 1964. *Storage and Flow of Solids, Bulletin No. 123.* Utah Engineering Experiment Station.

Jong, Wiebren De, and J. Ruud Van Ommen. 2014. *Biomass as a Sustainable Energy Source for the Future: Fundamentals of Conversion Processes.* doi:10.1002/9781118916643.

Jose, Shibu, and Thallada Bhaskar. 2015. *Biomass and Biofuels: Advanced Biorefineries for Sustainable Production and Distribution.* doi:10.1201/b18398.

Kan, Tao, Vladimir Strezov, and Tim J. Evans. 2016. "Lignocellulosic Biomass Pyrolysis: A Review of Product Properties and Effects of Pyrolysis Parameters." *Renewable and Sustainable Energy Reviews.* doi:10.1016/j.rser.2015.12.185.

Karunanithy, C., K. Muthukumarappan, and A. Donepudi. 2013. "Moisture Sorption Characteristics of Corn Stover and Big Bluestem." *Journal of Renewable Energy.* doi:10.1155/2013/939504.

Kaur, Meera. 2009 *Medical Foods from Natural Sources.* doi:10.1007/978-0-387-79378-8.

Kim, Seungdo, and Bruce E. Dale. 2004. "Global Potential Bioethanol Production from Wasted Crops and Crop Residues." *Biomass and Bioenergy.* doi:10.1016/j.biombioe.2003.08.002.

Kumar, Manish, and Kalyan Gayen. 2011. "Developments in Biobutanol Production: New Insights." *Applied Energy.* doi:10.1016/j.apenergy.2010.12.055.

Lam, P. S., S. Sokhansanj, X. Bi, S. Mani, C. J. Lim, A. R. Womac, M. Hoque, et al. 2007. "Physical Characterization of Wet and Dry Wheat Straw and Switchgrass – Bulk and Specific Density." In *2007 ASABE Annual International Meeting, Technical Papers.* doi:10.13031/2013.23552.

Laureano-Perez, Lizbeth, Farzaneh Teymouri, Hasan Alizadeh, and Bruce E. Dale. 2009. "Understanding Factors That Limit Enzymatic Hydrolysis of Biomass." In *Twenty-Sixth Symposium on Biotechnology for Fuels and Chemicals.* doi:10.1007/978-1-59259-991-2_91.

Limayem, Alya, and Steven C. Ricke. 2012. "Lignocellulosic Biomass for Bioethanol Production: Current Perspectives, Potential Issues and Future Prospects." *Progress in Energy and Combustion Science.* doi:10.1016/j.pecs.2012.03.002.

Lin, Fan, Christopher L. Waters, Richard G. Mallinson, Lance L. Lobban, and Laura E. Bartley. 2015. "Relationships between Biomass Composition and Liquid Products Formed via Pyrolysis." *Frontiers in Energy Research.* doi:10.3389/fenrg.2015.00045.

Lin, Guiying, Haiping Yang, Xianhua Wang, Yanyang Mei, Pan Li, Jingai Shao, and Hanping Chen. 2016. "The Moisture Sorption Characteristics and Modelling of Agricultural Biomass." *Biosystems Engineering.* doi:10.1016/j.biosystemseng.2016.08.006.

Loo, Sjaak van, and Jaap Koppejan. 2012. *The Handbook of Biomass Combustion and Co-Firing.* doi:10.4324/9781849773041.

Mabee, W. E., P. N. McFarlane, and J. N. Saddler. 2011. "Biomass Availability for Lignocellulosic Ethanol Production." *Biomass and Bioenergy.* doi:10.1016/j.biombioe.2011.06.026.

Man, Reinier de, and Laura German. 2017. "Certifying the Sustainability of Biofuels: Promise and Reality." *Energy Policy.* doi:10.1016/j.enpol.2017.05.047.

Mani, Sudhagar. 2015. "Fuel and Bulk Flow Properties of Coal and Torrefied Wood Mixtures for Co-Firing Application." In *International Bioenergy and Bioproducts Conference 2015, IBBC 2015: Leveraging Forest Biomass for Next Generation Fuels and Chemicals.*

Mann, Benjamin F., Hongmei Chen, Elizabeth M. Herndon, Rosalie K. Chu, Nikola Tolic, Evan F. Portier, Taniya Roy Chowdhury, et al. 2015. "Indexing Permafrost Soil Organic Matter Degradation Using High-Resolution Mass Spectrometry." *PLoS One.* doi:10.1371/journal.pone.0130557.

Mason, P. E., Darvell, L. I., Jones, J. M., and Williams, A. (2016) Comparative Study of the Thermal Conductivity of Solid Biomass Fuels. Energy and Fuels. 30. 2158–2163. https://doi.org/10.1021/acs.energyfuels.5b02261

Martín-Davison, Jessica San, Mercedes Ballesteros, Paloma Manzanares, Ximena Petit Breuilh Sepúlveda, and Alberto Vergara-Fernández. 2015. "Effects of Temperature on Steam Explosion Pretreatment of Poplar Hybrids with Different Lignin Contents in Bioethanol Production." *International Journal of Green Energy.* doi:10.1080/154350 75.2014.887569.

Maurya, Devendra Prasad, Ankit Singla, and Sangeeta Negi. 2015. "An Overview of Key Pretreatment Processes for Biological Conversion of Lignocellulosic Biomass to Bioethanol." *3 Biotech.* doi:10.1007/s13205-015-0279-4.

Mchenry, Mark P., David Doepel, and Karne de Boer. 2014. "Rural African Renewable Fuels and Fridges: Cassava Waste for Bioethanol, with Stillage Mixed with Manure for Biogas Digestion for Application with Dual-Fuel Absorption Refrigeration." *Biofuels, Bioproducts and Biorefining.* doi:10.1002/bbb.1433.

Nanda, Sonil, Ramin Azargohar, Janusz A. Kozinski, and Ajay K. Dalai. 2014. "Characteristic Studies on the Pyrolysis Products from Hydrolyzed Canadian Lignocellulosic Feedstocks." *Bioenergy Research.* doi:10.1007/s12155-013-9359-7.

Naylor, Rosamond L., Adam J. Liska, Marshall B. Burke, Walter P. Falcon, Joanne C. Gaskell, Scott D. Rozelle, and Kenneth G. Cassman. 2007. "The Ripple Effect Biofuels, Food Security, and the Environment." *Environment.* doi:10.3200/ENVT.49.9.30-43.

Pandey, Ashok. 2008. *Handbook of Plant-Based Biofuels.* doi:10.1201/9780789038746.

Panoutsou, Calliope, Ausilio Bauen, Hannes Böttcher, Efi Alexopoulou, Uwe Fritsche, Ayla Uslu, Joost N. P. van Stralen, et al. 2013. "Biomass Futures: An Integrated Approach for Estimating the Future Contribution of Biomass Value Chains to the European Energy System and Inform Future Policy Formation." *Biofuels, Bioproducts and Biorefining.* doi:10.1002/bbb.1367.

Patel, Madhumita, Xiaolei Zhang, and Amit Kumar. 2016. "Techno-Economic and Life Cycle Assessment on Lignocellulosic Biomass Thermochemical Conversion Technologies: A Review." *Renewable and Sustainable Energy Reviews.* doi:10.1016/j.rser.2015.09.070.

Pereira Ramos, Luiz. 2003. "The Chemistry Involved in the Steam Treatment of Lignocellulosic Materials." *Quimica Nova.* doi:10.1590/s0100-40422003000600015.

Perlack, R. D. 2005. *Biomass as Feedstock for a Bioenergy and Bioproducts Industry: The Technical Feasibility of a Billion-Ton Annual Supply.* Agriculture.

Pingali, Prabhu, Terri Raney, and Keith Wiebe. 2008. "Biofuels and Food Security: Missing the Point." *Review of Agricultural Economics.* doi:10.1111/j.1467-9353.2008.00425.x.

Pu, Yunqiao, Xianzhi Meng, Chang Geun Yoo, Mi Li, and Arthur J. Ragauskas. 2016. "Analytical Methods for Biomass Characterization during Pretreatment and Bioconversion?" In *Valorization of Lignocellulosic Biomass in a Biorefinery: From Logistics to Environmental and Performance Impact.* Nova Science Publishers.

Rabier, Fabienne, Michaël Temmerman, Thorsten Böhm, Hans Hartmann, Peter Daugbjerg Jensen, Josef Rathbauer, Juan Carrasco, and Miguel Fernández. 2006. "Particle Density Determination of Pellets and Briquettes." *Biomass and Bioenergy.* doi:10.1016/j.biombioe.2006.06.006.

Ropars, M., R. Marchal, J. Pourquié, and J. P. Vandecasteele. 1992. "Large-Scale Enzymatic Hydrolysis of Agricultural Lignocellulosic Biomass. Part 1: Pretreatment Procedures." *Bioresource Technology.* doi:10.1016/0960-8524(92)90023-Q.

Rosillo-Calle, Frank. 2012. *The Biomass Assessment Handbook.* doi:10.4324/9781849772884.

Rouse, Pascale C. 2014. "Comparison of Methods for the Measurement of the Angle of Repose of Granular Materials." *Geotechnical Testing Journal.* doi:10.1520/GTJ20120144.

Saba, N., M. Jawaid, K. R. Hakeem, M. T. Paridah, A. Khalina, and O. Y. Alothman. 2015. "Potential of Bioenergy Production from Industrial Kenaf (Hibiscus Cannabinus L.) Based on Malaysian Perspective." *Renewable and Sustainable Energy Reviews.* doi:10.1016/j.rser.2014.10.029.

Sadeghinezhad, E., S. N. Kazi, Foad Sadeghinejad, A. Badarudin, Mohammad Mehrali, Rad Sadri, and Mohammad Reza Safaei. 2014. "A Comprehensive Literature Review of Bio-Fuel Performance in Internal Combustion Engine and Relevant Costs Involvement." *Renewable and Sustainable Energy Reviews.* doi:10.1016/j.rser.2013.09.022.

Sansaniwal, S. K., M. A. Rosen, and S. K. Tyagi. 2017. "Global Challenges in the Sustainable Development of Biomass Gasification: An Overview." *Renewable and Sustainable Energy Reviews.* doi:10.1016/j.rser.2017.05.215.

Savy, Davide, and Alessandro Piccolo. 2014. "Physical-Chemical Characteristics of Lignins Separated from Biomasses for Second-Generation Ethanol." *Biomass and Bioenergy.* doi:10.1016/j.biombioe.2014.01.016.

Shastri, Yogendra, Alan Hansen, Luis Rodríguez, and K. C. Ting. 2014. *Engineering and Science of Biomass Feedstock Production and Provision.* doi:10.1007/978-1-4899-8014-4.

Sheng, Changdong, and J. L. T. Azevedo. 2005. "Estimating the Higher Heating Value of Biomass Fuels from Basic Analysis Data." *Biomass and Bioenergy.* doi:10.1016/j.biombioe.2004.11.008.

Sims, Ralph E. H., Astley Hastings, Bernhard Schlamadinger, Gail Taylor, and Pete Smith. 2006. "Energy Crops: Current Status and Future Prospects." *Global Change Biology.* doi:10.1111/j.1365-2486.2006.01163.x.

Singh, Anoop, Deepak Pant, Nicholas E. Korres, Abdul Sattar Nizami, Shiv Prasad, and Jerry D. Murphy. 2010. "Key Issues in Life Cycle Assessment of Ethanol Production from Lignocellulosic Biomass: Challenges and Perspectives." *Bioresource Technology.* doi:10.1016/j.biortech.2009.11.062.

Singh, Joginder, Meenakshi Suhag, and Anil Dhaka. 2015. "Augmented Digestion of Lignocellulose by Steam Explosion, Acid and Alkaline Pretreatment Methods: A Review." *Carbohydrate Polymers.* doi:10.1016/j.carbpol.2014.10.012.

Sluiter, Amie, Bonnie Hames, Raymond O. Ruiz, Christopher Scarlata, Justin Sluiter, David Templeton, and Department of Energy. 2004. "Determination of Structural Carbohydrates and Lignin in Biomass." *Biomass Analysis Technology Team, Laboratory Analytical Procedure.*

Sluiter, Justin B., Raymond O. Ruiz, Christopher J. Scarlata, Amie D. Sluiter, and David W. Templeton. 2010. "Compositional Analysis of Lignocellulosic Feedstocks. 1. Review and Description of Methods." *Journal of Agricultural and Food Chemistry.* doi:10.1021/jf1008023.

Soccol, Carlos Ricardo, Luciana Porto de Souza Vandenberghe, Adriane Bianchi Pedroni Medeiros, Susan Grace Karp, Marcos Buckeridge, Luiz Pereira Ramos, Ana Paula Pitarelo, et al. 2010. "Bioethanol from Lignocelluloses: Status and Perspectives in Brazil." *Bioresource Technology.* doi:10.1016/j.biortech.2009.11.067.

Steyermark, Al. 1984. "Kirk-Othmer Encyclopedia of Chemical Technology." *Microchemical Journal.* doi:10.1016/0026-265x(84)90073-0.

Szalay, Annamária, András Kelemen, and Klára Pintye-Hódi. 2015. "The Influence of the Cohesion Coefficient (C) on the Flowability of Different Sorbitol Types." *Chemical Engineering Research and Design.* doi:10.1016/j.cherd.2014.05.008.

Taherzadeh, Mohammad J., and Keikhosro Karimi. 2008. "Pretreatment of Lignocellulosic Wastes to Improve Ethanol and Biogas Production: A Review." *International Journal of Molecular Sciences*. doi:10.3390/ijms9091621.

Templeton, David W., Christopher J. Scarlata, Justin B. Sluiter, and Edward J. Wolfrum. 2010. "Compositional Analysis of Lignocellulosic Feedstocks. 2. Method Uncertainties." *Journal of Agricultural and Food Chemistry*. doi:10.1021/jf100807b.

Timilsina, Govinda R., and Ashish Shrestha. 2011. "How Much Hope Should We Have for Biofuels?" *Energy*. doi:10.1016/j.energy.2010.08.023.

Ueki, Yasuaki, Takashi Torigoe, Hirofumi Ono, Ryo Yoshiie, Joseph H. Kihedu, and Ichiro Naruse. 2011. "Gasification Characteristics of Woody Biomass in the Packed Bed Reactor." *Proceedings of the Combustion Institute*. doi:10.1016/j.proci.2010.07.080.

Vassilev, Stanislav V., David Baxter, Lars K. Andersen, and Christina G. Vassileva. 2013. "An Overview of the Composition and Application of Biomass Ash. Part 1. Phase-Mineral and Chemical Composition and Classification." *Fuel*. doi:10.1016/j.fuel.2012.09.041.

Vassilev, Stanislav V., Christina G. Vassileva, and Vassil S. Vassilev. 2015. "Advantages and Disadvantages of Composition and Properties of Biomass in Comparison with Coal: An Overview." *Fuel*. doi:10.1016/j.fuel.2015.05.050.

Vidal, Bernardo C., Bruce S. Dien, K. C. Ting, and Vijay Singh. 2011. "Influence of Feedstock Particle Size on Lignocellulose Conversion – A Review." *Applied Biochemistry and Biotechnology*. doi:10.1007/s12010-011-9221-3.

Wang, Michael, May Wu, and Hong Huo. 2007. "Life-Cycle Energy and Greenhouse Gas Emission Impacts of Different Corn Ethanol Plant Types." *Environmental Research Letters*. doi:10.1088/1748-9326/2/2/024001.

Wheeler, Tim, and Joachim Von Braun. 2013. "Climate Change Impacts on Global Food Security." *Science*. doi:10.1126/science.1239402.

Williams, C. Luke, Tyler L. Westover, Rachel M. Emerson, Jaya Shankar Tumuluru, and Chenlin Li. 2016. "Sources of Biomass Feedstock Variability and the Potential Impact on Biofuels Production." *Bioenergy Research*. doi:10.1007/s12155-015-9694-y.

Wiloso, Edi Iswanto, Reinout Heijungs, and Geert R. De Snoo. 2012. "LCA of Second Generation Bioethanol: A Review and Some Issues to Be Resolved for Good LCA Practice." *Renewable and Sustainable Energy Reviews*. doi:10.1016/j.rser.2012.04.035.

Wolosker, Herman, Elena Dumin, Livia Balan, and Veronika N. Foltyn. 2008. "D-Amino Acids in the Brain: D-Serine in Neurotransmission and Neurodegeneration." *FEBS Journal*. doi:10.1111/j.1742-4658.2008.06515.x.

Wu, Weixuan, Yuanfei Mei, Le Zhang, Ronghou Liu, and Junmeng Cai. 2015. "Kinetics and Reaction Chemistry of Pyrolysis and Combustion of Tobacco Waste." *Fuel*. doi:10.1016/j.fuel.2015.04.016.

Xu, Feng, Jianming Yu, Tesfaye Tesso, Floyd Dowell, and Donghai Wang. 2013. "Qualitative and Quantitative Analysis of Lignocellulosic Biomass Using Infrared Techniques: A Mini-Review." *Applied Energy*. doi:10.1016/j.apenergy.2012.12.019.

Xue, Qingluan, and Rodney O. Fox. 2015. "Computational Modeling of Biomass Thermochemical Conversion in Fluidized Beds: Particle Density Variation and Size Distribution." *Industrial and Engineering Chemistry Research*. doi:10.1021/ie503806p.

Yu, Xi, Mohamed Hassan, Raffaella Ocone, and Yassir Makkawi. 2015. "A CFD Study of Biomass Pyrolysis in a Downer Reactor Equipped with a Novel Gas-Solid Separator-II Thermochemical Performance and Products." *Fuel Processing Technology*. doi:10.1016/j.fuproc.2015.01.002.

# 9 Lignocellulosic Biomass Conversion into Second- and Third-Generation Biofuels

## 9.1 INTRODUCTION

Biofuels are the fuels which are extracted from different sources of biomass which are available around us and fall under the category of renewable material. Also, biomass has high potential for the sustainable and economic growth of industrialized countries (Lee and Lavoie 2013). The biofuels are generally categorized as follows: (1) first-generation biofuels; (2) second-generation biofuels; (3) third-generation biofuels. This chapter mainly focuses on second-generation and third-generation biofuels. Biofuels extracted from a variety of feedstocks, ranging from non-edible lignocellulosic feedstock to municipal solid wastes, are known as second-generation biofuels. The biomass used for the production of second-generation biofuels is classified into three different categories as per Beauchet, Monteil-Rivera, and Lavoie (2012): (1) homogeneous biomass; (2) quasi-homogeneous biomass; (3) heterogeneous biomass.

Wood chips, with a cost ranging from US$100 to nearly US$120/metricton, fall under homogeneous biomass. Agricultural and forest residues, having prices in the range of US$60 to US$80/ton, are considered as quasi-homogeneous biomass. Low-value feedstock like municipal solid waste, priced between US$0 and US$60/metricton, are considered to be non-homogenous or heterogeneous biomass. As compared to the prices of vegetable oil, corn, and sugarcane, this biomass has very low prices, so the final yield of biofuel is considered to be cost-effective and profitable in the market. The conversion of biomass to biofuels mainly follows one of two pathways: (1) bio pathways; (2) thermo pathways. As per Brennan and Owende (2010), third-generation biofuels are extracted from algal biomass and have a distinct growth yield in comparison to lignocellulosic biomass.

The biofuel production from algae depends on the lipid content of microorganisms. Mostly, *Chlorella* species are highly used because of their 60–70% lipid content; for example, the species *Chlorella protothecoides* has a high yield around 7.4 g/L/d (Ding et al. 2014). There are so many challenges, both technical and geographical, associated with algal biomass. Generally, in ideal growth conditions, algae will yield 1 to 7 g/L/d of biomass (Haykiri-Acma and Yaman 2010). For industrial purposes, large volumes of water are needed, contributing major issues for cold countries like Canada during the large part of the year in which the temperature is below 0°C. Before extracting lipid via ultrafiltration, the water content in the alge needs

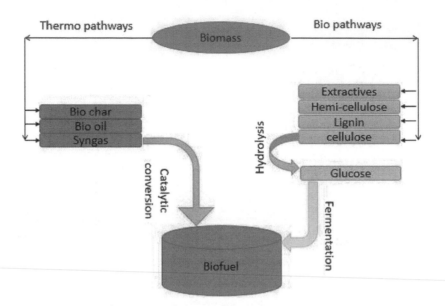

**FIGURE 9.1** Two different pathways for biomass conversion.

to removed because its creating a huge problem. Lipid extracted from algae is processed through transesterification to produce biodiesel or undergoes hydrogenolysis to produce kerosene-grade alkane, which is suitable for using in drop-in aviation fuels as discussed by Cui et al. (2010). Figure 9.1 shows two different pathways for biomass conversion.

### 9.1.1 ENERGY SECURITY AND GREENHOUSE EMISSION VS BIOFUEL

Energy is a fundamental need for society that supports human life by improving and developing wealth and the quality of life. Over a century, abundant amounts of conventional power from fossil fuels have played a supporting role for most of the industrialized countries and has brought great improvements in leading a wonderful life. The use of such fossil energy creates significant challenges and problems for the upcoming generation. The continuous use of fossil energy leads to a shortage in stock and also causes ecological destruction including water, air, and soil contamination. The uncertainty of future reserves of fossil fuels, known oil reserves, and the sustainable demand for energy cause a serious energy security problem, and these factors are also contributing highly to climate change globally (Cui et al. 2010).

To avoid vital damages from the anthropogenic point of view in regard to climate change, the level of $CO_2$ accumulation needs to stabilize to about 500 ppm. For stabilizing $CO_2$ accumulation to this level, emissions should be controlled to the present-day level. But this is one of the big challenges, as the expected $CO_2$ emission is going to be double by 2050 as per the current scenario. Some other action is required to control $CO_2$ emissions. Emission control is attained through less energy demand, low carbon energy supply, and energy efficiency. Bauen (2006) gives the concept of a series of seven triangular "wedges" which shows stabilization of emissions over the next 50 years. The main categories for achieving the series of seven triangular "wedges" are as follows: (1) energy conservation and efficiency, (2) the decarbonization of fuels and electricity supply, and (3) biological storage in soil and forest.

The improvement in efficiency and reduction in energy supply help a lot to stabilize emissions. They mainly depend on innovative energy supply for stationary and transport end uses. The conservation of energy by making behavioral changes like change of mode in transport, smaller vehicles, and less travel may provide a significant contribution to the stabilization of emissions. For efficient use of resources over a long period of time, energy efficiency and conservation have proven to be a major contributor to emission reduction (Bauen 2006).

In regard to energy supply, energy security is a huge challenge involving the provision of a consistent and acceptable energy supply at an affordable cost. Organisation for Economic Co-operation and Development (OECD) countries' energy imports are rising; in particular, some European countries' dependency on energy imports is expected to increase from 50% to 70% by 2030. Like environmental impact, energy insecurities also hinder the growth of society, as energy is a necessary input parameter for every activity happening around modern society. A major drop in the energy supply system can produce severe social and economic complications (Bauen 2006). Energy insecurities are defined as is a situation of elevated risks of a physical, perceived or real, and supply disruption.

The sudden rise in the price of energy causes inflation, which affects the performance of the economic system and thus results in loss of GNP and other effects such as unemployment rates. The abrupt rise in prices of energy supply is a sign of a high level of insecurity. The requirements of public policy, along with the market, may be capable of optimizing the level of energy security for society. To reduce the energy use problems that humans are facing in the present and for future generations, societies are shifting in terms of fossil energy to renewable energy wherever possible (Obeng-Odoom 2013). Out of the various renewable energy sources, biofuel extracted from biomass has high potential to replace fossil energy source to some extent. In both stationary and transport sectors, biofuels may play a crucial role in the upcoming generation's energy needs (Obeng-Odoom 2013).

Greenhouse gas (GHG) emission is one of the serious problems which is coming into the picture, whether fossil fuel or biofuel energy is being used. The main objective here is to find an energy source with ample ability to handle the greenhouse gas emissions efficiently (Faaij 2006). Let us consider the case of biofuel; if the biofuels are extracted in the right way, then greenhouse gas emission will be highly reduced, but if the method of extraction is wrong, then emission may be even more than that for fossil fuels. Plants utilize $CO_2$, a major greenhouse gas emission, for proper growth and food production (Menten et al. 2013). So, plants have the capacity to reduce the level of carbon dioxide in the environment and consequently reduce global warming. If fuels are obtained from plants, when partial combustion produces $CO_2$ then it can be utilized by the plants themselves for their nourishment. So, a continuous cycle having the idea of a one-to-one relationship will be developed. As a result, the gas produced will be taken by the plant and thus there will be zero impact on global warming. But development of such an ideal plant may not be possible (Menten et al. 2013).

For starters, a high amount of energy investment is required for growing the crop in the form of seed planting, ground preparation, and supplying water and nutrients. So, in the end, net negative energy is obtained (Van Eijck et al. 2014). Some companies are focusing on investment of energy in the form of sunlight so as to control the GHG emissions in the production phase. The algae are more suitable for this option where in spite of net energy input, there is no production of GHG (Van Eijck et al. 2014). The other problem coming into the picture is the use of land. The elimination of existing plant life is required for increasing the availability of land for growing biofuels. This is one big disadvantage associated with biofuel: carbon is produced at the cost of plant life. So, again algae are proved to be more beneficial. If somehow the above-mentioned problem is solved, then the impact of biofuels on the environment can be controlled. In such a case, the GHG emission and global warming effect will be lower with biofuels than with fossil fuels (Van Eijck et al. 2014).

## 9.1.2 Bioenergy around the Globe

Bioenergy contributes around 10% of the total primary energy supply globally by 2005 (47.2 EJ of bioenergy out of 479 EJ), mainly for heating and cooking purposes in the residential sector. Bioenergy contributed 78% of total renewable energy in 2005 (Purohit and Chaturvedi 2018). A total of 97% of biofuel is available in the form of solid biomass, out of which 71% is utilized for residential utilities. Recently, around 85% of bioenergy used as a solid fuel for lighting, cooking, and heating purposes with low efficiency (Guo, Song, and Buhain 2015). In most developing countries, traditional bioenergy dominates bioenergy consumption based on biomass. A variety of biomass sources are utilized for the production of bioenergy in various forms. Some of the biomass applicable for bioenergy production and utilized as electricity generation, heat and CHP (combined heat and power) are following below (Wright 2006).

1. Residues obtained during the processing of food, fiber, and wood in the industrial sector.
2. Short-rotation crops are grown in the form of energy plantation.
3. Agricultural residues from the farm sector.
4. Forest residues and agroforest residues.

The traditional forms of biomass like fuelwood, animal dung, and charcoal are crucial sources of bioenergy worldwide. In the past, wood fuels are considered as the most common sources all over the world and gave energy security service to a huge part of society. The purpose of the modern bioenergy system is to develop efficient conversion technology which can utilize the available biomass and convert it into the best form for households, cottage industry, and large-scale industry (Bhutto, Bazmi, and Zahedi 2011). Based on the state of the energy carrier, biofuels are broadly categorized into three different forms, namely, solid fuels (for example wood chips, firewood, pellets, charcoal, briquettes, etc.), liquid fuels (for example bio-oil, biodiesel, and bio-ethanol), and gaseous fuels (for example bio-gas, hydrogen, and synthesis gas) (Speight 2010).

A review of the recent status of bioenergy consumption in G8 + 5 countries is given in this section. This section mainly focuses on production, consumption, and trading of bioenergy in the different countries of the world. The statistics are given by IEA 2007, which provides detailed, complete, and consistent information on energy matters. The statistical analysis, taken for ten years from 1995 to 2005, shows how supply and consumption trends vary in G8 + 5 countries (Dam, Junginger, and Faaij 2010).

## 9.2 BIOMASS GASIFICATION

The conversion of solid and liquid organic compounds into the gaseous and solid states is the primary objective of the gasification process. The gaseous form of the product is generally termed as "syngas," while the solid form is "char," which consists of an inert and unconverted fraction of organic material present in biomass. The biomass gasification process is partial oxidation of biomass at high temperature, so the process is also called a thermochemical conversion process. The product gas has high calorific value and can generate power. This process helps in proper utilization of biomass available for all-around production of energy. It has high potential to be used similarly to fossil fuels and can control the continuous increase in the price of oil and fossil fuel products. This section mainly deals with the chemistry of gasification, gasification media, the equivalence ratio (ER), temperature, the pressure inside the gasifier, and the steam-to-biomass ratio (Molino, Chianese, and Musmarra 2016).

### 9.2.1 GASIFICATION CHEMISTRY

Generally, the chemical reaction that takes place during gasification is endothermic. The main steps of the gasification process are as follows (Molino, Chianese, and Musmarra 2016):

- Oxidation (exothermic process)
- Drying (endothermic process)
- Pyrolysis (endothermic process)
- Reduction (endothermic process)

*Oxidation steps*: The oxidation of biomass is carried out in the partial presence of oxygen in order to maintain the stoichiometric ratio for oxidizing only part of the fuel. The main reactions that take place during this process are the following:

(1) Char combustion

$$C + O_2 \rightarrow CO_2 \quad \Delta H = -394 \frac{kJ}{mole} \tag{9.1}$$

(2) Partial oxidation

$$C + \frac{1}{2}O_2 \rightarrow CO \quad \Delta H = -111\frac{kJ}{mole} \tag{9.2}$$

(3) Hydrogen combustion

$$H_2 + \frac{1}{2}O_2 \rightarrow H_2O \quad \Delta H = -242\frac{kJ}{mole} \tag{9.3}$$

These steps are mainly responsible for supplying the energy required for the entire process, and the product of combustion is a mixture of CO, $CO_2$, and $H_2O$; in this mixture, nitrogen from the air is used instead of pure oxygen.

**Drying steps:** In drying steps, the moisture present in feedstocks is mostly evaporated. The amount of heat needed for drying is proportional to the presence of moisture in the feedstock.

$$moist\ fuel \rightarrow fuels + H_2O \tag{9.4}$$

**Pyrolysis steps:** In these steps, the thermochemical decomposition of biomass takes place, and solids are converted into char and volatiles. The main reaction which shows the entire pyrolysis phenomenon is the following:

$$Biomass \leftrightarrow H_2 + CO + CO_2 + CH_4 + H_2O + tar + char\ (endothermic) \tag{9.5}$$

**Reduction steps:** In reduction steps, all the products of the above steps are involved; the gas mixture and char react with each other, which gives the final product: syngas. The reactions involved in these steps are the following:

(1) Boudouard reaction

$$C + CO_2 \leftrightarrow 2CO \quad \Delta H = 172\frac{kJ}{mole} \tag{9.6}$$

(2) Char reforming reaction

$$C + H_2O \leftrightarrow CO + H_2 \quad \Delta H = 131\frac{kJ}{mole} \tag{9.7}$$

(3) Water-gas shift reaction

$$CO + H_2O \leftrightarrow CO_2 + H_2 \quad \Delta H = -41\frac{kJ}{mole} \tag{9.8}$$

(4) Methanation reaction

$$C + 2H_2 \leftrightarrow CH_4 \quad \Delta H = -75\frac{kJ}{mole} \tag{9.9}$$

The Boudouard reaction and the char reforming reaction are endothermic, while the water–gas shift reaction and the methanation reaction are exothermic in nature; the overall reduction reaction is endothermic. The temperature at which the reduction reaction is carried out plays a significant role in determining the composition of syngas and their characteristics (Molino, Chianese, and Musmarra 2016).

## 9.2.2 GASIFYING MEDIUM

Gasifying agents react with solid carbon and higher hydrocarbon to reduce them into lower-molecular-weight compounds like $CO$, $H_2$, and $CH_4$. The main gasifying agents used for gasification are the following:

(1) Steam
(2) Oxygen
(3) Air

Steam and oxygen are the important gasifying media, which lead to the production of a mixture of syngas with a heating value ranging between 10 and 15 MJ/Nm$^3$, and thus are considered to be desirable properties for the synthesis processes (Ramalingam, Rajendiran, and Subramiyan 2020).

Steam helps in the steam-methane reforming reaction, so it is proved to be better for the adjustment of the $CO/H_2$ ratio in syngas. Also, steam carries heat as it enters the gasifier as a saturated state. If steam is the only gasifying medium, then it needs an external heat supply for maintaining the reactor temperature. Oxygen is used as a gasifying medium as it oxidizes and reacts with carbon-containing compounds via the exothermic route and supplies heat for the gasification process. Thus, with the addition of oxygen, the gasifier becomes autothermal. Air is also a good oxidizing agent, as it is cheap and abundantly available in the atmosphere, but the problem with the use of air is the presence of nitrogen, which reduces the heating value of the syngas (Ramalingam, Rajendiran, and Subramiyan 2020).

## 9.2.3 EQUIVALENCE RATIO

The equivalence ratio is a parameter that indicates quantitatively whether a fuel oxidizer mixture is lean, rich, or stoichiometric. The equivalence ratio is defined by the following equation (Jangsawang, Laohalidanond, and Kerdsuwan 2015)

$$\text{ER} = \frac{\left(\dfrac{\text{air}}{\text{fuel}}\right) \text{stoichiometric}}{\left(\dfrac{\text{air}}{\text{fuel}}\right) \text{actual}} = \frac{\left(\dfrac{\text{fuel}}{\text{air}}\right) \text{actual}}{\left(\dfrac{\text{fuel}}{\text{air}}\right) \text{stoichiometric}} \tag{9.10}$$

Three different cases that come into the picture are below.

(1) ER >1, fuel rich mixture
(2) ER <1, lean fuel mixture
(3) ER = 1, stoichiometric mixture

The main motive behind the gasification process is to produce synthesis gas by varying the equivalence ratio and gasifying agent temperature. For optimization of the gasification, process ER is critical. Two different cases of chemical equilibrium in gasification processes are identified. The first case is with an excess of carbon present in the gasification process, while the second is with an excess of a gasifying agent with all carbon gasified. In conclusion, the optimized values of the ER are given in the table below (Jangsawang, Laohalidanond, and Kerdsuwan 2015).

| Temperature range (K) | ER |
| --- | --- |
| 600–900 | 3.0 |
| 1000–1500 | 2.0 |
| 1600–2500 | 1.5 |

### 9.2.4 GASIFIER TEMPERATURE

The equivalence ratio is the main factor that may affect the gasifier temperature. With an increase in the equivalence ratio, the combustion process is enhanced, and thus the gasifier temperature increases in the combustion or oxidation zone. The increase in gasifier temperature helps in cracking of tar, which improves the endothermic char gasification, which results in the production of producer gas with a higher heating value. Mostly, the oxidation zone has the highest temperature as compared to other zones, and the temperature is always changing during gasification in all zones. This happens because of the self-regulating property of the gasifier (Guo et al. 2014).

## 9.3 GASIFIER DESIGN

Gasifiers are classified into two main categories, namely fixed bed gasifiers and fluidized bed gasifiers. Their designs are mainly based on certain parameters like gasification chemistry, gasifying media, equivalence ratio, steam-to-biomass ratio, gasifier temperature, and pressure. These parameters are also responsible for product yields. Further, fixed bed gasifiers are broadly classified into three main categories, namely (Prins, Ptasinski, and Janssen 2007)

1. Updraft gasifiers,
2. Downdraft gasifiers,
3. Cross draft gasifiers.

And fluidized bed gasifiers are classified into two main categories, namely

1. Circulating fluidized bed gasifiers,
2. Bubbling fluidized bed gasifiers.

## 9.3.1 Fixed Bed Gasifiers

The main parts of the fixed bed gasification system consist of a gasifier or reactor having a cleaning and gas cooling system. The gas and gasifying media may move in an upward or downward direction through the bed of solid fuel particles in the fixed bed gasifier. Its construction is simple, consisting of cylindrical space for fuel and a gasifying agent having a feeding unit, ash collection unit, and gas exit unit. The material used for the construction of a fixed bed gasifier system is concrete, steel, and/or firebricks. As gasification occurs in the fixed bed gasifier, the fuel bed moves slowly in a downward direction (Prins, Ptasinski, and Janssen 2007). The fixed bed gasifier generally operates with high conversion of carbon, high residence time, low gas velocity, and low ash flow as mentioned in Golden, Reed and Das (1988) and Carlos (2005). The major problem occurring in fixed bed gasifiers is with the removal of tar; however, current development in thermal and catalytic conversion of tar gives huge options. The fixed bed gasifier is more suitable for small-scale applications of heat and power. The main equipment used for the gas cleaning and cooling system is a cyclone separator, a dry filter, and wet scrubbers as mentioned in Ghosh, Sagar, and Kishore (2006).

### 9.3.1.1 Updraft Gasifier

The updraft gasifier, also called a counter-current gasifier, is a type of fixed bed gasifier which can use solid biomass having moisture content up to 60%, low volatile matter, and high ash content up to 25%. Some of the advantages of the updraft gasifier are as follows (Ayyadurai, Schoenmakers, and Hernández 2017).

1. Good thermal efficiency
2. Suitable for the high value of moisture in feedstocks
3. Low pressure drops across the reactor
4. Lower slag formation.

Direct firing, in which gases produced are directly utilized in a furnace or boiler, is used in the updraft gasifier. In the updraft gasifier, biomass is fed from the top and the gasifying medium is fed from the bottom of the gasifier. In this updraft gasifier product, gas comes out from the top and ash releases from the bottom. In designing the updraft gasifier, more attention is focused on the amount of tar in the final product (Ayyadurai, Schoenmakers, and Hernández 2017).

The biomass fed into the reactor is slowly burned with air. The upper layer of biomass is cracked first, followed by the next layer, and both layers are converted into char. The directional movement of tar is from bottom to top. In the pyrolysis zone, the temperature is around 600–800°C, where most of the tar was thermally braked, which results in tar-free product gas. The liquid fraction reaches a temperature of around 500°C, which shows that gas production was dominant because the liquid fraction is thermally cracked above 500°C. The main benefit which comes into the picture is minimization of tar content, which leads to improving the composition of the producer gas. Tar cracking is observed between 700 and 1250°C (Ciferno and Marano 2002).

The gasifier design focuses on the process and hardware. The factors under process include the following parameters.

1. Gasifier types,
2. Producer gas yield,
3. Operating conditions,
4. Reactor size.

According to Ciferno and Marano (2002) hardware consists of the following mechanical components:

1. Grate,
2. Reactor body,
3. Insulation,
4. Other components that are specific to reactors.

During the design of the gasifier, the specification of the plant plays a significant role; specification factors include fuel, gasification agent, and product gases. Generally, fuel specification consists of ultimate and proximate analysis. The specification for gasifying agents includes the selection of steam, oxygen, or air and their proportions (Sansaniwal, Rosen, and Tyagi 2017).

The parameters that drastically affect the design of updraft gasifiers are the following:

1. Gasifying agent: if air is used as a gasifying agent, the product gas lower heating value is in the range of 45 MJ/m³ (Roy, Datta, and Chakraborty 2013). High moisture and oxygen content in biomass results in lower heating value.
2. The equivalence ratio is the major parameter in conversion efficiency.
3. Capital cost; capital cost is least for air, followed by steam and pure oxygen.

Table 9.1 lists geometric, operating, and performance output parameters for the designed process.

According to Basu (2010) gasifier design generally begins with mass and energy balance. The mass and energy balance is common to all types of gasifiers. It involves the calculation of producer gas flow rate and biomass feed rate. The power output

**TABLE 9.1**

**Performance Parameters for Optimum Output**

| Geometric parameter | Operating parameter | Performance parameters |
|---|---|---|
| Reactor configuration | Preheat temperature of the air | Carbon conversion efficiency |
| Height | Reactor temperature | Cold-gas efficiency |
| Cross-sectional area | Amount and approximate quantity of gasifying medium | |

required for gasification is a primary input parameter denoted by Q in megawatts (MW). The volume flow rate of the producer gas, $V_g$ (Nm³/s), can be found by the following relationship.

$$V_g = \left( \frac{Q}{LHVg} \right) \tag{9.11}$$

The net heating value can be calculated by using the composition of the gas. The equation for finding the biomass feed rate $M_f$ is as follows:

$$M_f = \frac{Q}{LHV*\text{gasifier efficiency}} \tag{9.12}$$

The equation that is used to relate *LHV* and *HHV* is given by the following relationship.

$$LHV = HHV - 20,300*H - 2260*M \tag{9.13}$$

Where, $H$ = hydrogen mass fraction in the fuel, $M$ = moisture mass fraction, and *HHV* is in kJ/kg on a dry moisture ash-free basis. The relationship between *HHV* on a moisture ash-free basis and that on a dry basis is given by the following equation (Basu 2010).

$$HHV = HHV_d * \frac{(1-M)}{(1-ASH-M)} \tag{9.14}$$

Where subscript $d$ refers to dry, *ASH*=ash fraction in fuel on a raw fuel basis, and $M$=moisture fraction. $HHV_d$ is generally taken as 1821 MJ/kg. Finding $HHV_d$ from the elemental composition of biomass is done by the following equation:

$$HHV_d = 0.3491C + 1.1783H + 0.1005S$$
$$-0.015N - 0.1034O - 0.0211ASH \tag{9.15}$$

Where $C$, $H$, $S$, $N$, $O$, and *ASH* represent the fractions of carbon, hydrogen, sulfur, nitrogen, oxygen, and ash, respectively, in the fuel on a dry basis.

The stoichiometric air requirement of fuel for complete combustion is related as follows:

$$M_a = M_{th} * ER \tag{9.16}$$

Where $M_a$ = amount of air required for unit mass of biomass, $M_{th}$ = theoretical amount of air for the unit mass of biomass, and $ER$ = equivalence ratio. For the gasification of biomass, 0.25 $ER$ is the initial guess.

### 9.3.1.1.1 The Calculation for Reactor Diameter (D)

The size of the reactor, especially the cross-section of a cylinder, is calculated in terms of the diameter in which the fuel is being burned. This is given in terms of fuel consumed per unit time to the specific gasification rate of the fuel ranging from 100 to 250 kg/m²-h. The reactor diameter is calculated using the following relationship:

$$D = \left\{ \left. (4*FCR) \middle/ (SGR*\pi) \right. \right\}^{0.5}$$

(9.17)

Where FCR = fuel consumption rate, and SGR = specific gasification rate given by the following:

$$SGR = \left( \text{weight of biomass fuel used} \left( kg \right) \right) / \left( \text{reactor area} \left( m^2 \right) \right.$$

$$*\left( \text{reactor diameter} - 0.15 \left( m \right) * \text{operating time} \left( h \right) \right) \right).$$

(9.18)

### 9.3.1.1.2   The Calculation for Height (H)

The distance between the top and bottom ends of the reactor is the height of the reactor. The height will determine how long the gasifier will operate with a single load of fuel. Generally, it is the function of the time required to operate the gasifier (T), the specific gasifier rate, and the density of the fuel. The following relationship can be used to calculate the height of the gasifier:

$$H = \frac{(SGR*T)}{\rho}$$

(9.19)

For an operating time of 2.5 h, the density of the fuel is assumed to be 300 kg/m³. Time is the total time required to completely convert fuel into gases inside the reactor. This is the combination of time for igniting the fuel, generating the gas, and completely burning the fuel inside the reactor. It is the function of the density of the fuel, reactor volume, and fuel consumption rate. The relationship used for calculation is given by equation 9.20 (also see Table 9.2):

$$T = \frac{\rho*V}{FCR}$$

(9.20)

Where symbols have the usual meaning.

**TABLE 9.2**
**Dimensions of Updraft Gasifier**

| D | 0.6 m |
|---|---|
| H | 1.0 m |
| T | 2.45 hour |
| FCR | 30 kg/hr |
| AFR | 0.347 m³/s |

### 9.3.1.1.3   The Calculation for Airflow Rate (AFR)

The rate of flow of air required to gasify the fuel is called the AFR. It helps in determining the size of blower required for a reactor in gasification of fuel. This can be determined using FCR, the stoichiometric air of fuel (SA), the density of air ($\rho$), and ER (for wood 0.3 to 0.5). The formula used for calculation is given by the following relationship (Basu 2010):

$$V = \frac{AFR}{area\ of\ reactor} \tag{9.21}$$

## 9.3.1.2   Downdraft Gasifier

On the basis of utility, downdraft gasifiers are more prominent in small-scale applications. Their capacities range between 10 kW and 1 MW. Their construction and operation are simple and they contain less tar in producer gas. However, some of their disadvantages are below:

1. Grate blocking
2. Feedstock should have low bulk density
3. Only suitable for feedstock with lower moisture content.

Improvement in the design of downdraft gasifiers can be based on feeding system modification, air supply, producer gas recirculating system, and discharge system. The downdraft or co-current gasifier operates from top to bottom from the drying zone followed by pyrolysis, oxidation, and reduction zones. The heat generated in the oxidation zone helps in drying or reducing the moisture content of biomass. Also, the excess heat helps in the pyrolysis and reduction process. During volatile pyrolysis, the matter is released. The producer gas obtained during the reduction process comes out through the gas outlet (Susastriawan, Saptoadi, and Purnomo 2017).

The parameters playing a crucial part in the design of downdraft gasifier, resulting in high performance in terms of the quality of the producer gas and gasification efficiency, are mentioned below:

1. Biomass characteristics such as composition, moisture content, and size
2. Air-fuel ratio
3. Gasification temperature
4. Feeding rate.

The equation for gasification efficiency estimation is given by the following relationship.

$$\eta = \frac{(HV_{pg}) * xV}{HV_b} * 100\ \% \tag{9.22}$$

where $HV_{pg}$ represents the heating value of producer gas (kJ/Nm³) and $HV_b$ represents the heating value of biomass feedstocks (kJ/kg).

The heating value of producer gas can be evaluated using the following relationship.

$$HV_{pg} = \frac{(x_1 * HV)_{CO} + (x_2 * HV)_{H_2} + (x_3 * HV)_{CH_4}}{100} \qquad (9.23)$$

Where $x_1$, $x_2$, and $x_3$ are the percentages of $CO$, $H_2$, and $CH_4$ respectively in producer gas. These are measured using a gas chromatograph. The heating values of $CO$, $H_2$, and $CH_4$ are 12.71, 12.78, and 39.76 MJ/m³, respectively (Susastriawan, Saptoadi, and Purnomo 2017).

### 9.3.1.3   Cross Draft Gasifier

The higher exit temperature and poor carbon dioxide reduction are the major disadvantage of the cross draft gasifier. The ash pin, fire, and reduction zones are separated in cross draft gasifiers, unlike downdraft and updraft gasifiers. The design characteristic of the cross draft gasifier is to use wood, charcoal, and coke as a biomass feedstock for gas generation. It uses charcoal gasification with a high temperature of 1500°C in the oxidation zone, which can lead to material problems. The gas composition obtained from the cross draft gasifier is of low hydrogen and methane content, with higher carbon monoxide content (Golden, Reed, and Das 1988). The disadvantages of cross draft gasifiers are minimal tar cracking and very high temperatures in the oxidation zone.

## 9.3.2   Fluidized Bed Gasifier

The fluidized bed gasifier technique is a recent development offering excellent mixing and a high reaction rate for solid-to-gas conversion. Isothermal operation is an advantage of the fluidized bed gasifier. The operating temperature of the gasifier is around 800–850°C. To keep operating temperature in a state of suspension, air is blown through a bed of solid particles at a particular velocity. After reaching the optimum temperature, the biomass feedstock is fed into the gasifier (Siedlecki, de Jong, and Verkooijen 2011).

### 9.3.2.1   Circulating Fluid Bed Gasifier

To entrain large amounts of solids, the fluidizing velocity in the circulating fluid bed is high enough to convert the extensively developed wood waste conversion in pulp and paper mills for firing lime and cement kilns and steam generation for electricity (Raskin, Palonen, and Nieminen 2001).

### 9.3.2.2   Twin Fluid Bed Gasifier

To obtain a high heating value of gas, twin fluid bed gasifier is employed. It consists of two defined zones. In the upper zone, drying, low-temperature carbonization, and cracking of gases will take place. But in the lower zone, gasification of charcoal takes place. The temperatures of the gasifier's upper and lower zones are around 460 and 520°C, respectively. The pressure is around 30 mbar. To encourage carbon-steam reactions and hydrogen enrichment, steam is added to the reactions. The quality of gas is very clean with high energy content (Butt et al. 1991).

### 9.3.1.3 Entrained Bed Gasifier

In entrained flow gasifiers, a finely reduced feedstock is required, but no inert material is present. The entrained bed gasifiers operate at a very high temperature of 1200°C, depending on gasification medium employed, and hence the producer gas obtained from gasifier has a low concentration of tar and condensable gases. However, due to the operation of higher temperatures, material selection and ash melting are major problems in this gasifier. A conversion efficiency of 100% is reported for this gasifier, but not with the biomass feedstock (Bridgwater 1995).

## 9.4 GAS CLEANING AND COOLING

### 9.4.1 CLEANING DUST FROM THE GAS

In the cyclone separator, small carbon particles are very difficult to remove from the gasifier. The particulates present in the cyclone separator were not reduced to less than 5–30 g/Nm³. Ceramic filters used in the hot gas filtration showed the best results compared to other filters. The particulates present in the fixed bed are very low compared to the fluidized bed. In the gasification of peat and coal, high-temperature ceramic filters and metal handle filters are used. Tests with wood-based gasification created a further problem of clogging by soot by thermal cracking of tars both in the gas phase and on the filter surface (Kumara and Shukla 2016). Cooling the gas to a temperature below 500°C will eliminate the problem. However, if the temperature is below 400°C, there is a chance of tar deposition. Smaller fuel particulates can generate excessive dust content in the producer gas when compared to larger fuel blocks. Hardwood generates less dust compared to softwood (Kumara and Shukla 2016). Maize cob gasification leads to unembellished dust contamination. By properly designing the cyclone separator, 60–70% of the dust particles can be removed. The smaller dust particles can then be removed by scrubber or other means (Serrano et al. 2008).

During the Second World War, the gasifier used was installed with a filter containing wood wool, glass wool, sisal fiber, wood chips soaked in oil, and other types of material were used for removal of dust from the producer gas with a particle size below 60 μm, but results obtained were not favorable. By employing cloth filters, the success rate was higher. But the disadvantage is that the cloth filter was very sensitive to gas temperature. The dew point temperature of the gas is around 70°C when agriculture or wood gasification is done (Schneising et al. 2014). Water will condense below this temperature, and a pressure drop will be created in the filter section of the gasifier. At high temperatures, the cloth filter can eliminate dust, but the disadvantage is rapid build-up of dust if the cloth is not frequently cleaned. Smith et al. (2008) developed a woven glass wool filter that was better than cloth filters. The proposed woven glass wool filter can withstand 300°C by heating the insulating material by the hot gas stream coming from the gasifier; a temperature above 100°C can be maintained in the filter, thus avoiding the pressure drop and condensation problems. High-quality producer gas can be generated by using electrostatic filters. However, such filters are very costly and implemented only in 500-kW power generation plants.

### 9.4.2 Tar Cracking

Tars are a major problem in the gasifier. When the tar created in the gasifier starts condensing, fouling will be generated on the surface. The tar concentration mainly depends on the reactor type, processing conditions, and biomass feedstock. Coal gasification is much better than rice husk gasification in terms of tar content (Bridgwater 1995).

There are two basic ways of eliminating tar:

- Catalytic cracking
- Thermal cracking

#### 9.4.2.1 Catalytic Cracking

Catalytic cracking is an effective method for the elimination of tar. A tar conversion efficiency of 99% has been reported when using using nickel-based dolomite and another catalyst at a higher temperature of 800–900°C. Several researchers reported that by using a secondary reactor, tar elimination is better than that in the primary reactor. The dolomite catalyst has a negative aspect of catalyst deactivation and long-term operating experience. The settlement of carbon compounds on the catalyst creates a loss of activity. The metal catalyst is more susceptible to contamination (Bridgwater 1995).

#### 9.4.2.2 Thermal Cracking

The tar derived from biomass gasification is harder to crack by thermal treatment. The tar can be eliminated at the temperature around 800–1000°C by increasing the residence time, and contact surface area can also reduce the tar content. The overall efficiency of the gasifier will be reduced due to the energy supply to heat the surface. Due to the partial supply of oxygen, reduction of tar and the increase of $CO_2$ levels result. As a result, it will reduce the overall efficiency and increase the cost of oxygen at a temperature around 1300°C. In the case of fixed bed gasification, tar and oil are an important contribution to the energy content of gas delivered to a gas turbine combustor (Larson and Hughes 1996).

### 9.5 ALCOHOL PRODUCTION

#### 9.5.1 Thermodynamics of Bio-Methanol Synthesis

Syngas produced from biomass gasification can be used for producing methanol, hydrogen, synthetic natural gas, and dimethyl ether. Methanol can be produced by hydrogenation of CO and $CO_2$ in the presence of a Cu-based catalyst. Carbon monoxide chemically combines with hydrogen to form methanol. Carbon dioxide chemically combines with hydrogen to form methanol and water. Partial oxidation of methane can also be converted to methanol. Methane reacts with sulfuric acid in the presence of $PtCl_2$ catalyst to form methyl bisulfate, sulfur dioxide, and water. Methyl bisulfate hydrolyzes to methanol and sulfuric acid.

## 9.5.2 Unique Higher Alcohol Synthesis

Higher alcohol synthesis (HAS) is a series of exothermic reactions, where CO and $H_2$ (syngas) are converted into short alcohols over a catalyst (Basu 2010).

$$CO + 2H_2O \rightarrow CH_3OH \; \text{Methanol} \tag{9.24}$$

$$2CO + 4H_2O \rightarrow C_2H_5OH + H_2O \; \text{Ethanol} \tag{9.25}$$

$$3CO + 6H_2O \rightarrow C_3H_7OH + 2H_2O \; \text{Propanol} \tag{9.26}$$

$$4CO + 8H_2O \rightarrow C_4H_9OH + 3H_2O \; \text{Butaanol} \tag{9.27}$$

$$nCO + 2nH_2O \rightarrow C_nH_{2n}OH + (n-1)H_2O \; \text{any alcohol} \tag{9.28}$$

The main side reactions are the formation of hydrocarbons, normally dominated by methane together with short paraffin and olefins. Oxygenated by-products such as aldehydes, esters, and ethers might also be formed depending on the catalyst and operation conditions used.

$$CO + 3H_2O \rightarrow CH_4 + H_2O \tag{9.29}$$

The water–gas shift (WGS) reaction occurs simultaneously with catalysts having water–gas shift activity (Basu 2010).

$$CO + H_2O \rightarrow CO_2 + H_2 \tag{9.30}$$

### 9.5.2.1 Biobutanol Production

Due to the rapid rise in energy demand, creative technologies are needed for the production of biobutanol, which can be used as an alternative for fossil fuels. This biofuel can be produced from the use of syngas integrated with anaerobic micro-organisms. Syngas can be generated from the gasification of biomass. In this chapter, the syngas will be produced through downdraft gasification using biomass waste and fruit shells. In this work, $Fe_2O_3$ was selected and prepared from tropical soil for catalytic gasification. Out of 14 phases of iron oxide, $\gamma$-$Fe_2O_3$ would be the preferred phase of the iron oxide due to its advantages like rigid structure, easy availability, and magnetic recoverability. A fermenter with a capacity of 7 liters will be designed for generating 1g/L biobutanol from purified syngas using C1-fixing homoacetogenic and C4-producing butyrate microorganisms or butyrogens (Nanda et al. 2014).

### 9.5.2.2  Green Diesel Production

Oil will be extracted from fruit seeds by using an advanced ultrasonic and microwave-assisted technique to reduce solvent consumption and extraction time. In this proposed project, a hydrocracking reactor made up of a vertical tube stainless steel reactor with 20.46-mm inside diameter, 11.48-mm thickness, and 1016-mm length will be designed and fabricated. A novel bi-functional metal-based catalyst will be used for hydrodeoxygenation and hydrocracking of fruit seed oil. Halloysite-supported metal catalysts will be prepared using an incipient wetness impregnation method, and characterization will be done. The optimization of hydrocracking process parameters and its effects on reaction parameters on green diesel conversion will be analyzed using an artificial neural network (ANN) method (Khanal et al. 2010).

## 9.6  APPLICATION OF BIOFUEL IN FUEL CELLS

Biofuel cells have a variety of purposes, and the most significant uses are discussed below.

### 9.6.1  Transport and Energy Generation

Currently, fossil fuel is the primary supplier of energy in the world. Due to the limited availability of these resources, it is not possible to use fossil fuel indefinitely, and they pose environmental issues. To overcome these problems, biofuel cells using carbohydrates as their source of power can be developed and utilized. A report says that a fuel cell filled with a carbohydrate solution of 1 liter can drive a car for 16 to 19 miles. Similarly, the car with 50-liter tank capacity can power the car for 800 to 850 miles without any refueling. The benefits are not just environmental; biofuel cells would also eliminate the vapor risk factors corresponding to transport with a high load of volatile, flammable fuels, which have the additional risk of fire and road accidents (Rodionova et al. 2017).

### 9.6.2  Implantable Power Sources

Biofuel cells have the capacity to work in living organisms because the required oxygen and fuel can be taken from their living environment for their operation and can be used as a power source for medical implants for a particular range. For example, a glucose biosensor generates its electrical current using glucose oxidase as the anode and cytochrome C as the cathode. This method can be configured with a biosensor format to obtain a glucose level within a range of 1 to 80 mm. Relatedly, a lactate sensor has been developed. Developments have been already made in miniature fuel cells, and these cells can be used as the energy source for drug delivery systems as they are very small (Davis and Higson 2007).

### 9.6.3 Wastewater Treatment

It is evident that various fuel cells have generated power by oxidizing the chemical compounds present in the wastewater. This method has two valuable purposes: (1) elimination of organic compound present in the waste and (2) electrical power generation. A recent study on the topic estimated that an area of wastewater generation for 150,000 people is capable of producing 2.3 MW power (assuming output as 100%), even though a capacity of 0.5 MW could be practical (Davis and Higson 2007).

### 9.6.4 Robots

Robots can "live off the land" and fuel cells can be used as a source of power supply; the various challenges of the robots have already been discussed. The "Slugbot" is an early example; as its name implies, it hunts slugs. The "Slugbot" initially gains its power from a rechargeable battery, but after the slugs have been caught, they are stored in a storage tank before the battery begins running down. After this operation, the robot switches to a microbial fuel cell in which the captured slugs are used for microbial digestion and energy is produced to recharge the robot's battery (Gupta and Tuohy 2014).

## 9.7  LCA ON BIOFUEL PRODUCTION

Life cycle assessment (LCA) is widely used in order to quantify the direct and indirect environmental impacts and other factors related to life cycle energy balance. It involves accounting, assessing, and interpreting the potential environmental impacts generated by the production chain from a life cycle perspective, including raw materials, production, consumption, and waste utilization. Biofuel's life cycle assessment demonstrates the additional quantity of energy required to turn the energy present in the raw materials into useable energy of the fuel and provides an evaluation of the sustainability of the biofuel's production (Sikarwar et al. 2017).

## 9.8  CONCLUSION

Biofuels are more suitable for meeting the high energy demand of current and future generations, especially the second- and third-generation biofuels. Second-generation biofuel extraction has been made possible from a variety of feedstocks, even including municipal solid wastes. These are cost-effective and easily available all around us. Third-generation biofuels are extracted from algal biomass and have different yields with respect to lignocellulosic biomass. Biomass gasification is a suitable technology for converting solid wastes into combustible and noncombustible products. Syngas produced from gasifiers is used to produce methanol and higher alcohol using biological technology. The significant use of biofuels in fuel cells contributes to a wide range of applications.

## REFERENCES

Ayyadurai, Saravanakumar, Loek Schoenmakers, and Juan José Hernández. 2017. "Mass and Energy Analysis of a 60 KWth Updraft Gasifier Using Large Size Biomass." *Fuel.* doi:10.1016/j.fuel.2016.09.080.

Basu, Prabir. 2010. "Design of Biomass Gasifiers." In *Biomass Gasification Design Handbook.* doi:10.1016/b978-0-12-374988-8.00006-4.

Bauen, Ausilio. 2006. "Future Energy Sources and Systems-Acting on Climate Change and Energy Security." *Journal of Power Sources.* doi:10.1016/j.jpowsour.2006.03.034.

Beauchet, R., F. Monteil-Rivera, and J.M. Lavoie. 2012. "Conversion of Lignin to Aromatic-Based Chemicals (L-Chems) and Biofuels (L-Fuels)." *Bioresource Technology.* doi:10.1016/j.biortech.2012.06.061.

Bhutto, Abdul Waheed, Aqeel Ahmed Bazmi, and Gholamreza Zahedi. 2011. "Greener Energy: Issues and Challenges for Pakistan – Biomass Energy Prospective." *Renewable and Sustainable Energy Reviews.* doi:10.1016/j.rser.2011.04.015.

Brennan, Liam, and Philip Owende. 2010. "Biofuels from Microalgae – A Review of Technologies for Production, Processing, and Extractions of Biofuels and Co-Products." *Renewable and Sustainable Energy Reviews.* doi:10.1016/j.rser.2009.10.009.

Bridgwater, A.V. 1995. "The Technical and Economic Feasibility of Biomass Gasification for Power Generation." *Fuel.* doi:10.1016/0016-2361(95)00001-L.

Butt, A.R., M. Jepson, C. Milner, J. Moodie, and P. Annesley. 1991. "Scale-up and Transputer Modelling of the British Coal Twin-Bed Pyrolyser/Combustor System." In *Proceedings of the International Conference on Fluidized Bed Combustion.*

Ciferno, Jared P., and John J. Marano. 2002. *Benchmarking Biomass Gasification Technologies for Fuels, Chemicals and Hydrogen Production.* US Department of Energy, National Energy.

Cui, Hong, Scott Q. Turn, Vheissu Keffer, Donald Evans, Thai Tran, and Michael Foley. 2010. "Contaminant Estimates and Removal in Product Gas from Biomass Steam Gasification." *Energy and Fuels.* doi:10.1021/ef9010109.

Dam, J. Van, M. Junginger, and A.P.C. Faaij. 2010. "From the Global Efforts on Certification of Bioenergy towards an Integrated Approach Based on Sustainable Land Use Planning." *Renewable and Sustainable Energy Reviews.* doi:10.1016/j.rser.2010.07.010.

Davis, Frank, and Séamus P.J. Higson. 2007. "Biofuel Cells-Recent Advances and Applications." *Biosensors and Bioelectronics.* doi:10.1016/j.bios.2006.04.029.

Ding, Liang, Yongqi Zhang, Zhiqing Wang, Jiejie Huang, and Yitian Fang. 2014. "Interaction and Its Induced Inhibiting or Synergistic Effects during Co-Gasification of Coal Char and Biomass Char." *Bioresource Technology.* doi:10.1016/j.biortech.2014.09.007.

Eijck, Janske Van, Henny Romijn, Annelies Balkema, and André Faaij. 2014. "Global Experience with Jatropha Cultivation for Bioenergy: An Assessment of Socio-Economic and Environmental Aspects." *Renewable and Sustainable Energy Reviews.* doi:10.1016/j.rser.2014.01.028.

Faaij, André P.C. 2006. "Bio-Energy in Europe: Changing Technology Choices." *Energy Policy.* doi:10.1016/j.enpol.2004.03.026.

Ghosh, Debyani, Ambuj D. Sagar, and V.V.N. Kishore. 2006. "Scaling up Biomass Gasifier Use: An Application-Specific Approach." *Energy Policy.* doi:10.1016/j.enpol.2004.11.014.

Golden, T., B. Reed, and A. Das. 1988. *Handbook of Biomass Downdraft Gasifier Engine Systems.* SERI. U.S. Department of Energy. doi:10.2172/5206099.

Guo, Feiqiang, Yuping Dong, Lei Dong, and Chenlong Guo. 2014. "Effect of Design and Operating Parameters on the Gasification Process of Biomass in a Downdraft Fixed Bed: An Experimental Study." In *International Journal of Hydrogen Energy.* doi:10.1016/j.ijhydene.2014.01.130.

Guo, Mingxin, Weiping Song, and Jeremy Buhain. 2015. "Bioenergy and Biofuels: History, Status, and Perspective." *Renewable and Sustainable Energy Reviews.* doi:10.1016/j.rser.2014.10.013.

Gupta, Vijai K., and Maria G. Tuohy. 2014. *Biofuel Technologies: Recent Developments. Biofuel Technologies: Recent Developments.* doi:10.1007/978-3-642-34519-7.

Haykiri-Acma, H., and S. Yaman. 2010. "Interaction between Biomass and Different Rank Coals during Co-Pyrolysis." *Renewable Energy.* doi:10.1016/j.renene.2009.08.001.

Jangsawang, Woranuch, Krongkaew Laohalidanond, and Somrat Kerdsuwan. 2015. "Optimum Equivalence Ratio of Biomass Gasification Process Based on Thermodynamic Equilibrium Model." In *Energy Procedia.* doi:10.1016/j.egypro.2015.11.528.

Khanal, Samir K., Rao Y. Surampalli, Tian C. Zhang, Buddhi P. Lamsal, R.D. Tyagi, and C.M. Kao. 2010. *Bioenergy and Biofuel from Biowastes and Biomass. Bioenergy and Biofuel from Biowastes and Biomass.* doi:10.1061/9780784410899.

Kumara, Sunil, and S.K. Shukla. 2016. "Performance of Cyclone Separator for Syngas Production in Downdraft Gasifier." *Advances in Energy Research.* doi:10.12989/eri.2016.4.3.223.

Larson, Eric D., and Wendy E.M. Hughes. 1996. "Performance Modeling of Aeroderivative Steam-Injected Gas Turbines and Combined Cycles Fueled from Fixed or Fluid-Bed Biomass Gasifiers." In *ASME 1996 International Gas Turbine and Aeroengine Congress and Exhibition, GT 1996.* doi:10.1115/96-GT-089.

Lee, Roland Arthur, and Jean-Michel Lavoie. 2013. "From First- to Third-Generation Biofuels: Challenges of Producing a Commodity from a Biomass of Increasing Complexity." *Animal Frontiers.* doi:10.2527/af.2013-0010.

Menten, Fabio, Benoît Chèze, Laure Patouillard, and Frédérique Bouvart. 2013. "A Review of LCA Greenhouse Gas Emissions Results for Advanced Biofuels: The Use of Meta-Regression Analysis." *Renewable and Sustainable Energy Reviews.* doi:10.1016/j.rser.2013.04.021.

Molino, Antonio, Simeone Chianese, and Dino Musmarra. 2016. "Biomass Gasification Technology: The State of the Art Overview." *Journal of Energy Chemistry.* doi:10.1016/j.jechem.2015.11.005.

Nanda, Sonil, Javeed Mohammad, Sivamohan N. Reddy, Janusz A. Kozinski, and Ajay K. Dalai. 2014. "Pathways of Lignocellulosic Biomass Conversion to Renewable Fuels." *Biomass Conversion and Biorefinery.* doi:10.1007/s13399-013-0097-z.

Obeng-Odoom, Franklin. 2013. "The State of African Cities 2010: Governance, Inequality and Urban Land Markets." *Cities.* doi:10.1016/j.cities.2012.07.007.

Prins, Mark J., Krzysztof J. Ptasinski, and Frans J.J.G. Janssen. 2007. "From Coal to Biomass Gasification: Comparison of Thermodynamic Efficiency." *Energy.* doi:10.1016/j.energy.2006.07.017.

Purohit, Pallav, and Vaibhav Chaturvedi. 2018. "Biomass Pellets for Power Generation in India: A Techno-Economic Evaluation." *Environmental Science and Pollution Research.* doi:10.1007/s11356-018-2960-8.

Ramalingam, Senthil, Balamurugan Rajendiran, and Sudagar Subramiyan. 2020. "Recent Advances in the Performance of Co-Current Gasification Technology: A Review." *International Journal of Hydrogen Energy.* doi:10.1016/j.ijhydene.2019.10.185.

Raskin, Neil, Juha Palonen, and Jorma Nieminen. 2001. "Power Boiler Fuel Augmentation with a Biomass Fired Atmospheric Circulating Fluid-Bed Gasifier." *Biomass and Bioenergy.* doi:10.1016/S0961-9534(00)00056-8.

Rodionova, M.V., R.S. Poudyal, I. Tiwari, R.A. Voloshin, S.K. Zharmukhamedov, H.G. Nam, B.K. Zayadan, B.D. Bruce, H.J.M. Hou, and S.I. Allakhverdiev. 2017. "Biofuel Production: Challenges and Opportunities." *International Journal of Hydrogen Energy.* doi:10.1016/j.ijhydene.2016.11.125.

Roy, P.C., A. Datta, and N. Chakraborty. 2013. "An Assessment of Different Biomass Feedstocks in a Downdraft Gasifier for Engine Application." *Fuel.* doi:10.1016/j.fuel.2012.12.053.

Sansaniwal, S.K., M.A. Rosen, and S.K. Tyagi. 2017. "Global Challenges in the Sustainable Development of Biomass Gasification: An Overview." *Renewable and Sustainable Energy Reviews.* doi:10.1016/j.rser.2017.05.215.

Schneising, Oliver, John P. Burrows, Russell R. Dickerson, Michael Buchwitz, Maximilian Reuter, and Heinrich Bovensmann. 2014. "Remote Sensing of Fugitive Methane Emissions from Oil and Gas Production in North American Tight Geologic Formations." *Earth's Future*. doi:10.1002/2014ef000265.

Serrano, C., J.J. Hernández, C. Mandilas, C.G.W. Sheppard, and R. Woolley. 2008. "Laminar Burning Behaviour of Biomass Gasification-Derived Producer Gas." *International Journal of Hydrogen Energy*. doi:10.1016/j.ijhydene.2007.10.050.

Siedlecki, Marcin, Wiebren de Jong, and Adrian H.M. Verkooijen. 2011. "Fluidized Bed Gasification as a Mature and Reliable Technology for the Production of Bio-Syngas and Applied in the Production of Liquid Transportation Fuels – A Review." *Energies*. doi:10.3390/en4030389.

Sikarwar, Vineet Singh, Ming Zhao, Paul S. Fennell, Nilay Shah, and Edward J. Anthony. 2017. "Progress in Biofuel Production from Gasification." *Progress in Energy and Combustion Science*. doi:10.1016/j.pecs.2017.04.001.

Smith, Pete, Daniel Martino, Zucong Cai, Daniel Gwary, Henry Janzen, Pushpam Kumar, Bruce McCarl, et al. 2008. "Greenhouse Gas Mitigation in Agriculture." *Philosophical Transactions of the Royal Society B: Biological Sciences*. doi:10.1098/rstb.2007.2184.

Speight, J.G. 2010. *Biofuels Handbook*. No. 5. Royal Society of Chemistry. doi:10.1039/9781849731027.

Susastriawan, A.A.P., Harwin Saptoadi, and Purnomo. 2017. "Small-Scale Downdraft Gasifiers for Biomass Gasification: A Review." *Renewable and Sustainable Energy Reviews*. doi:10.1016/j.rser.2017.03.112.

Wright, Lynn. 2006. "Worldwide Commercial Development of Bioenergy with a Focus on Energy Crop-Based Projects." *Biomass and Bioenergy*. doi:10.1016/j.biombioe.2005.08.008.

# 10 The Microbiology Associated with Biogas Production Process

## 10.1 INTRODUCTION

Anaerobic digestion (AD) is a well-known technology for producing renewable natural gas, usually called biogas, from organic waste material such as livestock manure, food waste, and sewage sludge. The output producer gas is a combination of methane ($CH_4$; 60%–70%) and carbon dioxide ($CO_2$) including traces of other compounds. This gas is considered as a biofuel as it will directly be utilized for the production of heat and electricity, and up-gradation of this gas to around 90%–99% of methane can make it a viable fuel source for vehicles. The output of the up-gradation process is either liquid biogas (LBG) or compressed biogas (CBG). Compressed biogas is considered a traditional alternative. However, interest in utilizing LBG is gradually increasing because of its higher energy content, which in turn helps in long-distance transportation. Liquid biogas can be utilized in almost all applications where the LNG is usually used nowadays. The advantage of the gaseous state is that the up-gradation cost is lower than the injection cost of the existing natural gas network (Abu El-Rub, Bramer, and Brem 2008).

The anaerobic digestion process was initially used as a treatment of manure and sewage sludge, but nowadays, a variety of waste products and residues generated in industries, agricultural activities, and household wastes are digested in the AD. The substrates which are challenging to degrade, such as various wood residues, pulp residues, lignocellulose materials, paper production residues, microalgae, seaweed, and other biomass substrates that grow specifically in aquatic atmospheres, are gateways for innovation and optimization in this area (Schenk et al. 2008; Ometto et al. 2018). This paves the way for anaerobic digestion to be considered a key process in the growth of biorefinery units and circular economy growth (Hagman et al. 2018). Biogas produced from the anaerobic digestion process contains primarily methane which can potentially by used in renewable energy sources, which in turn will subsequently reduce fossil fuel energy consumption (Yang and Li 2014).

Anaerobic output digestate products are nutrient-rich and they can be potential materials for soil amendment (Sheets et al. 2015). Therefore, utilizing anaerobic digestion can not only mitigate environmental problems related to organic waste, but also additionally enhance energy, water, and food security. In the United States, there are a total of 247 anaerobic digestion systems in process at viable livestock facilities (US EPA 2020). It is calculated that there are 8200 livestock services reaping the benefits of anaerobic digestion (US EPA and Change Division 2011).

The commercial use of anaerobic digestion is expected to increase because of its economic and environmental advantages over recent techniques for dealing with municipal solid waste (MSW) and other types of wastes, for example, incineration, composting, and landfill (Kalyuzhnyi, Veeken, and Hamelers 2000). One reason behind the underutilization of anaerobic digestion might be a lower cost of electricity, which diminishes the financial encouragement to utilize anaerobic digestion for the production of energy (Yiridoe, Gordon, and Brown 2009).

Despite the numerous benefits of anaerobic digestion and advances in system design, various aspects should be improved for commercializing this technology. This includes improving economic and production efficiency, particularly in areas where traditional methods, such as landfilling, cost less (Jenkins et al. 2008). Further developments in feedstock pre-treatment, stability control, digestate utilization, and reactor design are required. The purpose of this chapter is to deliver a comprehensive study on recent anaerobic digestion technologies concerning microbiology of gas, reactor design, operating parameters, and the economic and environmental impacts of the same.

## 10.2 THE MICROBIOLOGY ASSOCIATED WITH THE BIOGAS PRODUCTION PROCESS

The effective generation of biogas involves a complicated biological process. Numerous types of organisms need to be active to generate biogas. Likewise, these microorganisms need to work intently for successful generation. Any interruption in this cooperation causes a decrease in the biogas generation and a breakdown of the whole process. Biogas process control requires information on the microbiology of the biogas production process and basic knowledge about the functioning of these microorganisms.

### 10.2.1 FUNCTIONING AND GROWTH OF MICROORGANISMS

The growth and functioning of an organism require a suitable nutrient base called a substrate. The substrate contains several different elements that provide food for the organism: an electron acceptor and a source of energy. These elements act as building blocks for constructing new cells, and they contain various trace elements such as metals and vitamins. Once the organisms access the substrate, then they will metabolize, generate new cells, and produce energy for its growth. For the biogas production process, different treated organic wastes serve as substrates for different microorganisms. The organic feedstock contains different components which facilitate the growth of microorganisms and lead to efficient biogas generation. Although bigger composition variations are not preferable, variations in time of growth are special because microorganisms grow faster only when specific substrates are present. Apart from the substrates, the development of microorganisms requires a favorable environment for them to bloom and function. Other factors influencing growth include oxygen content, temperature, pH, and salt concentration. Various organisms require different requirements to grow and function optimally. In a biogas production process, the activity of microorganisms increases when the environment of the reactor is conducive to the requirements of as many organisms as possible; this means that a particular atmosphere may not suitable for particular types of microorganisms, yet a few organisms will flourish in that particular condition. Whenever an organism uses the substrate, it will form new cells and different types of waste materials. This is what lies at the heart of a biogas production process: a series of various microorganisms use each other in the biological degradation of the product with respect to time. A few examples of microorganism waste constituents in the biogas production process are carbon dioxide, methane, hydrogen, and fatty acids; all are output products from the biogas generation process (Schnürer and Jarvis 2018).

### 10.2.1.1  Energy Source

Energy sources are the raw materials which the microorganisms utilize to obtain energy for their function and growth, for example, movement and admission of the substrate. It very well may be contrasted with diesel for a vehicle engine and solar energy in a plant. The source of energy for an organism can either be solar energy or chemical compounds. These compounds can be organic compounds like various fats, different types of sugars, and proteins. At a point when microorganisms utilize a chemical species as an energy source, that particular compound gets oxidized and protons/electrons will be moved to employ intermediate carriers into a final electron acceptor. Due to the electron exchange, energy is formed.

### 10.2.1.2  Electron Acceptors

A biological process relies upon the activities of the participating microorganisms. Hence, the way to improve the anaerobic digestion efficiency is to blend and accelerate these actions. Interspecies electron transport between the syntrophic accomplices assumes the significant job of oxidizing the higher natural organic content and reducing the carbon dioxide–to–methane ratio in an anaerobic digestion condition (Baek et al. 2018). Oxygen is considered as the final electron acceptor, called the "electron receiver," from an anaerobic respiration system (inhaling the oxygen). Respiration or fermentation occurs in an inert atmosphere. The fermentation process mainly utilizes different organic constituents as electron acceptors. The output product produced mainly contains various alcohols, acids, carbon dioxide, and

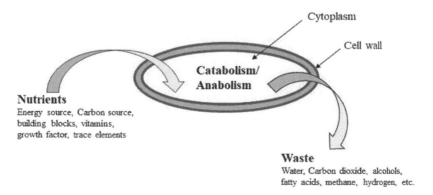

**FIGURE 10.1**   Cell metabolism (Schnürer and Jarvis 2018).

hydrogen. Anaerobic breathing mainly utilizes inorganic material as an electron acceptor. Materials that can be utilized for anaerobic process respiration include carbon dioxide ($CO_2$), iron ($Fe^{3+}$), nitrate ($NO_3^-$), sulfate ($SO_4^-$), and manganese ($Mn^{4+}$). A few microorganisms can utilize a solitary sort of acceptor; however, some electron acceptors are more beneficial than others due to the capacity to produce higher energy. The energy capacity of electron acceptors follows the following order: $O_2 > NO_3^- > Mn^{4+} > Fe^{3+} > SO_4^- > CO_2$. From this order, it is clear that carbon dioxide supplies the least energy and oxygen supplies the most energy. This implies that the microorganisms in an anaerobic digester mostly get lesser energy and subsequently lesser substrate than microorganisms developing in manure, where oxygen is used as an acceptor and they gain maximum energy from the substrate in this way (Schnürer and Jarvis 2018).

### 10.2.1.3   Building Blocks

The primary building block in anaerobic digestion is carbon, which constitutes nearly 50% of the dry weight. The percentages of hydrogen, nitrogen, oxygen are also given in Table 10.1. Additional significant building blocks in AD are phosphorous, sulfur, potassium, sodium chloride, and magnesium. When the energy source is natural, it is considered as a root for building blocks. When the source of energy is inorganic, the source of energy is carbon dioxide, which produces ammonia and carbon. A new cell is formed due to the energy produced by oxidation. The cell structure is also used as a guide for the rough composition of an optimized substrate. The growth of microbes will be partial if the concentration of any of these following building blocks is low.

### 10.2.1.4   Trace Elements and Vitamins

In anaerobic digestion, methanogenesis is the last and the most basic phase which generates biogas, and it is accomplished by a set of anaerobic bacteria called methanogens which belongs domain archaea. Other than the primary elements, such as C, H, N, S, and P, organisms greatly rely upon a component of trace elements for surviving and carrying out different cell metabolisms. Hence, it is essential to have an anaerobic digestion process with an optimal amount of trace elements for stable performance (Aarti, Arasu, and Agastian 2015; Choong et al. 2016). The stability of the process is a significant worry in a commercial anaerobic digestion system because inadequate process stability generally leads to unstable methane production. Furthermore, long instability brings process failure. Trace elements collected from the surrounding environment, in an anaerobic digestion system, such as vitamins and trace elements of organisms are supplied with the help of a substrate.

---

**TABLE 10.1**

**Bacterial Cell Approximate Composition (Schnürer and Jarvis 2018)**

| Component | C | H | N | O | S | P | Na | K | Ca | Fe | Mg | Other |
|---|---|---|---|---|---|---|---|---|---|---|---|---|
| % of dry weight | 50 | 14 | 14 | 20 | 1 | 3 | 1 | 1 | 0.5 | 0.5 | 0.5 | 0.5 |

---

Substance concentrations may vary greatly between different types of substrates. Hence, on some occasions, it is necessary to provide them individually with the anaerobic digestion process (Schmidt et al. 2014; Romero-Güiza et al. 2016). In an anaerobic digestion system, the most essential trace metals are transition elements like Ni, Co, and Fe (Oleszkiewicz and Sharma 1990). The dosage level of trace metals must be less than the toxic or inhibitory level, as an extreme metal concentration in the feed will be the basis for environmental problems and limit the quality of anaerobic products (Thanh et al. 2016).

## 10.2.2 Environmental Factors

The process associated with anaerobic digestion is very complex and involves various degradation processes. The microorganisms particularly involved in specific metabolic processes and they have various environmental requirements.

### 10.2.2.1 Temperature

The optimum temperature in anaerobic digestion will increase the efficiency and the speed of growth of the microorganisms. Organisms can be divided into various groups depending on their optimum growth temperature: mesophilic, thermophilic, hyperthermophilic (Mesbah and Wiegel 2008). The low temperature through the process leads to a decrease in microorganism's growth, biogas generation rate, and substrate utilization rate (Kim et al. 2006). Moreover, low temperatures result in the collapse of cell energy and lowering of the biogas yield, because of the formation of volatile gases (Kashyap, Dadhich, and Sharma 2003). Commonly, the anaerobic digestion process is performed at mesophilic temperatures (El-Mashad, van Loon, and Zeeman 2003). The anaerobic digestion performance at mesophilic temperature is quite stable and it needs a very small energy input (Fernández, Pérez, and Romero 2008). Chae et al. found that a better operation period is 18 days at a temperature of 35°C. A small change in temperature from 35°C to 30°C caused biogas production to decrease (Chae et al. 2008).

The overall change in temperature in the range of 35°C to 37°C is the most suitable for methane production and changes from the mesophilic temperature range to the thermophilic temperature range can cause a sudden slowdown in the production of biogas until the required population has increased (Briški et al. 2007). The biodegradation temperature of the substrate must be below 65°C since temperatures above 65°C lead to the destruction of its shape. However, the thermophilic circumstance has a certain set of advantages, like immediate degradation of organic wastes, higher biogas production, high pathogen destruction, and less viscosity of effluents (Zhu et al. 2009).

The optimal increases in temperature for the methanogenic bacteria are from 37°C to 45°C for *Methanobacterium*, from 35°C to 40°C for *Methanolobus*, from 37°C to 40°C for *Methanobrevibacter* and *Methanoculleus*, from 30°C to 40°C for *Methanobus*, and from 50°C to 55°C *Methanohalobium* and *Methanosarcina* (Ward et al. 2008). As stated above, an increase in temperature causes an increase in metabolic activity and higher degradation of organic products occurs. Hence, an optimum range of temperatures is required for the effective conversion of solid waste into high-quality biogas (Theuerl, Klang, and Prochnow 2019).

### 10.2.2.2  Oxygen

The significance of oxygen concentration differs incredibly for various microorganism networks involved in the biogas generation process. One type of organism, for example, one producing methane, might be extremely sensitive to the oxygen, and it interacts with air, it will immediately die. Other organisms can remain alive at very low concentrations of oxygen, at the same time others develop well if oxygen is available. The oxygen-free radicals are considered strong oxidizing specialists that will decimate cells by oxidizing different cell parts. Organisms that can survive within oxygen contain diverse guard frameworks, such as the various catalysts that will ensure cell formation against the oxygen. The microorganisms which are very sensitive to changes in oxygen do not have a protective enzymatic framework and are demolished due to an increase in oxygen concentrations (Schnürer and Jarvis 2018).

Usually, oxygen will act as an inhibitory agent for anaerobic digestion (Chu et al. 2005). In anaerobic digestion, the occurrence of the process is because of the involvement of anaerobic organisms and a group of methanogens and acetogens. It is believed that instability of the reactor, low methane yield, slow start-up, and sometimes overall reactor failures might occur because of oxygen entrainment in an anaerobic digester. Hence, the inoculum used in an anaerobic digester is deaerated before the operation of the reactor, and in some cases, oxygen scavenging occurs. Furthermore, hydrolysis of the particular matter in the digestion system is detected in the presence of oxygen (Botheju 2011).

### 10.2.2.3  pH

A range of optimum pH levels is suitable for an anaerobic process. For instance, methanogenesis prefers a neutral pH, around 7.0 (Huber et al. 1982). However, a few microorganisms are capable of surviving in lower and higher pH values, including the extremes. For a few microorganisms in the anaerobic digestion process, their pH values vary significantly. Microorganisms that produce acids during the fermentation process live under moderately acidic conditions (pH around 5) (Schnürer and Jarvis 2010). It is recommended that the production of hydrogen gas be maximum if the pH level of the anaerobic system is to be maintained at 9 (Dong et al. 2009). The substrate pH value can be adjusted by mixing in sodium carbonate, sodium bicarbonate, and potassium hydroxide. The maximum biogas yield for cow dung feedstock is obtained when the pH value is 7. Chaban, Ng, and Jarrell 2006), examined seven pH parameters in the range from 5 to 8

### 10.2.2.4  Salts

Salts are important to the function of microorganisms. Salts have necessary building blocks for an organism's growth, such as potassium, sodium, and chlorine. All organisms require a sufficient amount of salts for functioning. These elements are available in the various substrates and it is not necessary to add them separately for the biogas process. However, a few wastes that have higher salt concentrations cause excess air release, which prevents the growth of microorganisms in the biogas production process. Sugars and salts in large concentrations have a preserving impact that can restrain bacterial development. An excessive amount of salt also makes the cell siphon out water and lose both structure and capacity.

A few life forms can adjust to high salt fixations if they are permitted to modify gradually. The most general halophile grows much better in a salt concentration range of 20–30% of sodium chloride (Sambo, Garba, and Danshehu 1995). The few examples of materials that help to increase the salt level in biogases are waste materials like food waste and various protein-enriched materials. These may cause the release of ammonia (Chen, Cheng, and Creamer 2008).

## 10.3 BREAKDOWN OF ORGANIC COMPOUNDS

The anaerobic digestion process consists of various steps, with a huge number of microbial interactions and dependencies. Due to the presence of a few distinctive microorganisms required for the production of biogas, these activities of the microorganisms are required to form biogas as an output product (Schn, Bohn, and Moestedt 2016).

### 10.3.1 HYDROLYSIS

The anaerobic digestion process involves complex feedstocks, and the term "hydrolysis" is utilized to depict a wide scope of solubilization and depolymerization processes in which very complex natural polymer mixtures are decomposed into solvent-mixable monomers. The majority of these reactions, for example, triglyceride, polypeptide, and carbohydrate can be considered as hydrolysis in the strict sense, while others are reductive or oxidative bio-transformations (Sevier and Kaiser 2002). There are three types of main substrates for the hydrolysis process. They are proteins, lipids, and carbohydrates, which hydrolyze to various monosaccharides and glycerol, amino acids, and long-chain fatty acids. In the waste stream to the digester, the feedstock can be a mixture of three different substrates, for example, agricultural waste, manure, primary sludge, etc. (Batstone and Keller 2003).

During the hydrolysis stage, carbohydrates are separated into sugars, whole protein is converted into amino acids, and fats are converted into long unsaturated forms. A portion of the products of hydrolysis that include hydrogen and acetate acid might be utilized by methanogens in the final stage of the anaerobic digestion process (Matheri et al. 2017).

The hydrolysis is normally viewed as a rate-limiting process to degrade organic matter, for example, sewage sludge, crop residue, manure, etc. Hence, the rate in which the overall process occurs is determined by hydrolysis (Angelidaki et al. 2011).

### 10.3.2 FERMENTATION

The fermentation process in the anaerobic system is a conversion of input feedstock without inorganic electron acceptors, for example, sulfate, oxygen, and nitrate. Decrease of protons to shape hydrogen may happen; however, this is commonly facultative. Rather than the degradation of butyrate to acidic acid derivatives and hydrogen, this process is referred to as anaerobic oxidation (Angelidaki et al. 2011). The fermentation process in the biogas process comprises, similar to the hydrolysis reaction, not a single reaction but multiple. The biogas yield is depends on living

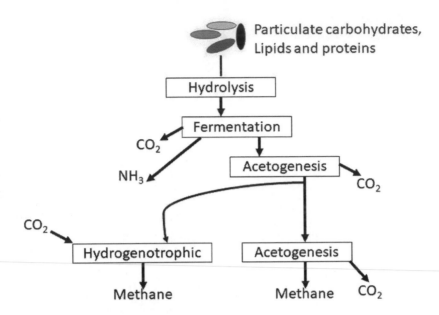

**FIGURE 10.2**    Steps involved in the anaerobic digestion process (Angelidaki et al. 2011).

organisms availability and feedstock treatment during this process. Numerous different microorganisms can live during this stage (Angelidaki et al. 2011).

Though the most available substrate for the fermentation process and main carbon flux are the amino acids and monosaccharides, both of these fermentation processes are different; however, utilization of this substrate is extensive, and the huge range of organisms, frequently clostridia and some other organisms, can be used for both of these substrates (Jabari et al. 2016; Ramsay and Pullammanappallil 2001). The fermentation of monosaccharides and amino acids has mutual elements, for both processes are comparatively rapid and energy-rich, and they have a various range of operating parameters for pH, reduction potential, and oxidation potential (Ramsay and Pullammanappallil 2001; Batstone, Picioreanu, and van Loosdrecht 2006).

### 10.3.3   Anaerobic Oxidation

The output product produced during the fermentation period is further degraded by different oxidation reactions. The output product produced during the fermentation process is additionally broken down into many anaerobic oxidation reactions. In an alternative option to methanogenesis, Valentine (2002) proposed two methods of achieving a sulfate-related methane oxidation process, which is energetically much more favorable and reliable with results based on previous cultivation studies. Studies show that few methylotrophic methanogens produced lipids which are highly isotopically depleted as a comparison with both product and substrate which in turn enhancing the methane oxidization. The methane-oxidizing mechanism produces hydrogen and acetic acids from both molecules of methane that are consumed by sulfate-reducing bacteria (SRB). The reaction between sulfate and methane provides double the amount of energy of methanogenesis reaction. Moreover, this model helps in explaining why a fraction of the lipids in the sample is depleted. Another mechanism suggested that acetate is formed from carbon dioxide and that methane consequently consumes acetate (Hinrichs et al. 2000; Pancost et al. 2000; Nauhaus et al. 2002; Thiel et al. 1999).

### 10.3.4 METHANE FORMATION

Methanogenic microorganisms are responsible for the generation of methane $CH_4$ from carbon dioxide, acetate hydrogen alcohols (Thauer et al. 2008). Methanogens are organisms that help to produce methane as the output product. All the methanogenic organisms belong to the Euryarchaeota. These are very numerous and large groups, all of which can produce methane, and they produce most of the energy in methanogenesis. Methanogenesis is very complex; it requires a large number of single coenzymes.

Methanogens are cultivated from the various operating conditions of anaerobic digestion. They are also common in environments which have extreme temperature, pH, and salinity. The usual methanogenic environments include freshwater sediments, marine sediments, flooded soils, animal and human gastrointestinal systems, anaerobic digestors, termites, landfills, the heartwood of the tree, and geothermal systems (Liu and Whitman 2008). Methanogenesis from marine environments is an important process that generates around 75 to 320 Tg of methane per year, but almost all of the methane is oxidized to $CO_2$; otherwise, it would escape into the atmosphere (Valentine 2002; Liu and Whitman 2008). Though methanogens are diverse and utilize various substrates, they are restricted to three important types: carbon dioxide, methyl group compounds, and acetate (Liu and Whitman 2008).

## 10.4   THE IMPORTANCE OF TECHNOLOGY TO MICROBIOLOGY

As already discussed, various microorganisms' interactions control the biogas production process. This is to attain a functioning and steady-state process with the ability to generate a high quantity of methane. This is very important to preserve an environment favorable for these microorganisms. In this situation, only technology can come into the picture. With the help of recent technologies, we can control and alter the condition of microorganisms. In this chapter, a few important conditions like biogas process start-up, process design, and important operating parameters are discussed.

### 10.4.1 START-UP OF A BIOGAS PROCESS

The biogas production process depends on microorganisms that are living, while the degradation of the substrate occurs normally in an 1 bar atmospheric condition This is especially true in a situation where biogas is generated naturally in wetlands, rice paddies, and lake deposits. Hence, when a manmade biogas process is started, it is convenient to use something like cow manure as an inoculum. On a basic level, the cow rumen works similar to a little biogas reactor, and entire essential organisms for methane generation are identified in the cow rumen. The temperature level in cow rumen is a little higher than the atmospheric temperature (around 39°C) of the residue where microorganisms live. The cow rumen's temperature is adjusted to a temperature in the mesophilic range, which is very similar to that of the man-made biogas reactor (Sun et al. 2015).

Whenever a biogas-producing reactor is started up, organisms present in the inoculum require some time to align themselves to the substrate. In an anaerobic digestion plant, both the environment and substrate vary from the actual environment and it requires some time (residence time) to stabilize the process. Through this period, the microorganisms can live in this new environment and grow. Hence, it is better to start the anaerobic digestion process with higher microbial activity. The environment in which the inoculum was initially taken for the anaerobic digestion process is different from that of the inoculum in the anaerobic digestion tank. The optimal condition for biogas production requires a longer start-up period (De Vrieze et al. 2012; Müller et al. 2016).

## 10.4.2 PROCESS DESIGN

A digestion system can be set with different configurations. For instance, a process can be run in a continuous or batch mode and with individual or multi-stage systems. The design choice sometimes depends on substrate composition and water content. This will give information as to whether it needs pumping or not. The anaerobic digestion process occurs in a closed tank in an inert atmosphere. However, few organisms in this process can utilize oxygen for their metabolic functions. This means that only a limited quantity of oxygen can enter the digestion tank without discontinuing the process. Though oxygen leakage permits non-methane-generating microorganisms to utilize organic substances for their development, a small amount of carbon is converted into methane (Schnürer and Jarvis 2018).

The commercial possibility of anaerobic digestion is largely dependent on its process steadiness as well as on its capability to manage various heterogeneous materials. The various operating factors like pH, mixing rate, temperature, retention time, loading rate, and availability of micro or macronutrients are very important for optimum performance of the anaerobic digestion system. In this regard, to achieve higher process efficiency, the above-mentioned parameters should be controlled effectively in order to allow these characteristics to play a significant role in the performance of the anaerobic digestion system (Jouzani 2018).

## 10.4.3 IMPORTANT OPERATING PARAMETERS

The following section explains some major operating parameters and their effects on the biogas production process.

### 10.4.3.1 Feedstock Composition

The anaerobic digestion process is used to deal with different types of solid wastes, industrial wastes, municipal solid wastes, and garden wastes, using a single type of feedstock. When a variety of feedstocks are present, it is called a co-digestion process. Both of these feedstocks have their characterization and composition that will produce an effect on the biogas production process (Rocamora et al. 2020). The methane potential for various feedstocks is very important to select good material and characterize it to optimize the efficiency of the process. The substrate composition also controls the process parameters and helps select the suitable substrate for the anaerobic digestion process to attain process stability and increase the process efficiency. A study conducted on the co-anaerobic digestion of fruit waste along with cattle manure, chicken manure, and vegetable waste indicated stable operation when cattle manure was around 50% of the feed. However, the addition of vegetable waste and fruit at around 30% in the raw feed increases the instability, does not change the methane production percentage, and decreases the pH level from 7.7 to 7.2. This result indicates the risk when acidification potential is present in the feedstock (André et al. 2019).

### 10.4.3.2   C/N Ratio

Input feedstock properties, for example, total organic carbon, total nitrogen, and C/N ratio are some important parameters in the anaerobic digestion process. Adding co-substrate to match some components is a common way to achieve the goal of stability in the anaerobic process. The nitrogen obtained from the anaerobic digestion reactor derived by using proteins requires microbial growth, and a low carbon-to-nitrogen ratio in the reactor can generate an accumulation of ammonia in the anaerobic reactor, which increases the toxicity, which in turn affects the methane and biogas yields and sometimes results in process failure (Jokela and Rintala 2003).

### 10.4.3.3   Particle Size

Particle size reduction is a general pre-treatment of feedstock for better results, and it produces the organic substance and enhances the feedstock kinetics by giving greater surface area. Regardless of the general application of the reduction in size for pre-treatment, this is important for better biogas yield. Various studies focus on a very small particle size of feedstock usually associated with better biogas generation, which will enhance the kinetics; some studies reported that size drop from 100 to 2 mm increases yield up to 24% with sisal fiber treatment. Understanding the benefits of reduction in particle size will be comparatively difficult and time consuming (Mshandete et al. 2006).

The particle size effect in biogas production is investigated in the landfill, where various literature shows that reduction in size led to reducing the yield of biogas because of higher fatty acid formation, which sometimes generates a movement of methanogenic activity (Barlaz, Ham, and Schaefer 1990).

## 10.5   SUBSTRATES

The feedstock supply to anaerobic digestion is called a substrate (food) for the microorganism's growth. Their properties have a major influence on process efficiency and stability. The composition of the substrate is necessary for both the quality and quantity of biogas formed.

### 10.5.1  SELECTION OF SUBSTRATES

Various types of organic feedstocks can be potentially utilized for the production of biogas, probably a lot more than used nowadays. The main raw source of materials used for the production of biogas today in Sweden is the municipal sewage sludge from treatment plants. Some other commonly used substrates for the production of biogas are waste from slaughter household, food waste and industries feed, and animal manure. Some examples of other materials used include fryer fat, grease trap waste, waste generated from paramedical industries and dairy farms, distillation waste, and grass silage. Soon, waste from the agricultural sector and different crop wastes will also become an important possible substrate for the production of biogas. Some other less general feedstocks that are being currently evaluated for the production of biogas include microalgae, feathers, grass, and woody biomass. The net amount of biogas production today concerning output energy is around 1.3 TWh/year, though theoretical energy potential in the domestic waste that excludes forest residue is approximately 15 TWh/year (Nordberg,).

### 10.5.2  PRE-TREATMENTS

Lignocellulose wastes, for example, forestry by-products and waste from meat, poultry, fish industries, and sewage sludge, have higher biogas potential. However, a very complex flow structure of the microbial cell from the sewage sludge and lignocellulose makes the hydrolysis process a rate-controlling step in the digestion system. The pre-treatment process is a well-established method to enhance the degradation of the above-mentioned wastes.

General pre-treatment approaches include physical (e.g., ultrasonic, irradiation, heat, and pressure), chemical (e.g., oxidation, ozonation, acids, and alkali treatments), and biological (e.g., the addition of bacteria/fungi/enzymes) treatments. The pre-treatment will enhance the availability of the feedstock to microbial degradation by breaking down the physical structure, reducing the particle size, changing the porosity of biomass, and increasing the surface area. Various studies that explored the impact on the yield of methane due to pre-treatment of different feedstocks and different methods showed improved efficiency of the process. Even though some studies have investigated the effect of pre-treatment on the microorganism communities and its relationship to enhanced methane yields, most of the investigations conducted to date are on anaerobic digestion of waste sludge with various results (Westerholm and Schnürer 2019).

### 10.5.3  SANITATION

It is essential to sanitize (kill) some pathogens present in the substrate to prevent contamination from reaching the digestate and substrate materials. The most general method to disinfect the substrate in the biogas production process is to heat the substrate at 70°C for about 1 hour. This is the remedy that is essential for low-problem animal manure in the EU Regulation EEC 2002. Another alternative method to Appert's method is pasteurization that gives a consistent level of sanitation, which has been in operation since 2007. Preconditioning is one of the methods which reduces the concentration of bacteria by 1,000,000 times and its heat resistant capacity by a thousand times. The presence of infective microorganisms from the substrate can impact the usefulness and quality of anaerobic residue (Schnürer and Jarvis 2010).

### 10.5.4 THICKENING

The dry solid contents of the substrate can be increased by permitting the feed-stock to pass via a screw or press. This is necessary in order to reduce the digester's volumetric loading. The free volume might be used to expand the organic loading, which would increase the quantity of gas yield. The problem in removing a quantity of water is that it involves some risk to the primary nutrients; for example, salts and some kind of organic feedstock can be dissolved into the water, causing it to lose its substrate for biogas generation. The dehydration process also causes increased wear on the wheel, the mixer, and the grinders (Schnürer and Jarvis 2010).

### 10.5.5 REDUCTION OF PARTICLE SIZE/INCREASED SOLUBILITY

There are a variety of pre-treatment methods used for the substrate in the biogas production process that can help in enhancing the decomposition behavior. The most general is mechanical pre-treatment, which is used for screws, blenders, androtary cutter mill knives. Degradation can be achieved by biological, chemical, and thermal treatments, and/or by adding acids/bases, hydrolytic enzymes, electroporation, and ultrasound methods. The total surface area of feedstock can be increased by also using grinding. The organisms is widely used in most of the methods for degrading material because organisms are active in the hydrolytic process. The total surface area depends on particle size: the shorter the particle is, the higher the overall surface area present in it (Vázquez-Padín et al. 2009).

## 10.6 TOXICITY

Various substances prevent the biogas production process, and methane producers are typically very sensitive when disturbed. The unwanted substance will get into the anaerobic digestion system by poorly arranged or polluted feedstock or it can be generated with the decay of a precursor unwanted substance. The impact of the toxic material may change, and the process might differ depending on some parameter, for example, the concentration of a substance, temperature, pH, or type of organisms available. If anaerobic digestate substrate residue is used as a fertilizer in the landfill area, there will be a different toxic substance produced that may adversely impact the soil microorganisms.

### 10.6.1 INHIBITION LEVELS

The difference in the concentrations of different substances can lead to the stoppage of the process. These different results in the level of inhibition are affected by various factors such as antagonism, synergism, complex formation, and adaptation. The microorganisms present in the system can sometimes recover before a disturbance, although the inhibition process is often irreversible. This means that organisms are not able to restore the repressive effects, and even repressive producing material will disappear. This process will be started again and fresh microorganisms will be produced. In another way, the microorganisms might be undamaged but inhibited later during an biogas generation period that can be regenerated. At this time, it is imperative to discuss the lag period, i.e., the period when microorganisms stop developing or in which their growth rate is slowed down because of inhibitory effects (Zhou, Brown, and Wen 2019; Chen, Cheng, and Creamer 2008).

### 10.6.2 Inhibiting Substances

Various substances will be inhibitory to the generation of the biogas process. A material is determined to be inhibitory to the biogas generation process when it results in opposing shifts in the microorganism's population or prevents bacterial development. The level of inhibition is generally shown by the reduction of stable levels of accumulation of organic acids and methane production. Some common inhibitors are ammonia, sulfide, and light metal ions (i.e., aluminum, calcium, magnesium, potassium, sodium) (Chen, Cheng, and Creamer 2008). Based on the source, the waste contains inhibitory or even toxic materials. Because of variation in the microorganisms, the composition of the waste, and operating parameters, the level of the inhibitory substance may vary every consecutive batch in the digestor. Before going to the anaerobic digestion process, it is essential to choose the correct material for doing anaerobic process and also to incorporate new methods to remove toxins before the anaerobic digestion process will help in making the waste treatment method effective.

## 10.7  MONITORING AND EVALUATION OF THE BIOGAS PRODUCTION PROCESS

It is necessary to cautiously monitor the generation of the biogas process, thereby spotting problems earlier and correcting them before they deteriorate the process. A few organisms like methane production organisms are very sensitive and may stop developing or wash away from the process. One example is that the temperature of the process must be monitored closely, as microorganisms are sensitive to temperature fluctuation. pH, fatty acid concentration, and ammonia and carbon dioxide content of biogas are all essential parameters that need to be monitored through the biogas process.

### 10.7.1 Monitoring Involved in the Biogas Process

The biogas production process is continuous; hence, it requires regular maintenance and supervision. Apart from this, mixer, pumps, gas collecting facilities, etc., must be working properly, and hence a frequent status check is essential. In addition to this, it is necessary to monitor the digestate tank and substrate tank because some microbial processes occur there too. The substrate tank must be kept at atmospheric temperature to avoid the decomposition process from making an impact on foaming and low pH, or the anaerobic digestor tank covered to avoid unintentional methane emission and nitrous oxide. It is necessary to monitor the daily basis of the biogas production process, and the inspection staff should be trained well. It is quite beneficial to make monitoring schedules and reports on a weekly or monthly basis to ensure the stability of operating parameters; for example, pH, temperature, alkalinity, ammonium, gas flow, and fatty acids are to be monitored.

Some parameters will be measured for longer periods between the biogas process. There are some laboratory procedures and existing sampling techniques to be conducted frequently for sampling, and some new methods that will facilitate continuous monitoring of the process. The biological process will be examined at various locations and approximate sampling places within the biogas generation process, including the inlet and outlet of the anaerobic digester and the sanitation or substrate tank. The best way to take a sample is at the period of mixing and pumping the slurry. Samples should be tested at shorter intervals if a disparity or problem is suspected. The sampling method at the location is significant. The biological process will be evaluated at various locations. Samples should be taken in the same way at all times.

### 10.7.1.1 Loading and Retention Time

The loading rate is important to obtain a constant decomposition rate. The time available for biological degradation is decided based on the retention time. Based on the microbiological view, these two process parameters are important for process efficiency. The organisms grow in a uniform substrate in a particular period and they need sufficient time to break down the substrate at a larger rate. The retention time and load are governed at a relatively high level; the uniform loading of the anaerobic digestor will ensure the analyzing of the substrate composition from the tank. Now, dry solids and volatile solids are mixed completely in the tank. It is tough to predict the exact rate of organic loading required; in this case, trial and error will be used, especially when a start-up new substrate is provided. The retention period is governed by several factors, such as characteristics of substrate and process temperature. Usually, slowly degraded materials like cellulose material require longer retention times as compared to easily degraded materials, for example, food waste.

The thermophilic process often runs at slightly lower retention time as compared to mesophilic processes from bacterial activity increases by an increment in temperature. Generally, the mesophilic process needs a retention period of approximately 1 day and a thermophilic process requires 12 days (Schnürer and Jarvis 2010). Most of the co-anaerobic digestion facilities in Sweden work at quite a higher retention time; about 20 to 30 days is usual. Here two shorter and longer times are used for degradation of the materials with higher water concentration, which include digestate sludge and industrial wastewater; those must be re-introduced into the anaerobic digestion process to retain the organisms or otherwise they will be washed away. In this situation, the hydraulic retention period will be decreased and the solid retention time will increase. The retention time of the process is only a couple of days (Kroeker et al. 1979).

### 10.7.1.2 Substrate Composition

The anaerobic digestion rate is strongly affected by the composition, complexity, and availability of the substrate present in the system (Ghaniyari-Benis et al. 2009; Zhao, Wang, and Ren 2010). Various types of carbon sources will support diverse microbial groups. Before doing the anaerobic digestion process, we need to characterize the substrate for lipid, carbohydrate, fiber, and protein content (Lesteur et al. 2010). Apart from this, the substrate must also be characterized by the quality of methane that can be generated by using anaerobic circumstances.

The carbohydrates are an essential organic component obtained from MSW for the generation of biogas (Dong et al. 2009). However, starch can act as a low-cost feedstock for the production of biogas, in contrast to glucose and sucrose. Experimentally obtained results show that the concentration of initial and substrate total solid content in the anaerobic reactor can disturb the efficiency of the whole biogas production process and the methane production amount during the process (Fernández, Pérez, and Romero 2008).

### 10.7.1.3 Gas Quantity

The quantity of biogas is very important to measure the condition of the process. Gas production rate reduction does not relate to the loading of the fresh substrate if the system is working optimally. The interrelation between the amount of biogas generated and the quantity of natural content provided also helps to evaluate the performance of the whole process. The usual natural gas production process produces biogas at a magnitude of 1 to 4 $m^3$ per $m^3$ of digester tank volume per day. The biogas plant should be able to accumulate this quantity of biogas all day. It is useful to interact with some devices to measure the amount of biogas produced in the collection unit. There are different varieties of flow meters used nowadays for this resolution. The quantity of biogas is frequently represented in cubic meters ($m^3$); for example, the quantity of biogas at 0°C and at an outside environmental pressure of 1.01325 bar (Schnürer and Jarvis 2010).

### 10.7.1.4 Gas Composition

The composition in biogas is another essential parameter to determine the process condition. A small quantity of methane and an increased fraction of $CO_2$ suggests that methane $CH_4$ generation was reserved. This is the indication that a few problems occurred in this process. All the compounds of gases consisting of biogas are generated at the time of decomposition by a different substance with the assistance of microorganisms. Natural biogas contains mainly carbon dioxide and methane, with other gases in very small quantities (i.e., hydrogen, nitrogen, ammonia, and sulfite). Commonly, biogas is saturated along with water vapors (Murphy 1949).

The composition of biogas can be analyzed by passing the produced biogas continuously over an analyzer. Another method is to get the sample and segregate gas from the gaseous phase for consecutive investigations. This procedure is regularly used when this process is analyzed from the laboratory. Various analysis methods may be applied. Using an anaerobic fermentation tube called an Einhorn saccharometer is a very fast method to find out the concentration of $CO_2$. This has a strong (7M) solution of lye, and within it, an identified quantity of biogas sample was injected. Hence, $CO_2$ will dissolve in the lye, and at the same time, methane will generate a gas bubble inside a tube. The $CO_2$ concentration can be evaluated by measuring the entire quantity of gas and then comparing this to the initial injected volume. Based on this procedure, it is essential to know the truth that immediate variation in the pH will cause the discharge of salt of bicarbonate, which dissolves all the organic substances in the anaerobic digester tank, in the form of carbon dioxide. The measured $CO_2$ concentration becomes higher than normal (Schnürer and Jarvis 2010).

### 10.7.1.5 Process Efficiency

The anaerobic digestion process relies upon the effective conversion of a biological substance into value-added products collectively known as biogas, which contains methane as a main ignitable constituent. The anaerobically generated biogas could be used as a combustible source for home lighting, cooking, and heating and also some additional applications. The anaerobic digestion processes largely depend upon mutual interaction of microorganisms in decomposing complex and organic matters to solvable monomers, for example, fatty acids, amino acids, glycerol, and simple sugars. For anaerobic digestion, process efficiency is a vital component to understand the actual biological process and possible chemical reactions. Even though there are many benefits of anaerobic digestion, poor operation stability will delay the technology from being widely implemented (Dupla et al. 2004).

Many factors will affect the stability and performance of the anaerobic system. Some chemical processes which are related to hydrogen partial pressure, interspecies hydrogen transfer, and microbial electrochemical systems are implemented to enhance the process efficiency of the anaerobic system through enhancing the microbial interactions. The above factors accelerate the methanogenic process and enhance the performance of the process.

## 10.8   THE DIGESTED RESIDUES

The organic substance degradation inside the AD process will generate biogas apart from the residue which is digested. If it is found to be of better quality, it will be utilized as a fertilizer for soil amendment. The nutrition content present in an organic substrate is exposed and added to end products of the anaerobic digestion. If the anaerobic digestion is performed with a comparatively pure substrate like manure, plant material, and food waste removed from the source, these residues can be potentially utilized as bio-manure. This product is different from residue sludge obtained from wastewater plants. Due to their metal content, organic digestate are hardly ever suitable for agricultural land applications. The nutritional content and quality of the digestion rate will be determined by many factors, which include the method of pre-treatment, variety of substrate, conditions of the process (i.e., residence time, temperature, etc.), storage, and post-digestion processes.

There are many environmental benefits of using anaerobically digested as compared to commercially available mineral fertilizer. The main aim of this is to decrease the utilization of fossil fuel and to recycle nutrients into the soil. The manufacturing and transportation of mineral fertilizers are considered energy-intensive and involve nitrous oxide emissions, which is considered as the most dangerous greenhouse emission. It is beneficial to digest the manure and then to spread it on the fields instead of using undigested manure. The nutrient availability in fresh manure is mainly dependent on animal type, but not all nutrients can be absorbed by the plant's roots. The result of this increase in nutrients is leaching from soil and finally mixing with groundwater, which causes eutrophication.

## 10.9 CONCLUSION

The biogas production process is a biological process that needs unity between various microorganisms to function correctly. Anaerobic digestion is a multi-purpose technology which can stabilize solid waste and convert waste organic complex feedstock into biogas. Anaerobic digestion generates biogas as an environmentally friendly and renewable energy resource, and it favors lower odour production and greenhouse gas emissions. The bacterial decomposition process that occurs in AD is understood; the complex process is monitored using modeling packages like anaerobic digestion modeling no.1 (ADM1) to understand the mechanism clearly.

Anaerobic digestion has a great ability to solve waste management problems and lead to profit options such as biogas production. This AD process utilizes waste material like cow manure as a feedstock; since this feedstock is abundantly available throughout the year, it can provide both energy recovery and economic benefits. So, the application of an AD system is very significant to achieve environmental goals and set up a sustainable environment. Better information on the processes and microbiology associated with AD is important to optimize the process parameters and increase the process efficiency. The substrate pre-treatment process might reduce the degradation time, and biogas up-gradation for alternative utilization and transportation fuel may increase the profitability of the anaerobic digestion process.

## REFERENCES

Aarti, Chirom, Mariadhas Valan Arasu, and Paul Agastian. 2015. "Lignin Degradation:A Microbial Approach." *South Indian Journal of Biological Sciences* 1 (3): 119. doi:10.22205/sijbs/2015/v1/i3/100405.

Abu El-Rub, Z., E. A. Bramer, and G. Brem. 2008. "Experimental Comparison of Biomass Chars with Other Catalysts for Tar Reduction." *Fuel* 87 (10–11): 2243–52. doi:10.1016/j.fuel.2008.01.004.

André, L., I. Zdanevitch, C. Pineau, J. Lencauchez, A. Damiano, A. Pauss, and T. Ribeiro. 2019. "Dry Anaerobic Co-Digestion of Roadside Grass and Cattle Manure at a 60 L Batch Pilot Scale." *Bioresource Technology* 289 (July). Elsevier: 121737. doi:10.1016/j.biortech.2019.121737.

Angelidaki, Irini, Dimitar Karakashev, Damien J. Batstone, Caroline M. Plugge, and Alfons J. M. Stams. 2011. *Biomethanation and Its Potential. Methods in Enzymology.* 1st ed. Vol. 494. Elsevier Inc. doi:10.1016/B978-0-12-385112-3.00016-0.

Baek, Gahyun, Jaai Kim, Jinsu Kim, and Changsoo Lee. 2018. "Role and Potential of Direct Interspecies Electron Transfer in Anaerobic Digestion." *Energies* 11 (1). doi:10.3390/en11010107.

Barlaz, Morton A., Robert K. Ham, and Daniel M. Schaefer. 1990. "Methane Production from Municipal Refuse: A Review of Enhancement Techniques and Microbial Dynamics." *Critical Reviews in Environmental Control* 19 (6): 557–84. doi:10.1080/10643389009388384.

Batstone, D. J., C. Picioreanu, and M. C. M. van Loosdrecht. 2006. "Multidimensional Modelling to Investigate Interspecies Hydrogen Transfer in Anaerobic Biofilms." *Water Research* 40 (16): 3099–108. doi:10.1016/j.watres.2006.06.014.

Batstone, Damien J., and J. Keller. 2003. "Industrial Applications of the IWA Anaerobic Digestion Model No. 1 (ADM1)." *Water Science and Technology* 47 (12): 199–206. doi:10.2166/wst.2003.0647.

Botheju, Deshai. 2011. "Oxygen Effects in Anaerobic Digestion – A Review." *The Open Waste Management Journal* 4 (1): 1–19. doi:10.2174/1876400201104010001.

Briški, F., M. Vuković, K. Papa, Z. Gomzi, and T. Domanovac. 2007. "Modelling of Composting of Food Waste in a Column Reactor." *Chemical Papers* 61 (1): 24–29. doi:10.2478/s11696-006-0090-0.

Chaban, Bonnie, Sandy Y. M. Ng, and Ken F. Jarrell. 2006. "Archaeal Habitats – From the Extreme to the Ordinary." *Canadian Journal of Microbiology* 52 (2): 73–116. doi:10.1139/w05-147.

Chae, K. J., Am Jang, S. K. Yim, and In S. Kim. 2008. "The Effects of Digestion Temperature and Temperature Shock on the Biogas Yields from the Mesophilic Anaerobic Digestion of Swine Manure." *Bioresource Technology* 99 (1): 1–6. doi:10.1016/j.biortech.2006.11.063.

Chen, Ye, Jay J. Cheng, and Kurt S. Creamer. 2008. "Inhibition of Anaerobic Digestion Process: A Review." *Bioresource Technology* 99 (10): 4044–64. doi:10.1016/j.biortech.2007.01.057.

Choong, Yee Yaw, Ismail Norli, Ahmad Zuhairi Abdullah, and Mohd Firdaus Yhaya. 2016. "Impacts of Trace Element Supplementation on the Performance of Anaerobic Digestion Process: A Critical Review." *Bioresource Technology* 209. Elsevier Ltd: 369–79. doi:10.1016/j.biortech.2016.03.028.

Chu, Li Bing, Xing Wen Zhang, Xiaohui Li, and Feng Lin Yang. 2005. "Simultaneous Removal of Organic Substances and Nitrogen Using a Membrane Bioreactor Seeded with Anaerobic Granular Sludge under Oxygen-Limited Conditions." *Desalination* 172 (3): 271–80. doi:10.1016/j.desal.2004.07.040.

Dong, Li, Yuan Zhenhong, Sun Yongming, Kong Xiaoying, and Zhang Yu. 2009. "Hydrogen Production Characteristics of the Organic Fraction of Municipal Solid Wastes by Anaerobic Mixed Culture Fermentation." *International Journal of Hydrogen Energy* 34 (2). International Association for Hydrogen Energy: 812–20. doi:10.1016/j.ijhydene.2008.11.031.

Dupla, M., T. Conte, J. C. Bouvier, N. Bernet, and J. P. Steyer. 2004. "Dynamic Evaluation of a Fixed Bed Anaerobic Digestion Process in Response to Organic Overloads and Toxicant Shock Loads." *Water Science and Technology* 49 (1): 61–68. doi:10.2166/wst.2004.0019.

El-Mashad, Hamed M., Wilko K. P. van Loon, and Grietje Zeeman. 2003. "A Model of Solar Energy Utilisation in the Anaerobic Digestion of Cattle Manure." *Biosystems Engineering* 84 (2): 231–38. doi:10.1016/S1537-5110(02)00245-3.

Fernández, J., M. Pérez, and L. I. Romero. 2008. "Effect of Substrate Concentration on Dry Mesophilic Anaerobic Digestion of Organic Fraction of Municipal Solid Waste (OFMSW)." *Bioresource Technology* 99 (14): 6075–80. doi:10.1016/j.biortech.2007.12.048.

Ghaniyari-Benis, S., R. Borja, S. Ali Monemian, and V. Goodarzi. 2009. "Anaerobic Treatment of Synthetic Medium-Strength Wastewater Using a Multistage Biofilm Reactor." *Bioresource Technology* 100 (5). Elsevier Ltd: 1740–45. doi:10.1016/j.biortech.2008.09.046.

Hagman, Linda, Alyssa Blumenthal, Mats Eklund, and Niclas Svensson. 2018. "The Role of Biogas Solutions in Sustainable Biorefineries." *Journal of Cleaner Production* 172: 3982–89. doi:10.1016/j.jclepro.2017.03.180.

Hinrichs, Kai Uwe, Roger E. Summons, Victoria Orphan, Sean P. Sylva, and John M. Hayes. 2000. "Molecular and Isotopic Analysis of Anaerobic Methaneoxidizing Communities in Marine Sediments." *Organic Geochemistry* 31 (12): 1685–1701. doi:10.1016/S0146-6380(00)00106-6.

Huber, Harald, Michael Thomm, Helmut König, Gesa Thies, and Karl O. Stetter. 1982. "Methanococcus Thermolithotrophicus, a Novel Thermophilic Lithotrophic Methanogen." *Archives of Microbiology* 132 (1): 47–50. doi:10.1007/BF00690816.

Jabari, Linda, Hana Gannoun, Eltaief Khelifi, Jean Luc Cayol, Jean Jacques Godon, Moktar Hamdi, and Marie Laure Fardeaub. 2016. "Bacterial Ecology of Abattoir Wastewater Treated by an Anaerobic Digestor." *Brazilian Journal of Microbiology* 47 (1). Sociedade Brasileira de Microbiologia: 73–84. doi:10.1016/j.bjm.2015.11.029.

Jenkins, Bryan M., Robert B. Williams, Linda S. Adams, Cheryl Peace, Gary Petersen, and Mark Leary. 2008. *Current Anaerobic Digestion Technologies Used for Treatment of Municipal Organic Solid Waste*, March.

Jokela, J. P. Y., and J. A. Rintala. 2003. "Anaerobic Solubilisation of Nitrogen from Municipal Solid Waste (MSW)." *Reviews in Environmental Science and Biotechnology* 2 (1): 67–77. doi:10.1023/B:RESB.0000022830.62176.36.

Jouzani, Gholamreza Salehi. 2018. *Metadata of the Chapter That Will Be Visualized in SpringerLink*. doi:10.1007/978-3-319-77335-3.

Kalyuzhnyi, S., A. Veeken, and B. Hamelers. 2000. "Two-Particle Model of Anaerobic Solid State Fermentation." *Water Science and Technology* 41 (3): 43–50. doi:10.2166/wst.2000.0054.

Kashyap, D. R., K. S. Dadhich, and S. K. Sharma. 2003. "Biomethanation under Psychrophilic Conditions: A Review." *Bioresource Technology* 87 (2): 147–53. doi:10.1016/S0960-8524(02)00205-5.

Kim, Jung Kon, Baek Rock Oh, Young Nam Chun, and Si Wouk Kim. 2006. "Effects of Temperature and Hydraulic Retention Time on Anaerobic Digestion of Food Waste." *Journal of Bioscience and Bioengineering* 102 (4): 328–32. doi:10.1263/jbb.102.328.

Kroeker, E. J., D. D. Schulte, A. B. Sparling, and H. M. Lapp. 1979. "Anaerobic Treatment Process Stability." *Journal of the Water Pollution Control Federation* 51 (4): 718–27.

Lesteur, M., V. Bellon-Maurel, C. Gonzalez, E. Latrille, J. M. Roger, G. Junqua, and J. P. Steyer. 2010. "Alternative Methods for Determining Anaerobic Biodegradability: A Review." *Process Biochemistry* 45 (4). Elsevier Ltd: 431–40. doi:10.1016/j.procbio.2009.11.018.

Liu, Yuchen, and William B. Whitman. 2008. "Metabolic, Phylogenetic, and Ecological Diversity of the Methanogenic Archaea." *Annals of the New York Academy of Sciences* 1125: 171–89. doi:10.1196/annals.1419.019.

Matheri, A. N., S. N. Ndiweni, M. Belaid, E. Muzenda, and R. Hubert. 2017. "Optimising Biogas Production from Anaerobic Co-Digestion of Chicken Manure and Organic Fraction of Municipal Solid Waste." *Renewable and Sustainable Energy Reviews* 80 (May). Elsevier Ltd: 756–64. doi:10.1016/j.rser.2017.05.068.

Mesbah, Noha M., and Juergen Wiegel. 2008. "Life at Extreme Limits: The Anaerobic Halophilic Alkalithermophiles." *Annals of the New York Academy of Sciences* 1125: 44–57. doi:10.1196/annals.1419.028.

Mshandete, Anthony, Lovisa Björnsson, Amelia K. Kivaisi, M. S. T. Rubindamayugi, and Bo Mattiasson. 2006. "Effect of Particle Size on Biogas Yield from Sisal Fibre Waste." *Renewable Energy* 31 (14): 2385–92. doi:10.1016/j.renene.2005.10.015.

Müller, Bettina, Li Sun, Maria Westerholm, and Anna Schnürer. 2016. "Bacterial Community Composition and Fhs Profiles of Low- and High-Ammonia Biogas Digesters Reveal Novel Syntrophic Acetate-Oxidising Bacteria." *Biotechnology for Biofuels* 9 (1). BioMed Central: 1–18. doi:10.1186/s13068-016-0454-9.

Murphy, Walter J. 1949. "Basic Data." *Analytical Chemistry* 21 (6): 651. doi:10.1017/s0074180900129663.

Nauhaus, Katja, Antje Boetius, Martin Krüger, and Friedrich Widdel. 2002. "In Vitro Demonstration of Anaerobic Oxidation of Methane Coupled to Sulphate Reduction in Sediment from a Marine Gas Hydrate Area." *Environmental Microbiology* 4: 296–305.

Nordberg, Ulf. n.d. *Biogas – Nuläge Och Framtida Potential.*

Oleszkiewicz, J. A., and V. K. Sharma. 1990. "Stimulation and Inhibition of Anaerobic Processes by Heavy Metals – A Review." *Biological Wastes* 31 (1): 45–67. doi:10.1016/0269-7483(90)90043-R.

Ometto, F., A. Berg, A. Björn, L. Safaric, B. H. Svensson, A. Karlsson, and J. Ejlertsson. 2018. "Inclusion of Saccharina Latissima in Conventional Anaerobic Digestion Systems." *Environmental Technology (United Kingdom)* 39 (5): 628–39. doi:10.1080/09593330.2017.1309075.

Pancost, Richard D., Jaap S. Sinninghe Damsté, Saskia de Lint, Marc J. E. C. van der Maarel, and Jan C. Gottschal. 2000. "Biomarker Evidence for Widespread Anaerobic Methane Oxidation in Mediterranean Sediments by a Consortium of Methanogenic Archaea and Bacteria." *Applied and Environmental Microbiology* 66 (3): 1126–32. doi:10.1128/AEM.66.3.1126-1132.2000.

Ramsay, Ian R., and Pratap C. Pullammanappallil. 2001. "Protein Degradation during Anaerobic Wastewater Treatment: Derivation of Stoichiometry." *Biodegradation* 12 (4): 247–56. doi:10.1023/A:1013116728817.

Rocamora, Ildefonso, Stuart T. Wagland, Raffaella Villa, Edmon W. Simpson, Oliver Fernández, and Yadira Bajón-Fernández. 2020. "Dry Anaerobic Digestion of Organic Waste: A Review of Operational Parameters and Their Impact on Process Performance." *Bioresource Technology* 299 (December). Elsevier: 122681. doi:10.1016/j.biortech.2019.122681.

Romero-Güiza, M. S., J. Vila, J. Mata-Alvarez, J. M. Chimenos, and S. Astals. 2016. "The Role of Additives on Anaerobic Digestion: A Review." *Renewable and Sustainable Energy Reviews* 58: 1486–99. doi:10.1016/j.rser.2015.12.094.

Sambo, A. S., B. Garba, and B. G. Danshehu. 1995. "Effect of Some Operating Parameters on Biogas Production Rate." *Renewable Energy* 6 (3): 343–44. doi:10.1016/0960-1481(95)00027-H.

Schenk, Peer M., Skye R. Thomas-Hall, Evan Stephens, Ute C. Marx, Jan H. Mussgnug, Clemens Posten, Olaf Kruse, and Ben Hankamer. 2008. "Second Generation Biofuels: High-Efficiency Microalgae for Biodiesel Production." *BioEnergy Research* 1 (1): 20–43. doi:10.1007/s12155-008-9008-8.

Schmidt, Thomas, Michael Nelles, Frank Scholwin, and Jürgen Pröter. 2014. "Trace Element Supplementation in the Biogas Production from Wheat Stillage – Optimization of Metal Dosing." *Bioresource Technology* 168. Elsevier Ltd: 80–85. doi:10.1016/j.biortech.2014.02.124.

Schn, A., I. Bohn, and J. Moestedt. 2016. "Protocol for Start-Up and Operation of CSTR Biogas Processes." (June): 171–200. doi:10.1007/8623.

Schnürer, A., and Å. Jarvis. 2018. *Microbiology of the Biogas Process.* https://www.research gate.net/publication/327388476_Microbiology_of_the_biogas_process.

Schnürer, Anna, and Asa Jarvis. 2010. "Microbiological Handbook for Biogas Plants." *Swedish Gas Centre Report 207*, 138. https://www.google.com/url?sa=t&rct=j&q=&esrc=s&source=web&cd=1&cad=rja&uact=8&ved=0ahUKEwjUhuL-wf3aAhVKl CwKHY8tCwQQFgguMAA&url=https%3A%2F%2Fpdfs.semanticscholar.org%2F7 5ff%2F5e5d9aec893cb10d5e146bb669cf7175d53c.pdf&usg=AOvVaw1kTZGDDV Na6CshGEyEF6Bt.

Sevier, Carolyn S., and Chris A. Kaiser. 2002. "Formation and Transfer of Disulphide Bonds in Living Cells." *Nature Reviews Molecular Cell Biology* 3 (11): 836–47. doi:10.1038/nrm954.

Sheets, Johnathon P., Liangcheng Yang, Xumeng Ge, Zhiwu Wang, and Yebo Li. 2015. "Beyond Land Application: Emerging Technologies for the Treatment and Reuse of Anaerobically Digested Agricultural and Food Waste." *Waste Management* 44. Elsevier Ltd: 94–115. doi:10.1016/j.wasman.2015.07.037.

Sun, Li, Phillip B. Pope, Vincent G. H. Eijsink, and Anna Schnürer. 2015. "Characterization of Microbial Community Structure during Continuous Anaerobic Digestion of Straw and Cow Manure." *Microbial Biotechnology* 8 (5): 815–27. doi:10.1111/1751-7915.12298.

Thanh, Pham Minh, Balachandran Ketheesan, Zhou Yan, and David Stuckey. 2016. "Trace Metal Speciation and Bioavailability in Anaerobic Digestion: A Review." *Biotechnology Advances* 34 (2). Elsevier B. V.: 122–36. doi:10.1016/j.biotechadv.2015.12.006.

Thauer, Rudolf K., Anne Kristin Kaster, Henning Seedorf, Wolfgang Buckel, and Reiner Hedderich. 2008. "Methanogenic Archaea: Ecologically Relevant Differences in Energy Conservation." *Nature Reviews Microbiology* 6 (8): 579–91. doi:10.1038/nrmicro1931.

Theuerl, Susanne, Johanna Klang, and Annette Prochnow. 2019. "Process Disturbances in Agricultural Biogas Production—Causes, Mechanisms and Effects on the Biogas Microbiome: A Review." *Energies* 12 (3). doi:10.3390/en12030365.

Thiel, Volker, Jörn Peckmann, Richard Seifert, Patrick Wehrung, Joachim Reitner, and Walter Michaelis. 1999. "Highly Isotopically Depleted Isoprenoids: Molecular Markers for Ancient Methane Venting." *Geochimica et Cosmochimica Acta* 63 (23–24): 3959–66. doi:10.1016/s0016-7037(99)00177-5.

US EPA, and Climate Change Division. 2011. *Market Opportunities for AgSTAR Biogas Recovery Systems at U.S. Livestock Facilities.*

US EPA, OAR. 2020. *AgSTAR: Biogas Recovery in the Agriculture Sector.* Accessed April 4. https://www.epa.gov/agstar.

Valentine David L. 2002. "Biogeochemistry and Microbial Ecology of Methane Oxidation in Anoxic Environments: A Review." *Antonie van Leeuwenhoek, International Journal of General and Molecular Microbiology* 81 (1–4): 271–82. doi:10.1023/A:1020587206351.

Vázquez-Padín, J. R., M. Figueroa, I. Fernández, A. Mosquera-Corral, J. L. Campos, and R. Méndez. 2009. "Post-Treatment of Effluents from Anaerobic Digesters by the Anammox Process." *Water Science and Technology* 60 (5): 1135–43. doi:10.2166/wst.2009.421.

Vrieze, Jo de, Tom Hennebel, Nico Boon, and Willy Verstraete. 2012. "Methanosarcina: The Rediscovered Methanogen for Heavy Duty Biomethanation." *Bioresource Technology* 112. Elsevier Ltd: 1–9. doi:10.1016/j.biortech.2012.02.079.

Ward, Alastair J., Phil J. Hobbs, Peter J. Holliman, and David L. Jones. 2008. "Optimisation of the Anaerobic Digestion of Agricultural Resources." *Bioresource Technology* 99 (17): 7928–40. doi:10.1016/j.biortech.2008.02.044.

Westerholm, Maria, and Anna Schnürer. 2019. "Microbial Responses to Different Operating Practices for Biogas Production Systems." *Anaerobic Digestion.* doi:10.5772/intechopen.82815.

Yang, Liangcheng, and Yebo Li. 2014. "Anaerobic Digestion of Giant Reed for Methane Production." *Bioresource Technology* 171 (1). Elsevier Ltd: 233–39. doi:10.1016/j.biortech.2014.08.051.

Yiridoe, Emmanuel K., Robert Gordon, and Bettina B. Brown. 2009. "Nonmarket Cobenefits and Economic Feasibility of On-Farm Biogas Energy Production." *Energy Policy* 37 (3): 1170–79. doi:10.1016/j.enpol.2008.11.018.

Zhao, Yang Guo, Ai Jie Wang, and Nan Qi Ren. 2010. "Effect of Carbon Sources on Sulfidogenic Bacterial Communities during the Starting-up of Acidogenic Sulfate-Reducing Bioreactors." *Bioresource Technology* 101 (9). Elsevier Ltd: 2952–59. doi:10.1016/j.biortech.2009.11.098.

Zhou, Haoqin, Robert C. Brown, and Zhiyou Wen. 2019. "Anaerobic Digestion of Aqueous Phase from Pyrolysis of Biomass: Reducing Toxicity and Improving Microbial Tolerance." *Bioresource Technology* 292 (June). Elsevier: 121976. doi:10.1016/j.biortech.2019.121976.

Zhu, Baoning, Petros Gikas, Ruihong Zhang, James Lord, Bryan Jenkins, and Xiujin Li. 2009. "Characteristics and Biogas Production Potential of Municipal Solid Wastes Pretreated with a Rotary Drum Reactor." *Bioresource Technology* 100 (3). Elsevier Ltd: 1122–29. doi:10.1016/j.biortech.2008.08.024.

# 11 Current Status and Perspectives of Biogas Upgrading and Utilization

## 11.1 INTRODUCTION

Anaerobic digestion (AD) of animal wastes, organic plants, food wastes, sewage wastewater, and industrial wastes generates biogas. Anaerobic digestion is a bio-chemical way to covert wastes into value. Biodegradable waste alone can process through this technique (Dahiya 2015). The process of converting biodegradable wastes into biogas in an anaerobic digester is clearly shown in Figure 11.1. The anaerobic digestion process is a step-by-step processes in which bacterial break-down of complex heterogenous organic substances occurs in the absence of oxygen (Kadam and Panwar 2017). The formation of biogas from biodegradable wastes has four major steps (Ghavinati and Tabatabaei 2018):

- **Hydrolysis:** The first step is the breakdown of organic substances into long-chain compounds by microorganisms. For example, carbohydrates and pro-teins are decomposed into sugars and amino acids.
- **Acidogenesis:** In the second step, microorganisms break long-chain com-pounds into single molecules. In this step, sugars and amino acids are con-verted into fatty acids (long-chain hydrocarbon), ethanol, $H_2S$, and $CO_2$.
- **Acetogenesis:** Long-chain fatty acids are broken into short-chain acids. Here ethanol and fatty acids are decomposed into $CO_2$, $CH_3COOH$, and $H_2$.
- **Methanogenesis:** Finally acetic acid and hydrogen combine together and form $CH_4$ and $CO_2$.

Environmental benefits of biogas production:

a. Biogas is one source of renewable energy.
b. Reduces landfill and methane liberation to the atmosphere.
c. Upgraded biogas will be utilized in transport vehicles instead of petroleum products (Ullah Khan et al. 2017).
d. Anaerobic digester sludge is a good fertilizer for agriculture purposes, and it simultaneously produces biogas (Sahota, Shah, et al. 2018).

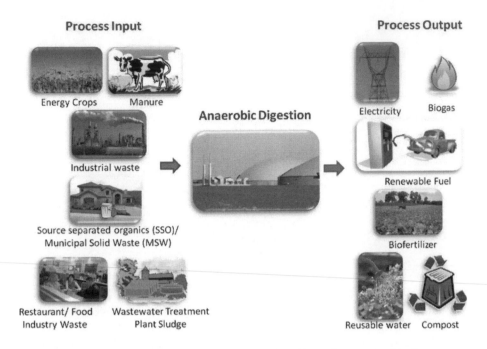

**FIGURE 11.1**   Conversion of biodegradable wastes into biogas in an anaerobic digester.

Biogas production is a conventional way to generate renewable energy while safely disposing of biodegradable wastes. Biogas is recognized as the best way to reduce the present energy demand and environmental impacts facing India (Starr et al. 2012).

The components present in the conventionally produced biogas are as follows:

- **Main components:** Methane (around 65%) and carbon dioxide (around 35%)
- **Trace components:** $H_2S$, water vapor, ammonia, oxygen, nitrogen, and halogenated volatile organic compounds.

Note: The percentage of trace components is always less than 2%. Percentages of trace components varies with respect to the source of biodegradable wastes fed into the anaerobic digester. The quantity of these trace compounds is very low compared to the major components (Kapoor et al. 2019). Traditionally generated biogas can be used for heating, cooking, power production, and lighting applications. Biogas is the best alternate to natural gas. The scope of biogas should be broadened to include facets such as transport, developing a substitute for natural gas network, and acting as the substrate for the production of chemicals and fuel cells. Hence, various countries are concentrating on upgrading biogas quality so that it is equal to natural gas. The main drawback of biogas is the presence of $H_2S$, $CO_2$, and other impurities. It must be treated before being used as transport fuel (Angelidaki et al. 2018).

Many upgrading technologies are being developed by researchers. Without purification, biomethane availability in raw plant biogas is only around 60%. Biogas purification techniques are mandatory for producing biomethane of more than 90% purity (Ullah Khan et al. 2017).

The operating principles involved in biogas upgradation can be categorized as follows:

i) Physicochemical processes
   a. Adsorption
   b. Absorption
   c. Cryogenic
   d. Membrane separations

ii) Biological processes
   a. *In situ*
   b. *Ex situ*
   c. Hybrid process (Angelidaki et al. 2018)

Each upgradation technology should be reviewed with respect to its operations, methane purity energy requirements, and cost savings. Deep analysis was conducted on technology gaps and implementation barriers, and detailed comparisons are given in this chapter. For a broad and successful implementation of biogas upgrading technology, research and development (R&D) trends such as the development of efficient biogas upgrading technologies, adsorbents, cost reduction, and methane loss were carefully evaluated.

This chapter provides a comprehensive description of the main principles of the various methodologies for biogas upgrading, the scientific and technical results related to their biomethanization efficiency, the challenges to further development, and the incentives and feasibility of valuation concepts.

The core objective of biogas upgrading technologies is to reduce $CO_2$ presence in the biogas and to enhance the biogas quality to be equivalent to natural gas. The volumetric energy density of upgraded biogas was high compared to that of raw biogas. That upgraded biogas can be used for vehicle's application in dual fuel mode. A comparison of the compositions of biogas, natural gas, and landfill gas is given in Table 11.1.

## 11.1.1 Need for Biogas Upgradation

i) Biogas contains $CH_4$, along with $CO_2$, $H_2S$, nitrogen, water vapor, oxygen, and hydrogen. Heating value is an appropriate tool to measure the energy density and quality of a particular fuel. Traditional raw biogas has a volumetric heating value of 21.5 MJ/m³, while the volumetric heating value of natural gas is 35.8 MJ/m³.

ii) Comparing raw biogas and natural gas heating values, raw biogas has poor energy density. To replace natural gas with biogas, we need to enhance energy content. Biogas contains a high volume of an incombustible constituent ($CO_2$). Eliminating $CO_2$ from raw biogas helps to increase the heating value of the gas and decrease the cost of compression and transport. With

**TABLE 11.1**

**Various Constituents Present in Biogas, Natural Gas, and Landfill Gas (Eriksson et al. 2016)**

| Component | Unit | Biogas | Natural gas | Landfill gas |
|-----------|------|--------|-------------|--------------|
| Methane ($CH_4$) | Volumetric % | Around 65 | 89 | Around 45 |
| Carbon dioxide ($CO_2$) | Volumetric % | Around 35 | 0.9 | Around 35 |
| Hydrogen sulfide ($H_2S$) | ppm | Around 3000 | 1-8 | 0-100 |
| Hydrogen ($H_2$) | Volumetric % | 0 | 0 | 0-3 |
| Nitrogen ($N_2$) | Volumetric % | 0.2 | 0.3 | 5-40 |
| Oxygen ($O_2$) | Volumetric % | 0 | 0 | 0-5 |
| Lower calorific value | MJ/kg | 20.2 | 48 | 12.3 |

supporting commercial and technical evidence, biogas has been suggested for electricity generation and transport applications (Awe et al. 2017).

iii) Other contaminants present in the biogas have negative impacts (corrosion and contamination) on IC engine components, steel chimneys, and boilers. Removing impurities from biogas allows its use without negative effects on downstream components.

iv) Greenhouse gas emissions were drastically reduced when upgraded biogas was used for combustion. Improved biogas combustion gives a lower quantity of carbon dioxide and nitrogen oxide than does combustion of gasoline or diesel (Song, Liu, Ji, Deng, Zhao, and Kitamura 2017).

v) The presence of $CO_2$ drops the energy yield from the burning of biogas (Ghatak and Mahanta 2016).

vi) During upgradation, separated $CO_2$ can be stored and utilized for other purposes (Morero, Groppelli, and Campanella 2017).

## 11.2   TECHNOLOGIES INVOLVED IN BIOGAS UPGRADING

Biogas upgrading is defined as removal of other gases except methane in biogas. The upgradation process of biogas consists of the following steps:

* Drying or condensation process: Removal of water vapor from raw biogas.
* Cleaning process: Removal of trace elements like $H_2S$, $N_2$, $O_2$, etc. (Ryckebosch, Drouillon, and Vervaeren 2011).

Recovery process: Improves the heat value (Pellegrini, De Guido, and Langé 2018).

Developed technologies for purification of biogas include absorption, adsorption, cryogenic, and membrane separation. The major objective of these technologies major objective is to separate $CO_2$ from raw biogas. Before $CO_2$ separation, some pre-processing is required to remove $H_2O$, $H_2S$, and siloxanes (Sahota, Vijay, et al. 2018).

## 11.2.1　Absorption

Absorption is the process in which constituents (absorbate) are dissolved by absorbent. Atoms or molecules cross the surface and enter the bulk volume of the material. During the absorption gas phase, impurities (mostly $CO_2$) present in biogas are dissolved by the liquid phase absorbent. Based on the absorbent usage, absorption is classified into physical and chemical absorption. Absorbent selection is the most challenging task for researchers.

### 11.2.1.1　Physical absorption

In the physical absorption process, there are no chemical reactions. Atoms or molecules only dissolve in absorbent, and there is no chemical reaction between absorbate and absorbent (Rotunno, Lanzini, and Leone 2017).

- Water scrubbing system – Water used as absorbent
- Organic absorbent – Organic solvent used as absorbent

### 11.2.1.2　Chemical absorption

In the chemical absorption process, chemical reactions take place between absorbate and absorbent. A chemical change exists after complete absorption.
- Amine solutions are mostly used as chemical absorbents

## 11.2.2　Physical Absorption Method Using Water Scrubbing System

Water scrubbing is most commonly used technique for biogas upgradation. The basic principle for physical absorption is Henry's law. Henry's law states that the mass of a dissolved gas in a given volume of solvent at equilibrium is proportional to the partial pressure of the gas (Sahota, Shah, et al. 2018). $CO_2$ and $H_2S$ absorbed from raw biogas based on the dissolvability of carbon dioxide and hydrogen sulfide in water but these gases are more soluble in methane (Cozma et al. 2013). A water scrubbing system will remove more than 90% of the $CO_2$ from raw biogas.

In a high-pressure water scrubbing system, the raw biogas is sent from the bottom of a high-pressure chamber as in Figure 11.2. At the same time, fresh water is

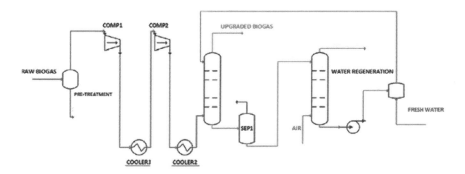

**FIGURE 11.2**　High-pressure water scrubbing system.

**TABLE 11.2**

**Advantages and Disadvantages of Water Scrubbing System (Rotunno, Lanzini, and Leone 2017)**

| Advantages: | Disadvantages: |
|---|---|
| • Simple principle and process | • Huge amount of fresh water is used in the water scrubbing process |
| • High-purity biomethane produced with minimum methane loss | • Requires external heat and high energy input for spraying water |
| • Chemicals are not required in entire process | |
| • Low operating and maintenance costs | |

showered from the top of the chamber. Raw biogas and water are allowed to flow in opposite directions, and the contact surface area between liquid and gases element is thus increased enormously. Several column arrangements are created inside the high-pressure chamber with packing materials. Biomethane of a maximum purity of 96% can be obtained using a water scrubbing system. Using a counter-flow water scrubbing system, up to 99% of the impurities from raw biogas will be removed, and this can also be done for landfill gas by altering the necessary parameters by using a suitable optimization tool (Cozma et al. 2014). The merits and demerits of water scrubbing system are given in Table 11.2.

### 11.2.3 Physical Absorption Method Using Organic Solvents

Physical absorption is based on the solubility of $CO_2$ in a chemical reaction–free solution, which in turn is based on Henry's law; therefore, high partial pressures of $CO_2$ and low temperatures are strongly recommended for its application. Performance of the physical absorption process is optimized in terms of absorption rates and solubility equilibrium of $CO_2$. Then, the rich ($CO_2$-loaded) solvent is regenerated in a desorption chamber. The working principles and system arrangements are very similar to those of the water scrubbing system. The performance of a physical solvent can be predicted. The solubility of an absorbate in the absorbent is directly proportional to the pre-processed raw biogas partial pressure. Hence, the performance of physical solvent processes are enhanced with rising gas pressure (Vega et al. 2018).

Some of the organic solvents commonly used for physical absorption are the following:

- N-methyl-2 pyrolidone
- Dimethyl ethers of polyethylene glycol
- Propylene carbonate
- Methanol

The merits and demerits of physical absorption using organic solvents are given in Table 11.3.

## TABLE 11.3
## Advantages and Disadvantages of Physical Absorption Using Organic Solvents

| Advantages: | Disadvantages: |
|---|---|
| • Highly recommended to separate $CO_2$ in pre-combustion processes<br>• Noticeable selectivity for hydrogen sulfide over carbon dioxide | • High capital and operating costs<br>• Chance of biological contamination |

### 11.2.4 CHEMICAL ABSORPTION METHOD USING AMINE SOLUTIONS

This section discusses biogas purification systems that utilize amine solution or soda lime to remove carbon dioxide from biogas by absorbing in it through chemical reaction (Ghatak and Mahanta 2016). Raw biogas from the biogas plant is sent into three-stage gas compressors with intercooling arrangements as shown in Figure 11.3. Inter air coolers help to reduce the compressed raw biogas temperature. Pressurized (above 20 bar) raw biogas passes into the absorption chamber from the bottom. Amine, soda lime, or NaOH solution passes into the absorption chamber from the top. Amine and pressurized raw biogas flow in opposite directions. This counter-current in the direction of the flow creates a large area of contact, and a chemical reaction takes place between the amine solution and raw biogas impurity ($CO_2$). Rich methane biogas was liberated at the top of the chamber after complete chemical absorption. For absorbent regeneration, biogas (20 bar, 48°C) enters into the pre-heater, and then the amine solution temperature is increased up to 100°C. Again, it is sent into the desorption chamber. Absorbent impurities are purified by desorbing agents. In the desorption chamber, water is used as desorption agent and vaporized, and it leaves along with $H_2S$ and carbon dioxide. In the desorption chamber, a small quantity of

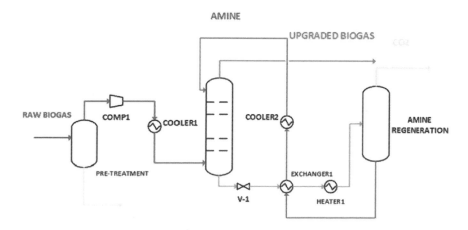

**FIGURE 11.3** Three-stage gas compressors.

**TABLE 11.4**

**Advantages and Disadvantages of Chemical Absorption Using Amine Solution**

| Advantages: | Disadvantages: |
|---|---|
| • Simple and easy to use in rural areas.<br>• High degree of biogas purity<br>• Preprocessing to remove hydrogen sulfide from raw biogas is not essential | • To maintain high methane concentration at the output, frequent replacement of absorbent is needed, which increases the operating cost<br>• Chemical absorbents are challenging to handle because of their corrosive nature |

water loss occurs in every cycle, and this will be replaced with externally provided fresh water (Cozma et al. 2014). The merits and demerits of physical absorption using organic solvents are given in Table 11.4 (Adnan et al. 2019).

Some of the chemical absorbents commonly used for biogas purification (Ramaraj and Dussadee 2015) are as follows:

- Diisopropanolamine (DIPA)
- Methyl diethanolamine (MDEA)
- Diglycol amine
- Diethanol amine
- Monoethanol amine

### 11.2.4.1  Adsorption

Adsorption is totally different from absorption. It is a surface deposition process, in which gaseous or liquid impurities are deposited on the surface of the adsorbing medium, which may be solid or liquid (Cozma et al. 2014). A surface attractive force acts between the adsorbent and the adsorbates.

- Physeorption:
  Physical binding (usually Van der Waals) surface attractive force between absorbent and adsorbate.
- Chemical absorption:
  Chemical binding (usually covalent bonds) involved between absorbate and absorbent.

Based on the process of adsorbent regeneration, biogas upgradation technologies can be classified into three different categories (Morero, Groppelli, and Campanella 2017)

- Vacuum swing adsorption
- Pressure swing adsorption
- Temperature swing adsorption (30–120°C) (Sahota, Vijay, et al. 2018).

The most commonly and commercially used adsorption technique is pressure swing adsorption, because of lower input energy needs, design flexibility, environmental safety, and reasonable performance output compared with other adsorption techniques (Cozma et al. 2014).

### 11.2.5 PRESSURE SWING ADSORPTION (PSA)

The surface deposition of adhesive elements on an adsorbing medium is known as adsorption. Raw biogas from an anaerobic digester is sent to a compressor to increase the pressure above atmospheric pressure. Then, $H_2S$ and water vapors are removed before it is sent into an adsorption column. Pre-processed and pressurized biogas is sent into adsorption columns. In the adsorption column, $CO_2$ molecules are attracted by the absorbent surfaces. Absorbed molecules are considered as a contaminant on the adsorbent surfaces.

In some rare cases, chemical reactions will take place between adsorbent and contaminants. In the case of such a chemical reaction, some non-hazardous components will be produced; these can be removed and utilized for secondary purposes. The adsorbing medium's efficiency is measured in terms of surface porosity. Because porous materials have large surface areas, porous adsorbent materials ultimately give the best-quality output (Ryckebosch, Drouillon, and Vervaeren 2011).

Various adsorbent materials commonly used in pressure swing adsorption systems are the following:

- Zeolites
- Molecular sieves
- Alkaline solids
- Iron sponge
- Silica gel
- Activated carbon

### 11.2.6 MEMBRANE SEPARATION

Membrane separation technology is a separation technology in which undesired constituents of the raw biogas are filtered by using various membranes. Membrane separation is most suitable for biomethane production from biogas. Membrane material selection is very challenging. The membrane separation technique has additional benefits that can be combined with other biogas upgrading techniques. Hybrid processes help to reduce the investment and operating costs compared with conventional processes (Song, Liu, Ji, Deng, Zhao, Li, et al. 2017). Organic, inorganic, and metal matrix membranes are commercially available in market. Advantages and drawbacks of each type of membrane are discussed in Table 11.5 (Ryckebosch, Drouillon, and Vervaeren 2011).

### 11.2.7 CRYOGENIC SEPARATION PROCESS

The basic principle for the cryogenic separation process is liquifying the gas state component into liquid phase species by reducing the temperature to as low as $-250°C$ (Hosseinipour and Mehrpooya 2019). Cryogenic separation also called

**TABLE 11.5**

**Organic Polymers and Inorganic Membrane Materials (Xia, Cheng, and Murphy 2016)**

| Organic polymers | Inorganic materials | Mixed matrix materials |
|---|---|---|
| **Merits:** | **Merits:** | **Merits:** |
| • Nowadays, most commonly used membrane materials for membrane gas separation<br>• Process ability good and surface area enhancing also possible<br>• Good choice compared with inorganic materials. | • Physical and chemical stability<br>• Quality of purified biogas was good while purifying through inorganic material membrane | • Introduced to overcome the demerits of inorganic membranes and organic membranes |
| **Demerits:** | **Demerits:** | |
| • Costly and poor availability<br>• Need to change simultaneously | • Worst machining properties and difficult to process<br>• Manufacturing cost is expensive and surface area enhancing is very complicated | |
| **Some organic membrane polymer materials:**<br>Polysulfone, polyethersulfone, cellulose acetate, polyimide, polyetherimide, polycarbonate (brominated), polyphenylene oxide, polymethylpentene, polydimethylsiloxane, polyvinyltrimethylsilane | **Some inorganic membrane materials:**<br>Nanoporous carbon, zeolites, alumina, cobalt, copper, iron, palladium, platinum, tantalum, vanadium, nickel, niobium | **Some mixed matrix membrane materials:**<br>Mixed conducting perovskites, metal organic frameworks, palladium alloys, ultra-microporous amorphous silica |

low-temperature upgradation technology. Liquid biogas (LBG) can be directly utilized as vehicular transport fuel. Liquid biogas production is only possible by cryogenic separation technique (Baena-Moreno et al. 2019). LPG is a fossil fuel which is derived from petroleum, while LBG is a renewable fuel which is derived from biogas. LBG is also known as a carbon-neutral fuel. The heating value of LBG is 2 times more than compressed natural gas (CNG).

The carbon dioxide liquification temperature differs from the methane liquification point. Therefore, if $CO_2$ is separated from raw biogas, pure liquid biomethane will be produced. Sometimes liquid biomethane also called liquid biogas. The energy requirement for the cryogenic separation technique is comparatively less than that for adsorption using amine solution. Liquid biomethane purity is very high compared with biomethane produced by other techniques. Low-temperature separation techniques also give liquid $CO_2$ as a by-product. Transportation and storage of liquid fuel are comparatively easier than compressed gaseous fuel storage. Transportation and storage space requirements are reduced significantly. The biogas impurity $CO_2$ is also collected in this technique; hence we can eliminate greenhouse gas emissions and global warming (Pellegrini, De Guido, and Langé 2018; Baena-Moreno et al. 2019).

## 11.2.8 CHEMICAL HYDROGENATION PROCESS

Before describing the chemical hydrogenation process, we need to know the definition of hydrogenation. Hydrogenation means conversion of unsaturated organic components into saturated organic components with the help of hydrogen molecules in the presence of a catalyst. Catalysts are substances that are used to speed up the rate of a reaction without being consumed during the process. Catalysts used for hydrogenation reactions are metals such as nickel, ruthenium, and platinum. We can also convert double- or triple-bond hydrocarbon to single-bond hydrocarbon by adding hydrogen atoms (Ramaraj and Dussadee 2015). In this process, hydrogen is added to carbon dioxide in the presence of a catalyst. The most commonly used catalysts for hydrogenation are nickel and ruthenium. The optimum condition for the hydrogenation reaction is 473 K temperature and 50–200 bar pressure. (Xia, Cheng, and Murphy 2016).

Biogas purification through biological methods are listed below:

- Chemoautotrophic method
- Photoautotrophic method
- Fermentation method
- Microbial electrochemical method

## 11.2.9 CHEMOAUTOTROPHIC METHODS

In the chemoautotrophic method, hydrogenotrophic methanogen microorganisms (bacteria) help to produce methane gas from carbon dioxide using hydrogen. Microorganisms are used for biogas upgradation, so it is eco-friendly and renewable in nature. It needs a continuous hydrogen gas supply that may be produced from renewable sources. Hydrolyzation of water is done with solar/wind plant electricity and can give hydrogen continuously. The investment costs of this technique are lower than those of other purification methods. The major merits of the chemoautotrophic method is that $CO_2$ is not liberated but rather is converted into methane (Wang et al. 2020). The process of hydrogenation of $CO_2$ to $CH_4$ with nickel catalyst is shown by the Sabatier reaction.

$$CO_2 + 4H_2 \rightarrow CH_4 + 2H_2O \tag{11.1}$$

Selecting the catalyst for hydrogenation is one of the challenging issues in enhancing the carbon dioxide conversion efficiency and purity of methane. While developing catalysts for hydrogenation, we need to focus on the operating temperature and the life of the catalyst. If catalyst reactivity is reduced, then methanol is produced from carbon dioxide. The formation of methanol, an exothermic reaction, is shown in Equation 11.2:

$$CO_2 + 3H_2 \rightarrow CH_3OH + H_2O \qquad (11.2)$$

It can be further classified into three types:

- *In situ* biogas upgrading
- *Ex situ* biogas upgrading
- Biogas upgrading systems using microbial communities

An overview of *in situ* and *ex situ* upgradation techniques is shown in Figure 11.4.

### 11.2.9.1 *In situ* Biological Biogas Upgrading

*In situ* systems directly send the hydrogen into an anaerobic digester. In an anaerobic digester, hydrogen combines with the carbon dioxide, and double bonds are broken into single-bond organic structures. At this stage, methane is formed from carbon dioxide by losing its double bonds by the action of microorganisms like autochthonous methanogenic archaea. There are two different types of methane formation from carbon dioxide: Wood–Ljungdahl and hydrogenotrophic methanogenesis (Voelklein, Rusmanis, and Murphy 2019).

- In hydrogenotrophic methanogenesis, the microorganism is directly added to the anaerobic digester unit; this causes interaction between carbon dioxide and hydrogen molecules to directly form pure saturated single-bond hydrocarbon (methane) in a single-step reaction as in Equation 11.1.

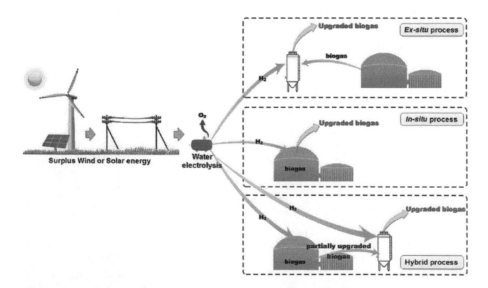

**FIGURE 11.4** *In situ* and *ex situ* upgradation techniques.

- At the same time Wood–Ljungdahl does not directly convert carbon dioxide into methane in a single step. This microorganism also converts carbon dioxide into methane indirectly. In the first step, homoacetogenic bacteria combine carbon dioxide with hydrogen to form acetic acid. In the second step, acetic acid converts into methane by acetoclastic methanogenic archaea bacteria. The first and second steps of the reaction are shown in Equations 11.3 and 11.4, respectively.

$$4H_2 + 2CO_2 \rightarrow CH_3COOH + 2H_2O \tag{11.3}$$

$$CH_3COOH \rightarrow CH_4 + CO_2 \tag{11.4}$$

### 11.2.9.2  *Ex situ* Biological Biogas Upgrading

Another biogas upgradation process is *ex situ* biological upgrading, in which a digester's externally provided $H_2$ and $CO_2$ are biologically transformed to $CH_4$ by the help of hydrogenotrophic methanogens. The $CH_4$ concentration in the product gas is more than 98%, permitting its utilization as a substitute for natural gas (Kougias et al. 2017). The *ex situ* biological upgradation system layout is shown in Figure 11.4.

### 11.2.9.3  Microbial Communities in Biological Biogas Upgrading Systems

Upgradation and purification of the biogas are essential because cleaned biogas offers reductions in harmful gas emissions and several other environmental advantages when it is utilized as a transportation fuel. Reducing carbon dioxide and hydrogen sulfide concentrations will considerably enhance the biogas quality. Several technologies are being developed and implemented for removing biogas impurities; these include physical absorption, absorption by chemical solvents, membrane separation, cryogenic separation, and chemical or biological methods. While the physicochemical method of removal is costly and environmentally harmful, the biological processes are considered as feasible and environmentally friendly. Moreover, algae biomass is plentiful and universal. Purification of biogas using algae involves the use of the photosynthetic capability of micro- or macroalgae to remove impurities existing in the biogas. Biological methods help to purify the biogas, improve the calorific value percentage of the gas, and make biogas with characteristics as close as possible to those of natural gas (Ramaraj and Dussadee 2015).

### 11.2.10  Photoautotrophic Methods

To produce biogas rich in methane concentration, the photoautotrophic technique is the most appropriate, since it gives the highest level of carbon dioxide sequestration. Apart from this, hydrogen sulfide contamination can also be eliminated in this method. Approximately 97% of $CH_4$ is recovered from this, with the recovery level based on the selected algae types. Algae are extensively used in the alteration process and are grown in mass in open ponds. Photosynthetic effectiveness is higher when the process is commenced in a closed pond. At the time of the recovery process, biogas is passed through the photo reactor for the effective conversion of gas to $CH_4$. The biggest downside is the high cost investment. This gas-processing technology has been utilized to instantaneously condense and disperse water and HC from the natural gas. Progress to allow for the high reduction of carbon dioxide and $H_2S$ is presently underway.

### 11.2.11  BIOGAS UPGRADING THROUGH OTHER FERMENTATION PROCESSES

In this process, biogas is upgraded through the biological conversion of $CO_2$ and transformed into valued liquid products such as acetate, ethanol, and butyrate (Omar et al. 2019). Different organisms are proficient in transforming carbon dioxide and hydrogen into liquid products. Most of these organisms are acetogens, which effectively ferment $C_1$ compounds using hydrogen as the electron donor and generate valuable chemicals and biofuels. However, different fermentation investigations have been studied using pure culture, which faces numerous limitations. Limited availability of adaptation measures for several substrates (e.g., toxic components) and high costs of operation to sustain the conversion conditions make these techniques less appropriate for large industries.

In general, $H_2$ needed for the fermentation process can be produced from large-scale sources such as petroleum refinery, soda manufacture, coal gasification, and petrochemical plants. The renewable electricity concept for the utilization of hydrogen generation is receiving great attention nowadays. This sustainable process, which also called conversion of power to biogas technology, depends on water electrolysis that utilizes surplus electricity generation by renewable energy resources, such as solar panels and windmills.

The upgradation of biogas of volatile fatty acids via mixed-values fermentation of carbon dioxide and hydrogen was described in an earlier study. This area of research is still in its infancy, and much progress concerning organisms and processing is necessary. Hence, a novel gas upgradation technique, involving the fermentation of carbon dioxide into valuable chemicals using various diverse culture acetogenic groups as biocatalysts and outside added hydrogen as a source of energy under various temperatures, has been suggested.

### 11.2.12  BIOGAS UPGRADING THROUGH MICROBIAL ELECTROCHEMICAL METHODS

A modern electrochemical separation cell was designed to *in situ* adsorb and regenerate $CO_2$ via alkali and acid regeneration alkali (Angelidaki et al. 2018). The electrochemical process is viewed to be the greatest and most cost-effective technique of biogas restoration for methane generation and $CO_2$ removal. One example is the electrochemical technique in the microbial electrolysis cell. Here, the oxidation of compounds by the help of bacteria issues electrons into the anode chamber, where they mix with protons in the cathode chamber to produce hydrogen that will be used for the recovery of biogas. Using biocathode in an electrolysis cell, $CH_4$ can be generated by reducing $CO_2$, for an attainment of 80% energy efficiency. The $CO_2$ reduction to $CH_4$ depends on the transfer of electrons and the $H_2$ generated. Based on the potential of the cathode, the reduction process happens. *In situ* and *ex situ* methods of biogas recovery using the electrolysis cell have been investigated experimentally. The results show that the performance of the *in situ* gas recovery technique is much better, having the greatest capacity to remove $CO_2$. Apart from this, it has been found that the removal of $CO_2$ is linked with both $CH_4$ production and $CO_2$ ionization. This ionization is because of the production of alkalinity from the cathode.

## 11.3 BIOGAS UPGRADATION TECHNOLOGIES UNDER DEVELOPMENT

### 11.3.1 INDUSTRIAL LUNG

The carboanhydrase enzyme is found in blood, where it catalyzes the breakdown of sources of carbon dioxide during cell metabolism. The carbonate-shaped dissolved carbon dioxide is then transferred to lungs, in which the same enzyme catalyzes the reverse reaction, by which carbon dioxide and water form.

The enzyme can also be used to dissolve the biogas dioxide from carbon and thus remove it from the gas. The production of the enzyme is expensive, and factors such as the life of the immobilized enzyme influence the feasibility of the operation. A research group from London and Sweden, researched the use of carboanhydrase for biogas valorization and developed the enzyme in a project with the addition of 6 histadines that were used to bind the enzyme to a solid base (Petersson, Holm-nielsen, and Baxter, n.d.). It was shown in the same analysis that biogas can be purified to 99% methane content. $CO_2$ Solutions Inc. is a Canadian company which created this technique and owns a bioreactor patent using a carbon dioxide–dissolving enzyme. They focus not only on biogas upgrading but also on technical ventilation, among other processes. Their research efforts are primarily based on enzyme immobilization and include work on mechanical bioreactors, cloning and the synthesis of enzymes, and scientific applications (Petersson and Wellinger 2009).

### 11.3.2 SUPERSONIC SEPARATION

The supersonic separation method is used for purifying biogas by condensing and removing water, heavy hydrocarbons, and other impurities. Supersonic separation technology consists of a lightweight cylindrical structure that essentially incorporates acceleration, removal of the gas/liquid cyclone, and recompression. A de Laval nozzle is used to increase the supersonic speed of the saturated feed gas, leading to low temperature and pressure. This causes droplets of hydrocarbon condensation to form water and fog. A high-vortex churn centrifuges the droplets against the surface, and a cyclone separator separates the liquids from the air.

## 11.4 COMPARATIVE ANALYSIS OF THE VARIOUS BIOGAS UPGRADATION TECHNOLOGIES

Summaries of various biogas upgradation technologies are given in Table 11.6 (Adnan et al. 2019; Angelidaki et al. 2018; Awe et al. 2017; Hjuler and Aryal 2016; Sahota, Shah, et al. 2018).

**TABLE 11.6**
**Summary of the Various Biogas Upgradation Technologies**

| Operating principle | Upgrading technique | Description |
|---|---|---|
| **Absorption** | Pressurized Water Scrubbing (PSW) | Carbon dioxide is dissolved in water under high-pressure conditions. Used water is purified by depressurization and then circulated again. |
|  | Organic physical absorption | Carbon dioxide is absorbed by organic solvent (polyethylene glycol). Used organic solvent is purified by heating/depressurizing and circulated again. |
|  | Chemical absorption using amine solution | Carbon dioxide is absorbed by amine solution. Used amine solution is purified by heating and circulated again. |
| **Adsorption** | Pressure swing adsorption (PSA) | Pressurized biogas passes through adsorption medium carbon dioxide deposit on adsorbent. Depressurization is the best way to regenerate adsorbent. |
| **Separation** | Membrane separation | Membrane helps to filter the raw biogas. Pressurized biogas passes into membrane and then carbon dioxide is separated from biogas. Biomethane is separated from $CO_2$. |
|  | Cryogenic separation | Low-temperature separation process in which biogas cooled until entire carbon dioxide condensation. Condensing point of carbon dioxide is greater than methane. |

## 11.4.1 Cost-Economics

Cost is considered as the key factor in determining the right solution; however, the best solution is not always the cheapest. Financially, water scrubbing and *in situ* are achievable, while technologies like cryogenic and chemical absorption give higher efficiency at higher capital and running costs (Cozma et al. 2013). However, in larger-scale applications, cryogenic technology is cost-effective in terms of maintenance costs. For membrane separation, the cost of investment is high and the running cost is also relatively high (Starr et al. 2012).

## 11.4.2 Technology

A technically strong procedure does not guarantee less complexity and low cost; hence, the optimal technology can differ according to requirements. A higher concentration of $CH_4$ is important for the use of biogas as a source of transport fuel (Awe et al. 2017). Biogas improving technologies such as chemical absorption and cryogenic separation maintain high purity of $CH_4$, but technologies such as pressure swing adsorption and membrane, along with $CO_2$, have the advantage of removing $O_2$ and $N_2$. Cryogenic separation reveals its benefits above other techniques, since it generates high-pressure liquid fuel and therefore needs no additional energy to compress the fuel to make it better for transport (Collet et al. 2011).

### 11.4.3 ENVIRONMENTAL SUSTAINABILITY

Biogas is one of the renewable energy sources. Anaerobic digestion mainly produces biogas, which is a carbon-neutral fuel and one of the most effective methods of energy storage. One of the best application examples is the use of biogas produced from household waste for cooking and heating in different countries (Branco, Serafim, and Xavier 2019). Production of biogas as a fuel does not lead to the build-up of greenhouse gases in the Earth's atmosphere, since the plants previously absorbed the carbon dioxide emitted during combustion (Collet et al. 2011).

Purity of methane and energy consumption (in the form of water and heat) are important factors for determining environmental sustainability. Techniques such as organic, water, and amine scrubbing techniques need more energy than PSA and membrane techniques. Compared to pressure swing adsorption, the demand for energy for cryogenic separation is quite high ($0.63$ kWh/m$^3$). Pressure swing adsorption can be recommended as an environmentally friendly solution because it also takes less energy to be a switch device, which thus eliminates instrumentation (Eriksson et al. 2016).

## 11.5 FUTURE PERSPECTIVES ON BIOGAS UPGRADATION

Currently, most of the countries focusing on generated biogas from AD plants utilize it directly as a transport vehicular fuel. Many biogas upgradation techniques are discussed in this chapter, along with their advantages and disadvantages (Kapoor et al. 2019). Many commercialization gaps still exist, and many researchers want to overcome the difficulties in commercialization and marketing. Biogas upgrading has the following prospects for future research (Sahota, Shah, et al. 2018).

- Moving towards hybrid upgradation technologies
- Proper utilization of methane content presenting off-gas
- Provision for economically feasible small-scale upgrading plants
- Supportive government policies
- Implementing biogas reforming technologies
- Commercial distribution of liquified biogas

### 11.5.1 MOVING TOWARDS HYBRID UPGRADATION TECHNOLOGIES

Good features of more than one upgradation technology combined together into a single system is known as a hybrid system. While moving to hybrid biogas upgradation techniques, the following factors need attention (Baena-Moreno et al. 2019).

- Efficient technology with high success rate,
- Minimized operational cost; techno-economically viable
- Increased $CH_4$ concentration and maximized $CO_2$ and $H_2S$ capture rate
- Low energy consumption

The following are some of the possible hybridized biogas upgradation systems (Song, Liu, Ji, Deng, Zhao, and Kitamura 2017; Song, Liu, Ji, Deng, Zhao, Li, et al. 2017).

- Hybrid membrane-absorption process
  - First, half of the $CO_2$ is removed by an absorber. Then, residual gas mixture is further purified by a membrane to achieve 90% total removal of $CO_2$.
- Hybrid membrane-cryogenic process (Song, Liu, Ji, Deng, Zhao, Li, et al. 2017)
- Hybrid membrane-adsorption process
  - Generates vehicular fuel–quality methane and improves methane recovery and process energy efficiency.
- Pressurized water scrubbing /Pressurized membrane technology
- Pressurized membrane technology /Cryogenic technology
- Cryogenic technology/Pressurized membrane technology /Temperature swing adsorption
- Temperature swing adsorption /Pressurized membrane technology
- *In situ/ex situ* biological upgradation

### 11.5.2   Utilization of Methane Available in Off-Gas

A major issue faced in upgrading plants is methane loss in the form of off-gas which is released to the environment. $CH_4$ is one of the greenhouse gases responsible for global warming. Off-gas treatment is essential before the off-gas leaves the plant. The percentage of methane in the off-gas depends on the upgrading technique (Kadam and Panwar 2017). Absorption, adsorption, and some membrane separation techniques have large quantities of methane in the off-gas. Removal of methane from off-gas can be done by direct combustion. Hence, methane will oxidize and give heat energy. This heat can either be utilized at the anaerobic digestion plant or be used for some other heating purposes. Otherwise, off-gas can be combined with raw biogas and fed into an existing combined heat and power gas engine. Future research needs to focus on effective capturing and utilization of methane in off-gas in biogas upgradation plants themselves (Kadam and Panwar 2017).

### 11.5.3   Making Small-Scale Upgrading Plants Economical

Usually, plants/system costs are related to the plant size/capacity. For rural applications, limited upgrading equipment is needed. However, costs of small-scale plants (below 200 Nm³/h) are very high, due to high cost incurred in the initial investment along with equipment upgrades; in addition, there are maintenance costs. Similar to large plants, plants operating at very low capacity require the same or greater number of devices like control valves and systems, analyzing instruments, and sensors for their operation. So, research will be needed in the coming years into realistic approaches to reduce the costs incurred in small-capacity plants. Upcoming challenges in the development of small-scale biogas production technology include making continuous output technically and economically viable, reducing methane in plant off-gas, and simplifying all operational factors. In addition, the methane generated can be directly utilized as transport fuel as an imminent relief to running small-scale units sustainably at low cost.

### 11.5.4 SUPPORT POLICIES

Massive development of biogas is only feasible in countries in which government policies promote it. Long-term steps include the use of alternative energy fuels for passenger traffic with the compressed biogas and heavy-duty transport with the liquified biogas. The output of rural biogas is used as a cooking fuel, as a heating source, and also for generation of electrical power at limited scale in rural places without grid links. Thus, the possibilities of transforming the waste of living organisms into biogas are still not completely known, and governments and policymakers must be involved in the framework to allow biogas upgrading research and development (Lima et al. 2018). Massive amounts of biogas can be produced from various renewable sources with favorable governmental support actions such as providing subsidies to producers and implementing administrative environments for market-level development of biogas-driven engine technology, offering alluring feed-in tariffs for the production of their own electricity with biogas technology, and providing seed funds to start-up companies working on the transformation of green energy sources to biogas. All these steps will enable societies to invest in the harvesting and transfer of bio-waste with the help of biogas plants, the development of centers for training skilled labor, and the provision of services for the development of biogas plants for a domestic purpose.

## 11.6 CONCLUSION

Biogas is generated from bio-digestible organic wastes through an anaerobic digester. Raw biogas contains impurities and low energy density. Hence its commercial usage is limited. That negative feedback can be rectified by biogas purification. In this chapter, clearly defined absorption, adsorption, separation, and biological upgrading techniques are thoroughly discussed. Different biogas upgradation techniques have evolved, and their respective merits and demerits are clearly stated. Upgraded biogas can be utilized as a vehicular fuel.

Biomethane production from raw biogas has a bright research scope. As of now, absorption, adsorption, and separation techniques have been commercialized and practically implemented in many biogas plants all around the world. However, biological techniques have not been practically implemented in real-world biogas plants. Many researchers are concentrating on 100% renewable energy biogas upgradation with green energy production.

## REFERENCES

Adnan, Amir Izzuddin, Mei Yin Ong, Saifuddin Nomanbhay, Kit Wayne Chew, and Pau Loke Show. 2019. "Technologies for Biogas Upgrading to Biomethane: A Review." *Bioengineering* 6 (4): 1–23. doi:10.3390/bioengineering6040092.

Angelidaki, Irini, Laura Treu, Panagiotis Tsapekos, Gang Luo, Stefano Campanaro, Henrik Wenzel, and Panagiotis G. Kougias. 2018. "Biogas Upgrading and Utilization: Current Status and Perspectives." *Biotechnology Advances* 36 (2): 452–66. doi:10.1016/j. biotechadv.2018.01.011.

Awe, Olumide Wesley, Yaqian Zhao, Ange Nzihou, Doan Pham Minh, and Nathalie Lyczko. 2017. "A Review of Biogas Utilisation, Purification and Upgrading Technologies." *Waste and Biomass Valorization* 8 (2): 267–83. doi:10.1007/s12649-016-9826-4.

Baena-Moreno, Francisco M., Mónica Rodríguez-Galán, Fernando Vega, Luis F. Vilches, Benito Navarrete, and Zhien Zhang. 2019. "Biogas Upgrading by Cryogenic Techniques." *Environmental Chemistry Letters* 17 (3): 1251–61. doi:10.1007/ s10311-019-00872-2.

Branco, Rita H.R., Luísa S. Serafim, and Ana M.R.B. Xavier. 2019. "Second Generation Bioethanol Production: On the Use of Pulp and Paper Industry Wastes as Feedstock." *Fermentation* 5 (1): 1–30. doi:10.3390/fermentation5010004.

Collet, Pierre, A. Hélias Arnaud, Laurent Lardon, Monique Ras, Romy Alice Goy, and Jean Philippe Steyer. 2011. "Life-Cycle Assessment of Microalgae Culture Coupled to Biogas Production." *Bioresource Technology* 102 (1): 207–14. doi:10.1016/j. biortech.2010.06.154.

Cozma, Petronela, Cristina Ghinea, Ioan Mămăligă, Walter Wukovits, Anton Friedl, and Maria Gavrilescu. 2013. "Environmental Impact Assessment of High Pressure Water Scrubbing Biogas Upgrading Technology." *Clean – Soil, Air, Water* 41 (9): 917–27. doi:10.1002/clen.201200303.

Cozma, Petronela, Walter Wukovits, Ioan Mămăligă, Anton Friedl, and Maria Gavrilescu. 2014. "Modeling and Simulation of High Pressure Water Scrubbing Technology Applied for Biogas Upgrading." *Clean Technologies and Environmental Policy* 17 (2): 373–91. doi:10.1007/s10098-014-0787-7.

Dahiya, Anju. 2015. *Bioenergy: Biomass to Biofuels. Bioenergy*. doi:10.1016/ B978-0-12-407909-0.00002-X.

Eriksson, Ola, Mattias Bisaillon, Mårten Haraldsson, and Johan Sundberg. 2016. "Enhancement of Biogas Production from Food Waste and Sewage Sludge – Environmental and Economic Life Cycle Performance." *Journal of Environmental Management* 175: 33–39. doi:10.1016/j.jenvman.2016.03.022.

Ghatak, Manjula Das, and P. Mahanta. 2016. "Biogas Purification Using Chemical Absorption." *International Journal of Engineering and Technology* 8 (3): 1600–05.

Ghavinati, H., and M. Tabatabaei. 2018. *Biogas: Fundamentals, Process, and Operation. Springer*. doi:10.1007/978-3-319-77335-3.

Hjuler, Klaus, and Nabin Aryal. 2016. *Review on Biogas Upgrading*. https://futuregas.dk/ wp-content/uploads/2018/06/FutureGas-WP1-Review-of-Biogas-Upgrading_revise d__final.pdf.

Hosseinipour, Sayed Amir, and Mehdi Mehrpooya. 2019. "Comparison of the Biogas Upgrading Methods as a Transportation Fuel." *Renewable Energy* 130: 641–55. doi:10.1016/j.renene.2018.06.089.

"In-Situ-Ex-Situ-and-Hybrid-Biological-Biogas-Upgrading-Technologies-Based-on-Hydroge n." n.d.

Kadam, Rahul, and N.L. Panwar. 2017. "Recent Advancement in Biogas Enrichment and Its Applications." *Renewable and Sustainable Energy Reviews* 73 (June): 892–903. doi:10.1016/j.rser.2017.01.167.

Kapoor, Rimika, Pooja Ghosh, Madan Kumar, and Virendra Kumar Vijay. 2019. *Evaluation of Biogas Upgrading Technologies and Future Perspectives: A Review. Environmental Science and Pollution Research*. Environmental Science and Pollution Research. doi:10.1007/s11356-019-04767-1.

Kougias, Panagiotis G., Laura Treu, Daniela Peñailillo Benavente, Kanokwan Boe, Stefano Campanaro, and Irini Angelidaki. 2017. "Ex-Situ Biogas Upgrading and Enhancement in Different Reactor Systems." *Bioresource Technology* 225: 429–37. doi:10.1016/j.biortech.2016.11.124.

Lima, Rodolfo M., Afonso H.M. Santos, Camilo R.S. Pereira, Bárbara K. Flauzino, Ana Cristina O.S. Pereira, Fábio J.H. Nogueira, and José Alfredo R. Valverde. 2018. "Spatially Distributed Potential of Landfill Biogas Production and Electric Power Generation in Brazil." *Waste Management* 74: 323–34. doi:10.1016/j.wasman.2017.12.011.

Morero, Betzabet, Eduardo S. Groppelli, and Enrique A. Campanella. 2017. "Evaluation of Biogas Upgrading Technologies Using a Response Surface Methodology for Process Simulation." *Journal of Cleaner Production* 141: 978–88. doi:10.1016/j.jclepro.2016.09.167.

Omar, Basma, Maie El-Gammal, Reda Abou-Shanab, Ioannis A. Fotidis, Irini Angelidaki, and Yifeng Zhang. 2019. "Biogas Upgrading and Biochemical Production from Gas Fermentation: Impact of Microbial Community and Gas Composition." *Bioresource Technology* 286 (May): 121413. doi:10.1016/j.biortech.2019.121413.

Pellegrini, Laura Annamaria, Giorgia De Guido, and Stefano Langé. 2018. "Biogas to Liquefied Biomethane via Cryogenic Upgrading Technologies." *Renewable Energy* 124: 75–83. doi:10.1016/j.renene.2017.08.007.

Petersson, Anneli, and Arthur Wellinger. 2009. "Biogas Upgrading Technologies – Developments and Innovations. IEA Bioenergy. Task 37. Report." *IEA Bioenergy Task 37-Energy from Biogass and Landfill Gas* (January 2009): 19. http://www.build-a-biogas-plant.com/PDF/IEA_Biogas_technologies.pdf.

Petersson, Anneli, Jens Bo Holm-Nielsen, and David Baxter. n.d. <Upgrading_Rz_Low.Final.Pdf>.

Ramaraj, Rameshprabu, and Natthawud Dussadee. 2015. "Biological Purification Processes for Biogas Using Algae Cultures: A Review." *International Journal of Sustainable and Green Energy* 4 (1): 20–32. doi:10.11648/j.ijrse.s.2015040101.14.

Rotunno, Paolo, Andrea Lanzini, and Pierluigi Leone. 2017. "Energy and Economic Analysis of a Water Scrubbing Based Biogas Upgrading Process for Biomethane Injection into the Gas Grid or Use as Transportation Fuel." *Renewable Energy* 102: 417–32. doi:10.1016/j.renene.2016.10.062.

Ryckebosch, E., M. Drouillon, and H. Vervaeren. 2011. "Techniques for Transformation of Biogas to Biomethane." *Biomass and Bioenergy* 35 (5): 1633–45. doi:10.1016/j.biombioe.2011.02.033.

Sahota, Shivali, Goldy Shah, Pooja Ghosh, Rimika Kapoor, Subhanjan Sengupta, Priyanka Singh, Vandit Vijay, Arunaditya Sahay, Virendra Kumar Vijay, and Indu Shekhar Thakur. 2018. "Review of Trends in Biogas Upgradation Technologies and Future Perspectives." *Bioresource Technology Reports* 1: 79–88. doi:10.1016/j.biteb.2018.01.002.

Sahota, Shivali, Virendra Kumar Vijay, P.M.V. Subbarao, Ram Chandra, Pooja Ghosh, Goldy Shah, Rimika Kapoor, Vandit Vijay, Vaibhav Koutu, and Indu Shekhar Thakur. 2018. "Characterization of Leaf Waste Based Biochar for Cost Effective Hydrogen Sulphide Removal from Biogas." *Bioresource Technology* 250: 635–41. doi:10.1016/j.biortech.2017.11.093.

Song, Chunfeng, Qingling Liu, Na Ji, Shuai Deng, Jun Zhao, and Yutaka Kitamura. 2017. "Advanced Cryogenic CO2 capture Process Based on Stirling Coolers by Heat Integration." *Applied Thermal Engineering* 114: 887–95. doi:10.1016/j.applthermaleng.2016.12.049.

Song, Chunfeng, Qingling Liu, Na Ji, Shuai Deng, Jun Zhao, Yang Li, and Yutaka Kitamura. 2017. "Reducing the Energy Consumption of Membrane-Cryogenic Hybrid CO2 Capture by Process Optimization." *Energy* 124: 29–39. doi:10.1016/j.energy.2017.02.054.

Starr, Katherine, Xavier Gabarrell, Gara Villalba, Laura Talens, and Lidia Lombardi. 2012. "Life Cycle Assessment of Biogas Upgrading Technologies." *Waste Management* 32 (5): 991–99. doi:10.1016/j.wasman.2011.12.016.

Ullah Khan, Imran, Mohd Hafiz Dzarfan Othman, Haslenda Hashim, Takeshi Matsuura, A.F. Ismail, M. Rezaei-DashtArzhandi, and I. Wan Azelee. 2017. "Biogas as a Renewable Energy Fuel – A Review of Biogas Upgrading, Utilisation and Storage." *Energy Conversion and Management* 150 (July): 277–94. doi:10.1016/j.enconman.2017.08.035.

Vega, Fernando, Mercedes Cano, Sara Camino, Luz M. Gallego Fernández, Esmeralda Portillo, and Benito Navarrete. 2018. "Solvents for Carbon Dioxide Capture." *Carbon Dioxide Chemistry, Capture and Oil Recovery*. doi:10.5772/intechopen.71443.

Voelklein, M.A., Davis Rusmanis, and J.D. Murphy. 2019. "Biological Methanation: Strategies for in-Situ and Ex-Situ Upgrading in Anaerobic Digestion." *Applied Energy* 235 (August 2018): 1061–71. doi:10.1016/j.apenergy.2018.11.006.

Wang, Han, Xinyu Zhu, Qun Yan, Yifeng Zhang, and Irini Angelidaki. 2020. "Microbial Community Response to Ammonia Levels in Hydrogen Assisted Biogas Production and Upgrading Process." *Bioresource Technology* 296 (October 2019): 122276. doi:10.1016/j.biortech.2019.122276.

Xia, Ao, Jun Cheng, and Jerry D. Murphy. 2016. "Innovation in Biological Production and Upgrading of Methane and Hydrogen for Use as Gaseous Transport Biofuel." *Biotechnology Advances* 34 (5): 451–72. doi:10.1016/j.biotechadv.2015.12.009.

# Index

Printed and bound by CPI Group (UK) Ltd, Croydon, CR0 4YY

24/10/2024

01778278-0009